D0504467

WITHDRAWN
FROM STOCK
QMUL LIBRARY

DATE DUE FOR RETURN

ENCOUNTER WITH MATHEMATICS

LARS GÅRDING

ENCOUNTER WITH MATHEMATICS

SPRINGER-VERLAG
NEW YORK   HEIDELBERG   BERLIN

181630

Lars Gårding
Mat. Inst. Fack 725
Lund 7, Sweden

Library of Congress Cataloging in Publication Data

Gårding, Lars, 1919–
   Encounter with mathematics.

   Includes index.
   1. Mathematics—1961- I. Title. II. Series.
QA37.2.G28   510   76–54765

All rights reserved.

No part of this book may be translated or reproduced in any
form without written permission from Springer-Verlag.

© 1977 by Springer-Verlag New York Inc.

Printed in the United States of America.

9 8 7 6 5 4 3 2 1

ISBN 0-387-90229-5 Springer-Verlag New York
ISBN 3-540-90229-5 Springer-Verlag Berlin Heidelberg

QUEEN MARY
COLLEGE
LIBRARY

PREFACE

Trying to make mathematics understandable to the general public is a very difficult task. The writer has to take into account that his reader has very little patience with unfamiliar concepts and intricate logic and this means that large parts of mathematics are out of bounds.

When planning this book, I set myself an easier goal. I wrote it for those who already know some mathematics, in particular those who study the subject the first year after high school. Its purpose is to provide a historical, scientific, and cultural frame for the parts of mathematics that meet the beginning student. Nine chapters ranging from number theory to applications are devoted to this program. Each one starts with a historical introduction, continues with a tight but complete account of some basic facts and proceeds to look at the present state of affairs including, if possible, some recent piece of research. Most of them end with one or two passages from historical mathematical papers, translated into English and edited so as to be understandable.

Sometimes the reader is referred back to earlier parts of the text, but the various chapters are to a large extent independent of each other. A reader who gets stuck in the middle of a chapter can still read large parts of the others. It should be said, however, that the book is not meant to be read straight through. It contains a wealth of material, some of it not considered elementary, e.g. Hilbert's Nullstellensatz in the algebra chapter and the Fourier inversion formula in the chapter on integration. These important items were included because in both cases it was possible to give simple and lucid proofs that fitted the context.

Three chapters are of a more general nature. The introductory one deals with models and reality and the final one with the sociology, psychology, and teaching of mathematics. There is also a chapter on the mathematics of the seventeenth century providing a fuller historical background to infinitesimal calculus. In an appendix there are a few words on terminology and notation and some advice on how to read and choose mathematical texts.

A preliminary draft of the book was read by Karl Gustav Andersson and Tomas Claesson, and Gunnar Blom read the probability chapter. The final draft has been read by William F. Donoghue Jr, Tore Herlestam, and Charles Halberg. I thank these six friends and critics for much valuable advice.

April, 1977   Lars Gårding
Lund

CONTENTS

Contents

6
THE HEROIC CENTURY   124

7
DIFFERENTIATION   134

8
INTEGRATION   168

9
SERIES   205

10
PROBABILITY   222

11
APPLICATIONS   240

12
THE SOCIOLOGY, PSYCHOLOGY, AND
TEACHING OF MATHEMATICS   253

APPENDIX   263

INDEX   265

1
MODELS AND REALITY

1.1 *Models*. The natural numbers. Celestial mechanics. Quantum mechanics. Economics. Language. 1.2 *Models and reality*. 1.3 *Mathematical models*.

Trying to understand the world around him, man organizes his observations and ideas into conceptual frames. These we shall call models. The insight gotten by applying logic to the concepts of a model will be called its theory. Mathematical models are logically coherent and have extensive theories. Others may be less strict but no less useful.

In the exact sciences, the validity of models is tested by logic and by experiment. This makes it necessary to distinguish very clearly between the model and the part of the outside world that it is supposed to represent. This principle is now current in many branches of science. When applied in a general way it puts human thought into an interesting perspective. Part of this chapter deals with man's relations to the models of the world that he himself has created. It starts with short descriptions and evaluations of some important models and ends with a survey of some mathematical models and their mutual relations.

1.1 Models

The natural numbers

The simplest mathematical model is the set of natural numbers 1, 2, 3, ... It is used for counting objects when all the properties of objects are disregarded except their number. The natural numbers appear in all languages. Some have names for more numbers than others, but there are always numbers that are so large that they have no names. This phenomenon is perhaps man's first encounter with infinity. In ancient times it led to serious questions like these: are there numbers so large that they cannot be counted? Or, in a more concrete setting: are the grains of sand on the earth uncountable? The second question was answered by Archimedes in a book called *The Sand Reckoner* (200 B.C.). He displayed a series of rapidly growing numbers and could show by some estimates of volume that some of these numbers were larger than the number of grains of sand on the earth and even in the solar system. We see here how the model of natural numbers answers a concrete question about the outside world. Abstraction has proved its worth. The situation is illustrated by the left part of Figure 1.1. The ragged contour on the left means that we have cut out a piece of the real world that can have properties without counterparts in the model. The straight lines and right angles of the model are supposed to illustrate its schematic character.

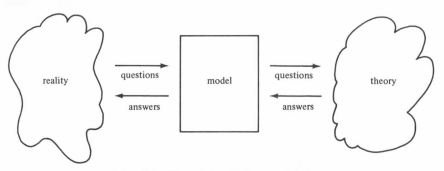

Figure 1.1   The triplet Reality-model-theory.

Let us now add the operation of multiplication to our model, the natural numbers. It then acquires a lot of very interesting properties. Some experimentation with multiplication shows that certain numbers are products of other numbers greater than 1, e.g., $20 = 2 \times 2 \times 5$, while others, e.g., 5, do not have this property. Those of the second kind are called *prime numbers* or *primes*. The first primes are 2, 3, 5, 7, 11, 13, 17. It is easy to continue the series but the amount of necessary checking increases very quickly when we come to large numbers. Under these circumstances, the following question presents itself: is there a finite or infinite number of primes? A simple and ingenious reasoning to be found in Euclid's *Elements* (270 B.C.) and explained in the next chapter provides the answer: the number of primes is infinite. We have here an example of a question put in the model that can be answered by theory, i.e., logical reasoning about the model. This is illustrated by the right part of Figure 1.1. The ragged contour of the theory indicates that it is not determined completely by the model. In our case it contains a host of theorems about the natural numbers, e.g., solvability and nonsolvability of certain equations, the distribution of primes, etc. The size and the power of the theory depends among other things on the ability of the mathematicians who created it.

Our three-part figure shall now serve to illustrate a number of important models other than the natural numbers.

Celestial mechanics

The part of reality that is to be analyzed consists of astronomical observations of the positions of the earth, the planets, and the sun at different times. In the model, these celestial bodies correspond to point-sized objects with masses that attract each other according to Newton's law of gravitation. Each object attracts every other one with a force whose size is proportional to the product of the two masses and inversely proportional to the square of the distance. This force is directed towards the attracting object. The movement is such that the product of the mass and the acceleration of an object equals the attractive force. The theory consists of mathematical propositions about the nature of such movements. It turns out in particular that the positions and the velocities of the objects at one time determine the subsequent motion uniquely. This model and its theory were created by Newton in the seventeenth century. It answers a multitude of astronomical questions. The orbits and masses of the planets can be computed with great accuracy. Small deviations from the predictions have led to the discovery of new planets. The orbits of artificial satellites are predicted with the aid of this model. The theory has grown continuously from the time of Newton to the present. Celestial mechanics has been an unparalleled success and made a deep impression on the philosophy of the eighteenth and nineteenth centuries.

Quantum mechanics

The problem is to analyze radiation from atoms, recorded as tracks and spectral lines on photographic plates. The model is a variant of celestial mechanics with objects corresponding to the atomic nuclei and the electrons, but the relatively close intuitive connection between the real objects and those of the model is lost. Certain objects of the model must be interpreted both as waves and as particles. The insight that led to the quantum-mechanical model comes, to a large extent, from the concepts that were fruitful in the theory of celestial mechanics, e.g., mass, energy, and momentum. The photographic plates do not by themselves provide much guidance. The theory of this model consists, among other things, of propositions about Hilbert space, an infinite-dimensional analogue of Euclidean space. Classical quantum mechanics has been very successful in the sense that the radiation frequencies and their dependence on electromagnetic fields are predictable from very few data. One property of the model, the complementarity principle, has gained a certain status in philosophy.

Small discrepancies between computed and observed frequencies have led to a more sophisticated model which takes into account Einstein's relativity theory, and in which the electromagnetic field is quantized in the same way as the model of the atom is a quantization of the model of planetary motion. This new model has been less successful. There were and still are unsolved difficulties with its theory. The embarrassing fact is that there is no model of this kind with a consistent theory that also has interesting applications. On the other hand, it is possible in the present models to predict certain facts from others. It is hard to guess what is going to happen next. Is the model going to change or is it the theory that will expand?

3

Economics

Let us consider the theory of prices in a market with perfect competition. The goal is to analyze the interplay between producers and consumers that determines prices. Producers and consumers appear also in the model but with very simplified motives. The producers want to maximize their profits, the consumers their utility. Profit and utility are assumed to be known for every state of the market. An equilibrium state is defined by, e.g., the property that any deviation from it is bad for at least somebody. The theory provides conditions that guarantee that equilibria exist and provides methods for computing the corresponding sets of prices. Since the mathematics of the theory is relatively simple, we might ask, why is it that grocery prices are not computed by some big computer? The answer is that utility functions are not so well-determined in practice that it makes sense to make such calculations. But this does not make the model worthless. Sometimes it permits qualitative conclusions and under all circumstances it is helpful when analyzing a market. The model serves as a coherent frame of concepts.

Very few economic models permit relevant numerical applications. This has led to a rather sharp division of economics into a theoretical branch, where people study more or less sophisticated models, and a descriptive branch dealing more directly with the real world.

Language

Classical grammar and parts of it can be thought of as models of language. It is only recently that one has distinguished between such models and their theory taken as a logical analysis of models. An important step in this direction was taken by Noam Chomsky in the fifties. He considered generative models, i.e., collections of rules and instructions leading to all well-formed sentences and no others. Using methods from mathematical logic Chomsky made it very plausible that a generative model must contain rules, the so-called transformations, that transcend the simple rules which are immediately deducible from the material. These were the only ones acceptable to the structuralists, who dismissed large parts of classical grammar as being infected by too much philosophical a priori. A transformation is, for instance, the passage from active to passive form. Chomsky sharpened the requirements on grammatical models, at the same time rejecting the idea that such models had to be constructed from language itself by some prescribed procedure.

With the above collection of examples in mind we can now make some general statements about the triplet reality-model-theory. In celestial mechanics there is a perfect balance and tight connections between the three parts, and theory has shown great vitality. Compared to this ideal state of affairs, the theory part may be insufficient as it is in quantum field theory. Many models are so schematic that they cannot make predictions and have to play the role of mere conceptual frames. Effective models are rare and cannot be deduced from observations by some automatic procedure. In general it is not reality itself but rather the critical points of contact between reality and a model that lead to improved models. When reality is confusing and difficult to observe, it may happen that the old model provides most of the intuition that is needed for a new one.

1.2 Models and reality

When we leave the exact sciences and mathematics, theory, in the sense that we have used it so far, becomes less important. It is then convenient to merge model and theory into one unit. We shall now restrict ourselves to model and reality. This pair plays an all-important part in human thought.

It is natural to consider, e.g., the theory of evolution as a model of how life developed on our planet. It is very convincing both by its inner coherence and the close fit with observation. Religions are models of man's position in the universe and the driving forces of history. Many of them assume the existence of gods and spirits whose relations to each other and to mankind mirror human society. Personifications of the sun, the moon, animals, strength, virtue, evil, etc., are common. The Christian confession of faith is a concise description of a model of the universe ruled by an almighty god who created it and governs over all life on earth, who punishes and rewards. A god of this kind is the answer given by many religions to man's wish to know his place in the universe and the purpose of his life.

Like the religions, the philosophical systems are models of man and his world. But they are more abstract and logic has a more important position in them. Plato, e.g., accepts that there is a god but the mainstay of his model of the world is a heaven of abstract ideas. Some of them correspond to things on earth, others represent goodness and beauty. The Greek philosophers drew a clear line between model and reality, and made logic the judge in philosophical controversies. Our scientific tradition starts with them. The philosophical systems were very important in the seventeenth and eighteenth centuries, with names like Descartes, Leibniz, Hume, Locke, Kant, and Hegel, but have now lost most of their attraction. An important offspring with great vitality is the marxist philosophy of history. It is a model of society where history is subordinated to economic forces and seen as a struggle between classes. One variant foresees the arrival of the classless society.

Most religions and many philosophical systems claim to combine universal validity with inner coherence and convincing ideas. But the flaws become apparent when they are confronted with reality or logic. Trying to cover too much they overreach themselves. Yet it is a fact that most people believe in at least one model of this kind. Models pretending to explain most of the world and human existence are certain to get followers. They

5

have been and are still very important in human society. For many people, the model that they believe in takes the place of the real world. Metaphorically speaking, they live in the model. A believing marxist interprets all phenomena in terms of his model, and a fanatic follower of a religion considers life as a play of shades and the model given by his religion as the only true world. Conflicts rooted in tensions between model and reality are a constant theme in literature. I give one example: the main character in Chekhov's short story, "The Man in the Case," is a teacher of Latin in a small town in southern Russia who believes very strongly in bourgeois morality of the strictest kind. A budding love affair between him and a young woman teacher collapses when he discovers that she rides a bicycle: a young woman on a bicycle, this is unthinkable.

I believe that man has a theoretical drive. His brain is a sorting machine where outside impulses are stored, ordered, and reworked into models of the world. Overwhelmed by a confusing and complicated reality, he sometimes takes refuge in the simple and safe world of the models. Most people have to content themselves with models constructed by others, but some can create new models or improve old ones. Our theoretical drive has given us both primitive ideas of the outside world and scientific theories.

1.3 Mathematical models

It may appear a bit pedantic to call the set of natural numbers a mathematical model separated from reality and logic. Most Greek philosophers would probably not have agreed to this division. But the Greek mathematicians created a model, Euclidean geometry, that fits exactly into the scheme. It is described in Euclid's *Elements* and deals with the straight lines, triangles, and other geometric objects that appear in our daily life, e.g., stretched cords and figures drawn in the sand. They were made into abstractions and their mutual relations were codified in a number of axioms of the type: "through any two distinct points there is exactly one straight line." With this, the model was fixed. The theory was then built with logic as the only tool. It contains a large number of results never contradicted by observation, e.g., the theorem that the sum of angles in a triangle is two right angles. Through Euclidean geometry, mathematics got its reputation as the strictest and purest of sciences.

The great advances in mathematics of the last four centuries have not been dominated by discussions about models. For the mathematicians it was evident that mathematical objects are abstractions and that logic had the final say about right or wrong. Those who applied mathematics to physics always had the experimental tests in mind. A detailed analysis of the ingredients of models did not have high priority under these circumstances. This was only done in the twentieth century with the systematization of mathematics. It led to mathematical models with names like set, group, ring, linear space, topological space. Their real world consists of mathematical objects rather than objects that appear in everyday life. In this sense they are second generation models. Set theory is an example of this (see Figure 1.2).

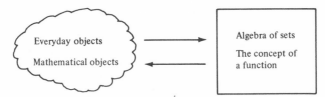

Figure 1.2   The real world of set theory consists of mathematical objects rather than everyday objects.

We let the model of set theory include the definition of a set, unions, intersections, set algebra, i.e., the rules for repeated unions and intersections, and the concept of a function. In elementary school, set theory is represented by collections of everyday objects. In mathematics it is a conceptual frame which is useful when applied to mathematical objects. This holds in particular for the concept of a function. Set theory as an independent subject does not exist any longer. It was created by Cantor around 1870 but has now been absorbed by logic and algebra and other parts of mathematics. Part of set theory can also serve as the real world of the natural numbers and this is done in mathematical logic (Figure 3).

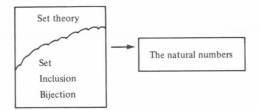

Figure 1.3   From sets to natural numbers.

Another example of a second generation model is the linear space, a model of some aspects of algebra, analysis, and geometry. The theory is called linear algebra. The concept of a group is a model of many different situations in all of mathematics. Group theory is much richer than linear algebra, whose main results can be explained and proved in a few pages (see Figure 1.4). Present international usage prescribes that linear algebra and group theory shall be presented axiomatically. After the recent re-organization of mathematics this seems the only reasonable way. The trouble is that children and most college students think in very concrete ways and find it difficult to come to terms with second generation mathematical models. In fact, they have difficulties even with the simplest models with close ties to reality. For this reason it has not been possible to

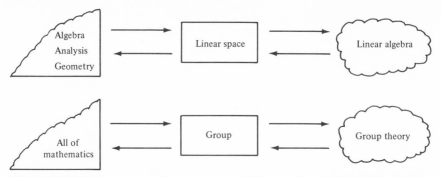

Figure 1.4   Linear spaces and groups are second generation mathematical models.

teach, e.g., the foundations of linear algebra and group theory in a meaningful way in the public schools. Although this stuff is so simple that everybody ought to understand it, most students do not. Those with mathematical ability and interest in the subject can get along very well in many mathematical models, but most people are really at home only among numbers and simple geometric figures. The half-way failure of the so-called new math is due to this fact. It was introduced with the purpose of teaching small children set theory illustrated by collections of everyday objects. The point was that this material is simpler than arithmetic or geometry and that afterwards the road was open to the numbers, to geometry, and to the terminology of mathematics including the concept of a function. But set theory is sterile, it has no interesting applications, and it is difficult to vary the exercises. On the contrary, the number model and the geometric model are much better playgrounds, with direct access to the infinitely varied outside world.

Literature

To consider mathematics as a collection of theories of mathematical models is an enlightening and useful point of view, but a good description of the subject should also give an idea of its richness and variety. The classic *What is Mathematics?* by R. Courant and H. Robbins (Oxford 1947) is perhaps the best effort in this direction written for the general public. It is a rambling and very readable account of some basic concepts and results in algebra, geometry, and analysis.

2
NUMBER THEORY

In this chapter we shall have a look at the numbers, in the simplest case the mathematical model that consists of the set of natural numbers $N = (1, 2, \ldots)$. Their basic properties are assumed, for instance the facts that given two such numbers m and n we have either $m < n$, $m = n$, or $m > n$, that every natural number n has an immediate successor $n + 1$ which also is the least number $> n$, and that every natural number is a successor of 1. They have some simple consequences, for instance that there is an infinite number of natural numbers and that a subset of natural numbers which contains 1 and, together with a number n, also contains its immediate successor $n + 1$, must be all of N. This last property is called the *principle of induction*. It shows that in order to verify the truth of an infinite sequence of propositions $P_1, P_2, \ldots$ it suffices to verify P_1 and that, for any n, $P_1, \ldots, P_n$ together imply P_{n+1}. This principle is used in mathematics as a matter of routine and we shall meet it right at the beginning of the chapter where P_n says that "the natural number $n + 1$ is either a prime or a product of primes." In most cases it is clear from the context what the propositions involved are. They are then not made explicit and as a rule the whole process is brought to the mind of the reader by simply mentioning the word induction.

We shall bring the operations of addition and multiplication of the natural numbers into our model. Already this leads to a very substantial theory. Later, we use subtraction and division and widen the horizon by imbedding the natural numbers into the complex numbers C. But mostly we shall be content with the integers $Z = (0, \pm 1, \pm 2, \ldots)$, the rational numbers Q, and the real numbers R.

2.1 The primes

Divisibility

The oldest and most important part of number theory deals with divisibility. That a natural number a is *divisible* by another natural number b means that the quotient a/b is an integer or, in other words, that there is a third natural number c such that $a = bc$. This is also expressed by saying that b *divides* a or is a *divisor* of a or that a is an *integral multiple* of b. The numbers b and c are also called *integral factors* of a. To give an example: 3 but not 5 divides 6. Since $a = a \cdot 1$ the number 1 divides every number and every number divides itself. These are the trivial divisors. Some experience of multiplication shows that divisibility is a complicated thing. There are numbers with many divisors, e.g., $120 = 2 \cdot 3 \cdot 4 \cdot 5$, and numbers with just trivial divisors, e.g., 3, 5, 17. Natural numbers p greater than 1 with only trivial divisors are called *primes*. The primes less than 100 are, in order, 2, 3, 5, 7, 11, 13, 17, 19, 23, 29, 31, 37, 41, 43, 47, 53, 59, 61, 67, 71, 73, 79, 83, 89, 97. As a first step in a theory of divisibility we shall now prove a simple result. To have a concise formulation we shall permit products with just one factor.

Theorem 1. *Every natural number greater than* 1 *is a product of primes.*

This theorem is stated and proved in Euclid's *Elements*, the great encyclopedia of Greek mathematics. Most of it deals with geometry; it was put together around 270 B.C. and has 13 parts traditionally called books. Not much is known about its author, Euclid, but his work kept its position as the most important textbook of mathematics until about 1850. If we include all kinds of editions of it, the *Elements* is actually one of history's major best sellers. We shall have more to say about it in the chapter on geometry and linear algebra. Some of our theorems can be found in the *Elements* in the sense that their content is there, although the form might be quite different. The Greeks did not, for instance, have our algebraic notation, and expressed themselves in many ways that seem strange to us.

We now turn to the proof of Theorem 1. Let a be the given number. If a is not itself a prime it is a product bc where b and c are natural numbers greater than 1 and hence smaller than a. If the theorem holds for all numbers smaller than a, in particular for b and c, it holds also for a. In fact, if b and c are products of primes, a is a product of primes. Now the theorem is true when $a = 2$ since 2 is a prime. Hence it must be true also when $a = 3, 4, \ldots$, i.e., for all natural numbers greater than 1. Here the principle of induction is used.

In spite of the rather formal proof we have given, Theorem 1 is an almost immediate consequence of the definition of a prime. Our next theorem is of another caliber. Stated and proved in the *Elements*, it is a marvel of intellectual curiosity, audacity, and ingenuity.

Theorem 2. *There are an infinite number of primes.*

In fact, let $p_1, \ldots, p_n$ be n primes. We shall see that in whatever way they are chosen and however many they may be, there is at least another prime. This will then prove the theorem. Consider the number $a = p_1 \ldots p_n + 1$. According to Theorem 1 there is a prime p and a natural number c, perhaps equal to 1, such that $a = pc$. Hence

$$p_1 \ldots p_n + 1 = pc.$$

This excludes that p is any of the numbers $p_1, \ldots, p_n$. If, for instance, $p = p_k$ and b is the product of the other numbers $p_1, \ldots$, then

$$pb + 1 = pc$$

so that $p(c - b) = 1$. But this is impossible since p is greater than 1 and $c - b$ at least equal to 1.

The prime number theorem

Theorem 2 shows that there is no catalogue listing all primes, but human curiosity has produced many large catalogues, nowadays done by computers. In principle, the computation is simple: deleting successively all proper multiples of the primes from the natural numbers leaves only the primes. One starts by deleting all the multiples of 2, then all the multiples of 3, and so on. The Greeks called this method the Sieve of Erathostenes. It shows that the primes get more and more rare among large natural numbers. One might ask: how rare? The prime number theorem gives a rough answer: For large integers n the number of primes less than n is about $n/\log n$ in the sense that the quotient between these two numbers tends to 1 as n tends to infinity. In a less precise manner we can also say that the "probability" that a large number n shall be a prime is about $1/\log n$. The prime number theorem was proved around 1890 independently by Hadamard and by de la Vallée-Poussin with advanced analytical tools. An elementary (but not simple) proof by Atle Selberg in 1948 was a mathematical event. The prime number theorem is a rather rough estimate and perhaps not very characteristic of the primes. It holds also for many kinds of artificial primes obtained from the natural numbers by procedures similar to the Sieve of Erathostenes. An as yet unsolved problem is to decide whether or not there is an infinite number of twins among the primes, i.e., primes whose difference is two. Example: 101 and 103.

Modules

We now leave the large primes and proceed with the theory of divisibility. A set $M = (u, v, \ldots)$ of integers is called a *module* if it has the property that $u - v$ belongs to M whenever u and v do. Iteration shows

11

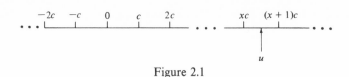

Figure 2.1

that this means that

$$u, v \in M \quad \Rightarrow \quad xu + yv \in M \tag{1}$$

for all integers x and y. The simplest example of a module is the set of all integral multiples xa of a fixed number a, i.e., $0, \pm a, \pm 2a, \ldots$. Let us denote it by $\mathbf{Z}a$. We get another example by letting M consist of all numbers $u = xa + yb$ where a and b are fixed and x and y are arbitrary integers. Let us denote this module by $\mathbf{Z}a + \mathbf{Z}b$. The point of the concept of a module is that already our first example is exhaustive: to every module $M \neq 0$ there is precisely one number $c > 0$ such that $M = \mathbf{Z}c$. To see this, it suffices to let c be the least positive number in M and mark out the module $\mathbf{Z}c$ on the real line (see Figure 2.1). Every integer u is then a member of the module $\mathbf{Z}c$ or else lies between two consecutive numbers xc and $(x + 1)c$ in it. If we have the latter and u belongs to M the number $u - xc$ is also in M. But it is > 0 and $< c$ and this contradicts the definition of c. Hence $M = \mathbf{Z}c$. With this we have proved the first part of

Theorem 3 (Greatest common divisor). *To every pair of natural numbers a and b there is a unique natural number c such that*

$$\mathbf{Z}a + \mathbf{Z}b = \mathbf{Z}c. \tag{2}$$

The number c divides a and b, and, moreover, every number with this property divides c.

Let us prove the last statement of the theorem. According to (2), both a and b are multiples of c. The formula also says that there are integers x and y such that $c = ax + by$ and hence a number dividing a and b must divide c, then obviously the largest of these common divisors. When $c = 1$ we say that the numbers a and b are *relatively prime* to each other.

Theorem 3 contains everything essential about the divisibility properties of integers. It occurs in another form in the *Elements* with a similar proof.

Let p be a prime and a an integer not a multiple of p. Then 1 is the largest common divisor of a and p and we have $\mathbf{Z}a + \mathbf{Z}p = \mathbf{Z}$, i.e., there are numbers x and y such that

$$1 = xa + yp. \tag{3}$$

From this follows the well-known fact that if p divides a product ab, it must divide either a or b. For if p does not divide a, (3) shows that $b = xab + byp$ is a multiple of p. We have shown

Theorem 4. *If a prime divides a product of integers, it divides at least one factor.*

Theorem 2 shows that to every natural number > 1 there is at least one set $(p_1, \ldots, p_k)$ of different primes and natural numbers $n_1, \ldots, n_k$ such that

$$a = p_1^{n_1} \ldots p_k^{n_k}. \tag{4}$$

According to Theorem 4, the set $(p_1, \ldots, p_k)$ consists of *all* primes dividing a. Hence it is uniquely determined by a. The same is true of the exponents $n_1, \ldots, n_k$. In fact, n_j is the largest number n such that p_j^n divides a. We state this as a theorem: the factorization of a natural number > 1 into primes is unique apart from the order of the factors. We shall see later that this need not be true in certain other situations, where one has analogues of Theorem 1 and Theorem 2 but not of Theorem 3.

2.2 The theorems of Fermat and Wilson

Let us start with the implication

$$a, b \text{ integers, } p \text{ a prime} \implies p \text{ divides } (a+b)^p - a^p - b^p. \tag{5}$$

This follows from the binomial theorem since

$$(a+b)^p - a^p - b^p = \sum_{k=1}^{p-1} \binom{p}{k} a^k b^{p-k}$$

where all the binomial coefficients

$$\binom{p}{k} = p!/k!(p-k)!, \qquad 0 < k < p,$$

are integers. Since p does not divide the denominator it must divide the quotient. Hence p divides $\binom{p}{k}$ when $0 < k < p$.

The following famous result was obtained by Fermat around 1640. It was not known to the Greeks.

Fermat's theorem. *The implication*

$$p \text{ a prime} \implies p \text{ divides } a^p - a, \tag{6}$$

is true for every integer a.

We prove the theorem the way Euler did it in 1736. Combining a special case of (5), namely

$$p \text{ divides } (b+1)^p - b^p - 1$$

with the hypothesis that (6) is true when $a = b$ gives the result that p divides the sum

$$(b+1)^p - b^p - 1 + b^p - b = (b+1)^p - (b+1).$$

Hence, since our implication (6) holds for $a = 1$, induction shows that it holds when a is any natural number. In order to show that the theorem is true also when a is an integer $\leqslant 0$, just note that if it is true for a, it holds for $a - p$.

We are now going to give a better proof that uses the concept of a module. Let m be a fixed integer. When $x - y$ belongs to the module $\mathbf{Z}m$, i.e., when $x - y$ is an integral multiple of m, we say that x is *congruent* to y *modulo* m and write

$$x \equiv y \bmod m \tag{7}$$

or, more simply, $x \equiv y$ when it is clear what number m we refer to. Mod 5 we have, for instance, $8 \equiv 3$, $4 \equiv -1$ and $10^{11} \equiv 0$, but $7 \not\equiv 1$ and $99 \not\equiv 66$. It is clear that, mod m, every integer is congruent to precisely one of the numbers $0, 1, \ldots, m - 1$. The point of using congruences is that they behave like equalities under addition and multiplication. We see immediately that

$$x \equiv y, \quad z \equiv u \quad \Rightarrow \quad x + z \equiv y + u \quad \text{and} \quad xz \equiv yu. \tag{8}$$

The last congruence is clear from the fact that $y = x + am$, $u = z + bm$ with integral a and b.

Given an integer a, the existence of an integer a' such that $aa' \equiv 1$ mod m means precisely that there is another integer m' such that $a'a + m'm = 1$, i.e., that a is relatively prime to m. We say that a' is an *inverse* of a mod m. In particular, when $m = p$ is a prime, every $a \not\equiv 0 \bmod p$ has an inverse mod p. Inverses mod 5 are, for instance, $1' = 1$, $2' = 3 = -2$, $3' = (-2)' = 2$, $4' = (-1)' = -1$. Note that when a has an inverse a' mod m, multiplication by a' proves the cancellation law $ax \equiv ay \bmod m \Rightarrow x \equiv y \bmod m$.

We can now reproduce Ivory's proof from 1806 of Fermat's theorem, rediscovered in 1828 by Dirichlet. Let $a \not\equiv 0 \bmod p$ and consider the numbers $a, 2a, \ldots, (p - 1)a$. They are all different mod p for if xa and ya are two of them, then $x \not\equiv y \bmod p$ so that also $xa \not\equiv ya \bmod p$. Hence, apart from the order they are congruent to $1, 2, \ldots, p - 1 \bmod p$. Multiplying them together and using (8) this gives $a^{p-1}(p - 1)! \equiv (p - 1)!$ and, by cancellation,

$$a \not\equiv 0 \bmod p \quad \Rightarrow \quad a^{p-1} \equiv 1 \bmod p, \tag{9}$$

which is Fermat's theorem expressed in another way. The same line of reasoning can be used if we replace p by an arbitrary natural integer m and the numbers $1, \ldots, p - 1$ by the set of numbers between 0 and m that are relatively prime to m. The product of any two such numbers has the same property and if there are $\varphi(m)$ of them, the proof above yields Euler's generalization of Fermat's theorem, namely

$$a \text{ relatively prime to } m \quad \Rightarrow \quad a^{\varphi(m)} \equiv 1 \bmod m.$$

Example: when $m = 6$, then 1 and 5 are the only numbers between 0 and 6 that are relatively prime to 6. Hence $\varphi(6) = 2$ and the theorem says that $1^2 \equiv 1$ and $5^2 \equiv 1 \bmod 6$.

Fermat's theorem has a companion, Wilson's theorem, proved by Waring in 1740 but probably known to Leibniz. It states that

$$p \text{ a prime} \implies (p-1)! \equiv -1 \bmod p. \tag{10}$$

This is true for $p = 2$. For $p > 2$ one proves (10) by collecting the numbers $1, 2, \ldots, p-1$ in pairs x, x' such that $xx' \equiv 1 \bmod p$. Since $x^2 \equiv 1 \implies (x-1)(x+1) \equiv 0 \implies x \equiv 1$ or -1, only the numbers $x = 1$ and $x = p-1 \equiv -1$ are their own inverses, and the rest occur in pairs with a product $\equiv 1$. Multiplying everything together, (10) follows from (8).

Squares

Writing Fermat's theorem for $p > 2$ as

$$a^{p-1} - 1 = (a^{(p-1)/2} + 1)(a^{(p-1)/2} - 1) \equiv 0$$

we see that for every $a \neq 0$ at least one of the factors on the right is $\equiv 0$. Which one it turns out to be depends on whether or not a is a square mod p, i.e., whether or not there is an integer b such that $a \equiv b^2 \bmod p$. We shall see that if $a \not\equiv 0 \bmod p$, then

$$a \text{ is a square mod } p \implies a^{(p-1)/2} \equiv 1 \bmod p \tag{11}$$

$$a \text{ is not a square mod } p \implies a^{(p-1)/2} \equiv -1 \bmod p. \tag{12}$$

In fact, if we put $a \equiv b^2$ in (11), this statement follows from Fermat's theorem. To prove (12), collect the numbers $1, 2, \ldots, p-1$ in pairs x, x^* such that $xx^* \equiv a$. Since a is not a square mod p, x and x^* are always different and we get $(p-1)/2$ pairs with the product $a^{(p-1)/2} \equiv (p-1)!$ so that (12) follows from Wilson's theorem.

Let the symbol

$$\left(\frac{a}{p} \right)$$

denote 1 when a is a square mod p and -1 when a is not a square mod p. One of the most famous theorems of number theory, first proved by Gauss (1801), is the quadratic reciprocity law. It says that

$$\left(\frac{p}{q} \right)\left(\frac{q}{p} \right) = (-1)^{(p-1)/2}(-1)^{(q-1)/2}$$

when p and q are odd primes.

2.3 The Gaussian integers

Norm and divisibility

To get some perspective on the concept of divisibility it is necessary to consider other numbers than just the natural ones. To begin with we shall consider *the Gaussian integers*, which are the complex numbers $\alpha = a + ib$

where a and b are ordinary integers, here referred to as rational integers. The set of Gaussian integers will be denoted by $\mathbf{Z}^*$. They form a quadratic lattice in the complex plane (see Figure 2.2). It is clear that sums and products of Gaussian integers are Gaussian integers. Divisibility is defined in a natural way. For instance, $(2 + i)(3 + i) = 5 + 5i$ means that $2 + i$ and $3 + i$ are divisors of $5 + 5i$ which in turn is an integral multiple of each of these numbers. On the other hand, the number 2 does not divide $1 + i$ since the quotient $(1 + i)/2$ is not a Gaussian integer. The prime number concept can be transferred to the Gaussian integers, but $5 = (2 + i)(2 - i)$ shows that not all rational primes are Gaussian primes. The *norm square* $|\alpha|^2 = a^2 + b^2$ of a Gaussian integer α is a rational integer ≥ 0 and will be used to measure the size of α. It has the property that $|\alpha\beta|^2 = |\alpha|^2 |\beta|^2$. The Gaussian integers of norm square 1, i.e., ± 1 and $\pm i$ are not considered to be primes. They are called units. If we go back and try to prove our previous theorems for the Gaussian integers, we have no difficulties with Theorem 1 and Theorem 2 provided we use induction with respect to the norm square. Every Gaussian integer of norm square > 1 is a prime or a product of primes, and there are infinitely many primes.

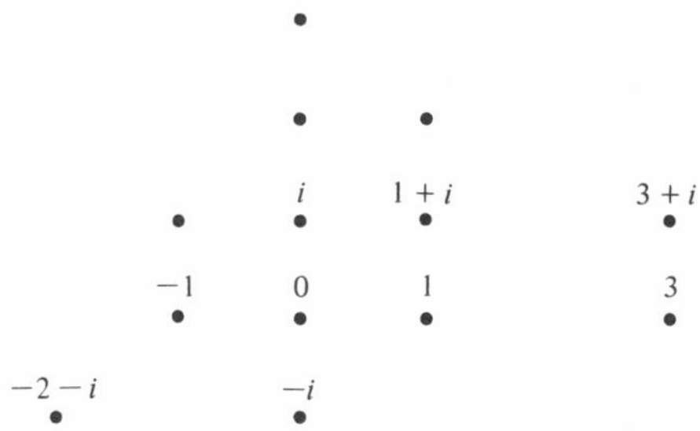

Figure 2.2

Ideals

For the Gaussian integers it is possible to define the concept of a module in two ways. Either one requires (1) for all rational integers x and y, or for all Gaussian integers x and y. It is customary to keep the word module in the first case and to use the word *ideal* in the second case. A generalized form of this concept plays a very important role in algebra and will come up in the next chapter. An example of an ideal in the Gaussian integers $\mathbf{Z}^*$ is the set $\mathbf{Z}^*\gamma$ of all Gaussian integral multiples $z\gamma$ of a fixed Gaussian integer γ. It can also be described as the set of numbers $x\gamma + iy\gamma$ where x and y are arbitrary rational integers. In the complex plane the ideal $\mathbf{Z}^*\gamma$ constitutes a quadratic lattice according to Figure 2.3. *Every ideal J in the Gaussian integers is of the form $\mathbf{Z}^*\gamma$.* In fact, if $J = 0$ take $\gamma = 0$. If $J \neq 0$ choose an element $\gamma \neq 0$ in J with least possible norm and consider the ideal $\mathbf{Z}^*\gamma$. It is evident from the figure that the distance from

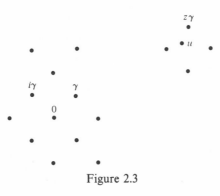

Figure 2.3

any complex number u to a suitably chosen element of $\mathbf{Z}^*\gamma$ is at most $|\gamma|/\sqrt{2}$. In other words, there is a Gaussian integer z such that $|u - z\gamma| \leqslant |\gamma|/\sqrt{2}$. Now if u belongs to J so does $u - z\gamma$. But its norm is less than $|\gamma|$ and this is possible only if $u = z\gamma$.

From what we have now proved it follows that Theorem 3 holds for the Gaussian integers with the difference that the equality $\mathbf{Z}^*\alpha + \mathbf{Z}^*\beta = \mathbf{Z}^*\gamma$ only determines γ up to multiplication by a unit ± 1 or $\pm i$. The theorem about unique factorization into primes is also true with the same reservation.

A theorem by Euler

Combining Fermat's theorem with what we know about the Gaussian integers we shall get a windfall in the form of a theorem stated by Fermat and proved by Euler in 1749. Note first that if an odd number m is the sum $a^2 + b^2$ of the squares of two rational integers, one must be even and the other one odd and hence m is of the form $4n + 1$ with integral n.

Theorem 5. *Every rational prime of the form $4n + 1$ is the sum of the squares of two rational integers.*

PROOF. In fact, let p be such a prime. According to (11) the number -1 is then a square mod p, i.e., p divides at least one number $x^2 + 1 = (x + i) \cdot (x - i)$ where x is a rational integer. If p were a Gaussian prime p would divide $x + i$ or $x - i$ which is impossible since the numbers $(x \pm i)/p$ are not Gaussian integers. Hence p is not a Gaussian prime so it must be the product of two Gaussian integers which are not units. If one factor is $a + bi$, the other one must be $a - bi$ and hence $p = a^2 + b^2$ is the sum of the squares of two rational integers. Finally, let us note that the factors $\alpha = a \pm bi$ are Gaussian primes. For if $\alpha = \beta\gamma$, then $p = |\alpha|^2 = |\beta|^2|\gamma|^2$ and hence β or γ is a unit. It is not difficult to see that the numbers α and all rational primes of the form $4n + 3$ when multiplied by units constitute all Gaussian primes.

Algebraic number theory

A real or complex number a is said to be *algebraic of degree n* if it is a root of an equation

$$x^n + a_{n-1}x^{n-1} + \cdots + a_0 = 0 \tag{13}$$

with rational coefficients $a_{n-1}, \ldots, a_0$, and *algebraic* if it is a root of such an equation for some n. When the coefficients are integers we say that a is an *algebraic integer*. When $n = 1$ we get back the ordinary rationals and integers but if $n > 1$ we have to deal with numbers like $\sqrt{2}$ (from $x^2 - 2 = 0$) or nth roots of unity (from $x^n - 1 = 0$). Let us remark in passing that if a rational number $x = b/c$ is an algebraic integer, satisfying (13) with integral coefficients, then x is an integer dividing a_0. In fact, multiplying (13) by c^n shows that c divides b^n and hence that $c = \pm 1$ when b and c are relatively prime.

When x is a root of (13), using the equation repeatedly we can express not only x^n but also the higher powers $x^{n+1}, x^{n+2}, \ldots$ in terms of $x_1 = 1, \ldots, x_n = x^{n-1}$. More precisely, using the language of modules, we have

$$x^m \in \mathbf{Q}x_1 + \cdots + \mathbf{Q}x_n \tag{14}$$

for all integers $m \geqslant 0$. It is an important fact that, whatever the numbers $x_1, \ldots, x_n$ are, a number x with these properties is algebraic. To see this we just have to use the simplest facts about systems of linear equations with rational coefficients (see 4.3). For according to (14), the numbers $1, x, \ldots, x^n$ are linear combinations of $x_1, \ldots, x_n$ with rational coefficients and hence, if $y_0, \ldots, y_n$ are any numbers, then $y_0 + y_1 x + \cdots + y_n x^n$ is a sum $f_1(y)x_1 + \cdots + f_n(y)x_n$ where $f_1, \ldots, f_n$ are linear combinations of $y_0, \ldots, y_n$ with rational coefficients. Putting $f_1(y) = 0, \ldots, f_n(y) = 0$ gives n linear equations for $n + 1$ unknowns $y_0, \ldots, y_n$ which, by the theorem of underdetermined systems, has a solution $y_0 = a_0, \ldots, y_n = a_n$ with rational $a_0, \ldots, a_n$ not all equal to zero. Hence x is algebraic.

Suppose now that all powers of x and y are rational linear combinations of, respectively, $x_1, \ldots, x_n$ and $y_1, \ldots, y_m$. Then, if a and b are rational, all powers of $ax + by$ and xy are rational linear combinations of the nm products $x_1 y_1, x_1 y_2, x_2 y_1, \ldots, x_n y_m$. Hence rational multiples, sums, and products of algebraic numbers are algebraic. Multiplying (13) by x^{-n} we see that if $x \neq 0$ is algebraic, so is $1/x$.

In the same way we can show that integral multiples, sums, and products of algebraic integers are algebraic integers. We then replace $\mathbf{Q}x_1 + \cdots + \mathbf{Q}x_n$ by $\mathbf{Z}x_1 + \cdots + \mathbf{Z}x_n$ and restrict $x_1, \ldots, x_n$ to such numbers that one of them is 1 and that all products $x_j x_k$ are linear combinations of $x_1, \ldots, x_n$ with integral coefficients, properties which hold when $x_1 = 1, \ldots, x_n = x^{n-1}$ and x satisfies (13) with integral coefficients. If x is a linear combination of $x_1, \ldots, x_n$ with integral coefficients, the same is true of all products $xx_1 = a_{11}x_1 + \cdots + a_{1n}x_n$, $xx_2 = a_{21}x_1 + \cdots + a_{2n}x_n$, and so on, and this we can write as a system of linear homogeneous equations with integral coefficients and the coefficient x,

$$(a_{11} - x)x_1 + \cdots + a_{1n}x_n = 0$$
$$\cdots$$
$$a_{n1}x_1 + \cdots + (a_{nn} - x)x_n = 0.$$

Hence, by the properties of determinants, $D(x)x_k = 0$ for all k, in particular $D(x) = 0$ where

$$D(x) = \det(A - xE)$$
$$= (-x)^n + (a_{11} + a_{22} + \cdots + a_{nn})(-x)^{n-1} + \cdots + \det A$$

with $A = (a_{jk})$, E the unit $n \times n$ matrix, is the determinant of the matrix $A - xE$ of the system (see the end of 4.3). Since $D(x)$ has integral coefficients, x is an algebraic integer. Reasoning as above we then conclude that integral multiples, sums, and products of algebraic integers are algebraic integers.

What we have proved so far is the point of departure of algebraic number theory which is an old, large, and much-cultivated part of mathematics. It is dominated by the fact that the theorem of unique factorization into primes holds only exceptionally. The theory deals with modules of the type $M = \mathbf{Z} + \mathbf{Z}a_1 + \cdots + \mathbf{Z}a_m$ where $a_1, \ldots, a_m$ are algebraic numbers and it is required that M is a ring, i.e., it shares with the Gaussian integers $\mathbf{Z} + \mathbf{Z}i$ the property that $u - v$ and uv belong to M when u and v belong to M. In most such rings there is no unique factorization into powers of different primes. Example: the numbers 2, 3, and $1 \pm i\sqrt{5}$ of the ring $\mathbf{Z} + \mathbf{Z}i\sqrt{5}$ turn out to be primes but $2 \cdot 3 = (1 + i\sqrt{5})(1 - i\sqrt{5})$. But the theory of divisibility can be saved if we replace numbers in M by ideals J in M, i.e., subsets of M such that $u - v$ and xu belong to J when u and v belong to J and x is any number in M. Ideals can be multiplied in a natural way and one can show that every ideal is an essentially unique product of prime ideals similar to those of the prime numbers. There are also generalizations of, e.g., Fermat's theorem and the quadratic reciprocity theorem.

2.4 Some problems and results

What has been said in this chapter so far is just a small sample of number theory. Its practical applications are not overwhelming but it has always attracted the best mathematicians like Gauss and Hilbert ($\sim$ 1900). The reason is, perhaps, that the field is full of unsolved problems which are important if put in the proper context and often seem to require new methods for their solution. Number theory has one advantage over most other parts of mathematics: many of the results (but as a rule not the proofs) can be understood with just a little knowledge of mathematics. Here are some examples.

Pythagorean numbers and Fermat's great theorem

Pythagorean numbers are natural numbers x, y, z such that $x^2 + y^2 = z^2$. Examples: 3, 4, 5 and 5, 12, 13. Apart from the order between x and y one gets all Pythagorean numbers by letting the numbers p, q, r in the formulas

$$z = (p^2 + q^2)r, \qquad x = (p^2 - q^2)r, \qquad y = 2pqr$$

run through all natural numbers, keeping p larger than q. This is proved in the *Elements* but the fact was probably known some centuries earlier to the Babylonians and Assyrians. The step from Pythagorean numbers to the search for natural numbers x, y, z such that $x^n + y^n = z^n$ where $n > 2$ does not require much imagination. Fermat stated without proof that such numbers do not exist, and this statement is called Fermat's great theorem. It has been proved for many values of n, for instance $n = 3$, but not for all. The problem still attracts amateurs. The first efforts to find a proof led, around 1800, to algebraic number theory, a field much more important than the problem that started it.

Sums of squares

Lagrange proved in 1770 that every natural number is the sum of at most 4 squares.

Additive number theory

If we try to write natural numbers as sums of primes we shall soon discover that we do not need many primes in each sum. The boldest conjecture was made by Goldbach around 1750: every even number > 2 is the sum of two primes. So far this has not been proved but it is known that, for instance, every sufficiently large integer is the sum of at most 20 primes.

Approximation of real numbers by rationals

It had been observed for a long time and it was shown by Dirichlet around 1840 that the inequality $|a - (p/q)| < 1/q^2$ has infinitely many integral solutions p, q when a is a real irrational number. He also proved that if a is algebraic of degree n, then the inequality

$$\left| a - \frac{p}{q} \right| < \frac{1}{q^m} \tag{13}$$

has only a finite number of integral solutions p, q when $m > n$. Since it is easy to construct irrational numbers a for which (13) has an infinite number of integral solutions p, q for every m, it follows among other things that there are irrational numbers that are not algebraic. These are the so-called *transcendental numbers*. Hermite proved in 1873 that e, the basis of the natural logarithms, is transcendental and Lindemann in 1882 that the number π is transcendental. This was the final (negative) answer to the problem of squaring the circle, put more than 2000 years before.

According to (13) an algebraic number cannot be approximated too well by rational numbers. It has been shown later that the condition $m > n$ can be sharpened to $m > (2n)^{1/2}$. The final result came in 1955 when Roth proved that the correct bound is 2, independently of n. In other words, if a is an algebraic number and $m > 2$, then the inequality (13) has only a finite number of integral solutions p, q. We might say that all such numbers are equally difficult to approximate by rationals.

2.5 Documents

Euclid on the infinity of primes

Here follows Euclid's proof that there is an infinite number of primes, translated by Heath. It is taken as a matter of course that it suffices to work with three primes, the reasoning in the general case being the same. The word "measures" means "divides" and was thought of geometrically, as the figure indicates.

"Prime numbers are more than any assigned multitude of primes.

Let A, B, C be the assigned prime numbers; I say that there are more prime numbers than A, B, C. For let the least number measured by A, B, C be taken and let it be DE; let the unit DF be added to DE. Then EF is either prime or not. First, let it be prime; then the prime numbers A, B, C, EF have been found which are more than A, B, C. Next, let EF not be prime; therefore it is measured by some prime number. Let it be measured by the prime number G. I say that G is not the same with any of the numbers A, B, C. For, if possible, let it be so. Now A, B, C measure DE; therefore G also will measure DE. But it also measures EF. Therefore G, being a number, will measure the remainder, the unit DF; which is absurd. Therefore G is not the same with any one of the numbers A, B, C. And by hypothesis it is a prime. Therefore the prime numbers A, B, C, G have been found which are more than the assigned multitude of A, B, C."

Dirichlet on Fermat's theorem

What we have said in connection with Fermat's theorem is all taken from a small article by Dirichlet from 1828, "New proofs of some results in number theory." Here, in an English verbatim translation, is his leisurely proof of Fermat's theorem.

"With a and p as before [p a prime, a not divisible by p] let us consider the following $p-1$ multiples of a,

$$a, 2a, 3a, \ldots, (p-1)a.$$

It is easy to see that two of them cannot give the same remainder when divided by p; for if the remainders coming from ma and na were equal, $ma - na = (m-n)a$ would be divisible by p which is impossible since a is not divisible by p and $m-n$ is $<p$ and not zero. The remainders which we obtain by dividing the $p-1$ multiples by p being all different and not zero, it is easy to see that these remainders must coincide with the numbers of the series $1, 2, 3, \ldots, p-1$ when one disregards the order between them. It follows from this that the product of the $p-1$ multiples of a should give the same remainder as the product $1 \cdot 2 \cdots (p-1)$.

The difference of these products is then a multiple of p. But this difference is easily put in the form

Pierre de Fermat   1608–1665

$$(a^{p-1} - 1) 1 \cdot 2 \cdot 3 \cdots (p-1)$$

and, $1 \cdot 2 \cdot 3 \cdots (p-1)$ not being divisible by p, we conclude that $a^{p-1} - 1$ is a multiple of p or, which is the same thing, a^{p-1} divided by p gives the remainder 1."

Literature

There are plenty of books on number theory. *An Introduction to Number Theory* by Hardy and Wright (Oxford, 1959) is nice to read and has a wealth of material including some algebraic number theory. Alan Baker's *Transcendental Number Theory* (Cambridge, 1975) is an up-to-date account of the theory of transcendental numbers. The chapter on number theory of *What is Mathematics?* by Courant and Robbins (Oxford, 1947) has an appendix with a nonstrict but enlightening treatment of the prime number theorem.

3
ALGEBRA

3.1 *The theory of equations*. Equations of degree less than five. Polynomials. Degree and divisibility. Prime polynomials. Polynomials in several variables. 3.2 *Rings, fields, modules, and ideals*. Laws of composition. Ring, field, module. Noncommutative rings, quaternions. Finite fields. Integral domains and quotient fields. Algebraic elements. Ideals. Noetherian rings, primary decomposition, maximal ideals. Algebraic manifolds, Hilbert's Nullstellensatz. 3.3 *Groups*. Bijections. Symmetries. Permutations. Groups of bijections. Affine groups. Abstract groups. Homomorphisms and isomorphisms. Group actions. Orbits. Invariant subgroups. Finitely generated abelian groups. Finite groups. Crystal groups and patterns. Literature. Solvable groups. Galois theory. Constructions by ruler and compass. Group representations. History. 3.4 *Documents*. Ars Magna. Galois on permutation groups. Cayley on matrices.

The word *algebra* is part of the title of an Arabic manuscript from about 800 A.D. giving rules for solving equations, and until about 100 years ago algebra was just the theory of equations. Nowadays it is best described as dealing with more or less formal mathematical operations and relations. Modern algebra is really a collection of second generation abstract models drawn from many parts of mathematics. Economy of notation prescribes that new symbols should be avoided unless absolutely necessary. For this reason familiar operational signs, e.g., those for addition and multiplication, are used again and again but acquire new meanings depending on the model. The objects of algebra are classified by the kind of operations that can be performed in them. They carry names like "ring" and "ideal," which may sound funny at first; but this feeling wears off with a closer acquaintance.

The content of a first course in algebra has shifted with time. It used to be only the theory of equations but has now taken an abstract turn toward the so-called algebraic structures like ring and group. We shall give a little of everything. In the first part we solve the equations of degree less than five and then proceed to the Fundamental theorem of algebra and the divisibility of polynomials. The second part, about rings, fields, modules, and ideals has the flavor of a typical algebra text crammed with definitions and simple illustrative examples. It is there just to give the reader an idea of what abstract algebra is about and to provide enough background for a presentation of the Galois fields and Hilbert's *Nullstellensatz* (theorem of zeros), extending the Fundamental theorem of algebra to more than one variable. The third part, about groups, should be easier to read and has a wide horizon. At the end I try to explain what Galois theory is about and how it is used to settle the question of the solvability of equations by radicals, the doubling of the cube, and the trisection of the angle.

3.1 The theory of equations

Equations of degree less than five

Let us write equations of the first, second, and third degree as

$$x + a = 0 \tag{1}$$

$$x^2 + ax + b = 0 \tag{2}$$

$$x^3 + ax^2 + bx + c = 0. \tag{3}$$

In all cases the coefficients a, b, c are given rational, real, or even complex numbers and the problem is to find the roots, i.e., all numbers x satisfying the equation. There is not much to be said about (1). It has a unique root $x = -a$. The equation (2) has precisely two roots

$$x = -\frac{a}{2} \pm \sqrt{\frac{a^2}{4} - b} \,, \tag{2*}$$

different unless the discriminant $a^2 - 4b$ vanishes. This is proved by rewriting (2) as

$$\left(x + \frac{a}{2}\right)^2 = \frac{a^2}{4} - b. \tag{2'}$$

When a, b are real and the discriminant is negative, the roots are complex. The solution (2*) of (2) was known in principle in the Babylonian empire (2000 B.C.). In the equations treated then, a and b were rational numbers and the discriminant a square so that the roots were also rational. Negative numbers or 0 were not known and in the texts of the time (tablets of burnt clay) both (2) and (2*) are described in words and with numerical values of a and b. Negative and complex numbers entered into the number model much later in connection with the efforts in Italy, around 1500, to solve the equation (3) of degree three and also the equation of degree four,

$$x^4 + ax^3 + bx^2 + cx + d = 0. \tag{4}$$

In both cases a preliminary simplification was introduced profiting from the fact that

$$\left(x + \frac{a}{3}\right)^3 = x^3 + ax^2 + \ldots, \qquad \left(x + \frac{a}{4}\right)^4 = x^4 + ax^3 + \ldots,$$

where the dots denote lower powers of x. Replacing x in (3) and (4) by $x - a/3$ and $x - a/4$, respectively, one gets the simpler but equivalent equations

$$x^3 + px + q = 0 \tag{3'}$$

$$x^4 + px^2 + qx + r = 0 \tag{4'}$$

where p, q, r are certain new coefficients. Both equations were solved by very ingenious methods. To solve (3') put $x = u + v$. Since

$$(u + v)^3 = u^3 + 3u^2v + 3uv^2 + v^3$$

we can then rewrite the equation as

$$u^3 + v^3 + (3uv + p)(u + v) + q = 0.$$

After that we try to determine u and v in such a way that the product of the two parentheses vanishes. One way to do this is to require the two equations $3uv + p = 0$ and $u^3 + v^3 + q = 0$. Then $u^3 + v^3 = -q$, $u^3v^3 = -p^3/27$ and this means that both u^3 and v^3 are roots of the equation of degree two,

$$w^2 + qw - \frac{p^3}{27} = 0.$$

Hence, if we are lucky, the formula

$$x = \sqrt[3]{-\frac{q}{2} + \sqrt{\frac{q^2}{4} + \frac{p^3}{27}}} + \sqrt[3]{-\frac{q}{2} - \sqrt{\frac{q^2}{4} + \frac{p^3}{27}}} \tag{3*}$$

should give all solutions of (3'). An insertion into (3') shows this to be true provided the arguments (as complex numbers) of the square roots and cube roots on the right side of (3*) are chosen suitably, but we do not go into the details of this. The equation (4') of degree four can also be solved in a similar way by extraction of square roots and cube roots. The expression that corresponds to (3*) is then so complicated that it is usually not written out in full. One method of solution is to try to write the left side of (4') as a product $(x^2 + 2kx + l)(x^2 - 2kx + m)$ of two quadratic factors. After some computation this leads to an equation of degree three for k^2. The coefficients l and m are then easy to get. The formula (3*) is called Cardano's formula after the mathematician who published it in 1545 in a book on algebra with the proud title of *Ars Magna* (*The Great Art*). In this work the equation of degree four is also solved.

The solution of the equations of degree three and four was a decisive step forward from Greek mathematics. It led to repeated attacks for 250 years on equations of degree $n > 4$,

$$x^n + a_{n-1}x^{n-1} + \cdots + a_0 = 0 \tag{5}$$

where the coefficients $a_{n-1}, \ldots, a_0$ are given numbers. To solve such an equation meant of course to find an expression for x involving rational combinations of the coefficients and repeated square and higher roots, in other words, a reduction to ordinary arithmetic combined with solutions of the equations $x^m = b$. At the end of the eighteenth century suspicions were aroused that this is impossible, and Abel proved in 1824 that equations of higher degree than four cannot be solved by root extractions except for special values of the coefficients. Abel's proof is now obsolete, but viewed in its context it was a fabulous achievement. In modern algebra the impossibility is a consequence of the so-called Galois theory, a standard item in many textbooks. Galois, who was a contemporary of Abel, discovered that the heart of the problem lies in group theory, but it took 50 years before this was generally realized. We shall return to Galois theory at the end of this chapter when the proper concepts, group and algebraic field, have been presented.

Even if (5) cannot be solved by extraction of roots, it is conceivable that there are solutions. Gauss proved in 1799 the following

Theorem 1 (The Fundamental theorem of algebra). *Every equation* (5) *with complex coefficients and $n > 0$ has at least one root, i.e., there is at least one complex number x satisfying the equality.*

This theorem got its name at a time when algebra was identical with the theory of equations and the name is no longer justified. But the theorem is fundamental for many fields of science and technology where complex numbers are used and it is of very great practical importance to compute good approximations of roots of algebraic equations.

From the fundamental theorem of algebra it follows by a simple argument given in the next section that the equation has not only one but n roots $c_1, \ldots, c_n$, some perhaps coinciding, such that the equality

$$x^n + a_{n-1}x^{n-1} + \cdots + a_0 = (x - c_1) \ldots (x - c_n) \qquad (6)$$

holds for all complex values of x.

Polynomials

Polynomials are expressions of the form

$$P = a_0 + a_1x + \cdots + a_nx^n. \qquad (7)$$

They appear, for instance, in the theory of equations. Here $a_0, \ldots, a_n$ are numbers called the *coefficients* of the polynomial and x is an *indeterminate symbol* or, if we want, *a variable* that can take all complex values. The polynomials $1, x, x^2, x^3, \ldots$ are called *powers* of x. Polynomials are added and multiplied as if x were a number. If P and

$$Q = b_0 + b_1x + \cdots + b_mx^m \qquad (8)$$

are two polynomials, one gets their sum

$$P + Q = a_0 + b_0 + (a_1 + b_1)x + \cdots \qquad (9)$$

by adding corresponding coefficients and their product

$$PQ = a_0b_0 + (a_1b_0 + a_0b_1)x + \cdots + a_nb_mx^{n+m} \qquad (10)$$

by multiplying the right sides of (7) and (8) and assembling the result as a new polynomial. Example: when $P = 1 + x + x^2$ and $Q = 2 + 3x$ then $P + Q = 3 + 4x + x^2$ and $PQ = 2 + 5x + 5x^2 + 3x^3$. The sets of all polynomials with rational, real, or complex coefficients will be denoted by, respectively, $\mathbf{Q}[x]$, $\mathbf{R}[x]$, and $\mathbf{C}[x]$. For simplicity we shall also use the notation $A[x]$ where $A = \mathbf{Q}, \mathbf{R}$, or $\mathbf{C}$. According to (9) and (10) every $A[x]$ is a *ring* in the sense that $P \pm Q$ and PQ are in $A[x]$ when P and Q are. When all the coefficients of Q vanish we write $Q = 0$ and call it *the null polynomial*. It has the property that $P + 0 = P$, $P0 = 0$ for every polynomial P.

When P is a polynomial (7), let $P(x)$ denote its value at x, i.e., the complex number one gets from the right side when x is a complex number. Note that the sum $P + Q$ and the product PQ have been defined so that $(P + Q)(x) = P(x) + Q(x)$ and $(PQ)(x) = P(x)Q(x)$ for all complex numbers x.

Degree and divisibility

When $a_n \neq 0$ in (7) we say that P *has degree* n and write $n = \deg P$. Hence the numbers $a \neq 0$ are the polynomials of degree 0, while the polynomials $ax + b$ with $a \neq 0$ are those of degree 1, and so on. It follows from (10) that

$$\deg PQ = \deg P + \deg Q \qquad (11)$$

when $P \neq 0$ and $Q \neq 0$. It is convenient to attach to the null polynomial a symbolic degree, $\deg 0 = -\infty$, with the property that $\deg 0 = \deg 0 + \deg P$ for all P. Then (11) holds without restriction. The polynomials in a given ring $A[x]$ have the following important property: if P and Q are polynomials and $Q \neq 0$, then there are polynomials S and R such that

$$P = SQ + R \quad \text{where} \quad \deg R < \deg Q. \qquad (12)$$

Here $\deg 0 = -\infty$ is of course < 0. In the proof one has to distinguish between several cases. If $\deg Q = 0$, Q is just a number $b \neq 0$ and we can take $R = 0$ and $S = b^{-1}P$ which is indeed a polynomial. Next let $\deg Q > 0$. If $\deg P < \deg Q$ take $S = 0$ and $R = P$. This leaves us with the only interesting case, namely when $\deg P \geqslant \deg Q > 0$. Let P and Q be given by (7) and (8) with $\deg P = n$ and $\deg Q = m$. The polynomial

$$x^{n-m}Q = b_0 x^{n-m} + b_1 x^{n-m+1} + \cdots + b_m x^n$$

then has degree n, and if we multiply it by $a_n b_m^{-1}$ the power x^n gets the same coefficient a_n as x^n in P. Hence, putting $S_1 = a_n b_m^{-1} x^{n-m}$, we get $P = S_1 Q + R_1$ where the degree of $R_1 = P - S_1 Q$ is less than the degree of P. If also $\deg R_1 < \deg Q$ we are done. If not, we can in the same manner write R_1 as $S_2 Q + R_2$ and hence P as $(S_1 + S_2)Q + R_2$ where $\deg R_2 < \deg R_1$. If R_2 still will not do, i.e., if $\deg R_2 \geqslant \deg Q$ we can rewrite R_2. Since the degrees diminish in every step we will finally end up with (12).

The proof is illustrated by the following computation where $Q = x - 1$ and $P = x^3 + x^2 - 3$,

$$P = x^2(x - 1) + 2x^2 - 3 = x^2(x - 1) + 2x(x - 1) + 2x - 3$$

$$= (x^2 + 2x)(x - 1) + 2(x - 1) - 1 = (x^2 + 2x + 2)(x - 1) - 1.$$

We shall now use (11) and (12) to prove (6). When deg $P = n > 0$ and $Q = x - c_1$ has degree one, there are, according to (12), a polynomial S and a number r such that $P = SQ + r$ and hence also $P(x) = S(x)(x - c_1) + r$ for all complex numbers x. If $P(c_1) = 0$ this gives $r = 0$ so that $P = SQ$ and, by virtue of (11), deg $S = $ deg $P - 1$. If deg $S > 0$ the equation $S(x) = 0$ has a root $x = c_2$ which gives a new factor $x - c_2$ and so on. After $n = $ deg P steps we get the desired result (6) also as an equality between polynomials. It follows from (6) that every root x of the equation $P(x) = 0$ is one of the numbers $c_1, \ldots, c_n$. They are called *zeros* of the polynomial P. It is clear that only the null polynomial $P = 0$ has more than deg P zeros. From this it follows, for instance, that two polynomials P and Q, of degree n at most, which take the same values at $n + 1$ different points are equal, i.e., have the same coefficients. In fact, $P - Q$ must be the null polynomial. Since (6) is an equality between polynomials, the polynomial P and the polynomial

$$(x - c_1) \ldots (x - c_n)$$

$$= x^n - (c_1 + \cdots + c_n)x^{n-1} + \cdots + (-1)^n c_1 \ldots c_n$$

have the same coefficients. This gives us the well-known connections between the coefficients and zeros of a polynomial, e.g., $c_1 + \cdots + c_n = -a_{n-1}$ and $c_1 \ldots c_n = (-1)^n a_0$.

Prime polynomials

From (11) and (12) follow some important facts about divisibility in a ring $A[x]$. When $PQ = R$ we say that P and Q are factors or divisors of R and also that P and Q divide R. According to (11), the polynomials dividing $R = 1$ are just the numbers $\neq 0$ in A. In this connection, we shall call them *units*. A polynomial of positive degree divisible only by units and unit multiples of itself is said to be *prime*. Using (11), and reasoning as in the proof of Theorem 1, Chapter 2, one sees that every polynomial of positive degree is a product of prime polynomials. A subset J of the ring $A[x]$ is said to be an *ideal* if $UP + VQ$ belongs to J when P and Q do and U and V are any polynomials in $A[x]$. According to (12) there is to every ideal $J \neq 0$ a polynomial $Q \neq 0$ such that $J = A[x]Q$, i.e., J consists of all PQ with P arbitrary in $A[x]$. In fact, choose a $Q \neq 0$ in J of least possible degree and consider the ideal $I = A[x]Q$ which is a part of J. To every P in J there is, according to (12), a polynomial S such that the degree of $R = P - SQ$ is less than the degree of Q. But R is in J and this is possible only when $R = 0$. Hence I is all of J. Combining what we have now proved with the line of reasoning used to prove Theorems 3 and 4 of Chapter 2, we get an important theorem about decomposition of polynomials into prime factors. Note that (12) has rendered us the same service as the figures of Chapter 2.

Theorem 2. *Every polynomial with complex (real, rational) coefficients and of positive degree is a product of prime polynomials with the same kind of coefficients. The prime polynomials are unique apart from the order and numerical factors.*

Whether a given polynomial is prime or not depends strongly on A, i.e., whether A is $\mathbf{Q}$, $\mathbf{R}$, or $\mathbf{C}$. Under all circumstances it follows from (11) that every polynomial of degree one, $ax + b$ with $a \neq 0$, is a prime polynomial. When $A = \mathbf{C}$, (6) shows that all prime polynomials are of this kind. When $A = \mathbf{R}$ it is easy to show that the only additional prime polynomials are those of degree two with complex zeros, $ax^2 + 2bx + c$ with $ac - b^2 > 0$. When $A = \mathbf{Q}$ there are prime polynomials of every degree $n > 0$, e.g., $x^n - 2$.

All these facts have been known since the time of Gauss and now belong to a general education in mathematics, but Theorem 2 is not the ultimate result. Looking through the proof we shall find that it works provided A is a field, a notion that will be explained in the next section. It means, roughly speaking, that the usual arithmetical laws hold in A. Since there are many more fields than $\mathbf{Q}$, $\mathbf{R}$, and $\mathbf{C}$, this remark gives Theorem 2 the status of a general algebraic result. In retrospect, the proof was not difficult. On the other hand, the theorem applies in many seemingly different situations.

Since there are fields A with a finite number of elements (see the next section), it is not without interest to remark that for any field A, the ring $A[x]$ has an infinite number of prime polynomials. In particular, given any polynomial P in $A[x]$ there is another polynomial Q relatively prime to it. To see this, we just have to repeat Euclid's proof: if $P_1, \ldots, P_n$ are prime polynomials, then $P = a + P_1 \ldots P_n$ where $a \neq 0$ is in A, is not divisible by any of them. For if, e.g., P is divisible by P_1, then P_1 would be a factor of a, contradicting that $\deg P_1 > 0$.

Polynomials in several variables

Having dealt with the set $A[x]$ of polynomials in one variable with coefficients in $A = \mathbf{Q}$, $\mathbf{R}$, or $\mathbf{C}$ we might as well introduce the set $A[x, y]$ of polynomials in two variables x and y with coefficients in A, i.e., finite sums

$$P = a + bx + cy + dx^2 + exy + fy^2 + gx^3 + hx^2y + \ldots \qquad (13)$$

with coefficients $a, b, \ldots$ in A. Such a polynomial is considered to be zero only when all its coefficients vanish; two polynomials are considered equal precisely when all corresponding coefficients are equal; and they are added

29

by adding corresponding coefficients, and multiplied by multiplication term by term and writing the result in the form (13), using the fact that $xy = yx$. Example: if $P = 2 + 3x + y$ and $Q = 3 + y^3$, then $P + Q = 5 + 3x + y + y^3$ and $PQ = 6 + 9x + 5y + 3xy^3 + y^4$. These rules extend in a natural way to the set $A[x_1, \ldots, x_n]$ of all polynomials in n variables $x_1, \ldots, x_n$ with coefficients in A. Its elements are finite sums

$$P = \sum a_{k_1 \ldots k_n} x_1^{k_1} \ldots x_n^{k_n}$$

where $k_1, \ldots, k_n$ are integers ≥ 0 and the coefficients $a_{k_1 \ldots k_n}$ are elements of A. Polynomials in several variables will be important in the next section.

3.2 Rings, fields, modules, and ideals

Laws of composition

Trying to formulate in a precise way the rules for computation that hold in the number models $\mathbf{Q}$, $\mathbf{R}$, and $\mathbf{C}$ and transplanting them to an unspecified set A with elements $a, b, c, \ldots$ we get the following list of laws of composition with specific properties.

1. *Addition.* To every pair of elements a, b of A there is a unique third element of A called the *sum* of a and b and denoted by $a + b$ such that

$$a + b = b + a \qquad \text{(the commutative law)} \qquad \text{(i)}$$

$$(a + b) + c = a + (b + c) \qquad \text{(the associative law)} \qquad \text{(ii)}$$

for all a, b, c in A.

2. *Subtraction.* There is precisely one element of A called *zero* or *null* and denoted by 0 such that

$$a + 0 = 0 + a \qquad \text{(iii)}$$

for all a. To every element a of A there is precisely one element b called *opposite* to a and denoted by $-a$ such that

$$a + (-a) = (-a) + a = 0. \qquad \text{(iv)}$$

(Note: The sum $a + (-b)$ is also written as $a - b$.)

3. *Multiplication.* To every pair of elements a, b of A there is a unique third element of A called the *product* of a and b and denoted by ab such that

$$(ab)c = a(bc) \qquad \text{(the associative law)} \qquad \text{(v)}$$

for all a, b, c in A.

4. *Division.* There is in A precisely one element called the *unit*, or simply *one*, and denoted by 1 such that

$$a1 = 1a \qquad \text{(vi)}$$

for all a. To every element $a \neq 0$ of A there is precisely one element called the *inverse* of a and denoted by a^{-1} such that

$$a^{-1}a = aa^{-1} = 1. \qquad \text{(vii)}$$

5. *Distributivity*. Addition and multiplication have the property that

$$a(b + c) = ab + ac \qquad (b + c)a = ba + ca \qquad \text{(the distributive law)}$$

(viii)

holds for all a, b, c in A.

(Note: The formula (iii) is said to express that null is the neutral element under addition, the formula (vi) that one is the neutral element under multiplication.)

Ring, field, module

Abstract algebra was born in the beginning of the nineteenth century when mathematicians began to study sets equipped with all or some of these laws of composition. A set with the properties listed under the headings of addition, subtraction, multiplication, and distributivity is called a *ring*. If A also has the properties listed under division, A is said to be a *division ring*. When

$$ab = ba \qquad \text{(the commutative law for multiplication)} \qquad \text{(ix)}$$

holds for all a and b in A, the ring is said to be *commutative*, and commutative division rings are called *fields*. Rings may or may not have a unit with the property (vi). We already have many examples of commutative rings and of fields. The integers $\mathbf{Z}$ form a ring with a unit, and the module $\mathbf{Z}a$ where $a \neq 1$ is an integer is a ring without unit. The rational numbers $\mathbf{Q}$, real numbers $\mathbf{R}$, and complex numbers $\mathbf{C}$ are fields. The corresponding sets of polynomials, $\mathbf{Q}[x]$, $\mathbf{R}[x]$, $\mathbf{C}[x]$ are rings with units, in all cases the number 1. Also $\mathbf{Z}[x]$, the set of polynomials with integral coefficients, is a ring with a unit. In none of these cases do we have a field since $PQ = 1$ implies deg $P = 0$, i.e., only polynomials of degree zero can have an inverse. When A is a ring, all polynomials (7) with coefficients $a_0, \dots, a_n$ in A constitute a new ring, the polynomial ring $A[x]$, in which addition, subtraction, and multiplication are defined by (9) and (10), x is supposed to commute with the elements of A, and two polynomials are considered to be equal if and only if all corresponding coefficients are equal. When A is commutative or has a unit, $A[x]$ has the same property. The same statements hold, e.g., for the ring $A[x, y]$ of polynomials in two variables x, y with coefficients in A where, of course, x and y commute with each other and with the elements of A. This ring can also be seen as the ring $B[y]$ of all polynomials in y with coefficients in $B = A[x]$ for, according to (13), every P in $A[x, y]$ is a sum $c_0 + c_1 y + c_2 y^2 + \dots$ where $c_0, c_1, \dots$ are polynomials in x with coefficients in A. The same remarks apply to the ring $A[x_1, \dots, x_n]$ of all polynomials in the variables $x_1, \dots, x_n$ with coefficients in A.

Let A with elements $a, b, \dots$ be a ring with a unit. A set M with elements $\xi, \eta, \dots$ having the properties listed under addition and subtraction is said to be an *A-module* when a product $a\xi$ of elements a in A and elements ξ in M is defined such that $a\xi$ belongs to M and, in addition,

$$(a + b)\xi = a\xi + b\xi, \quad a(\xi + \eta) = a\xi + b\eta, \quad a(b\xi) = (ab)\xi, \quad 1\xi = \xi \quad \text{(x)}$$

for all a, b in A and ξ, η in M. When $A = \mathbf{Z}$ consists of the integers, the products are $0\xi = 0, \pm \xi, \pm 2\xi, \ldots$ and the module property (x) follows since M has addition and subtraction. It is clear that every ring A is an A-module and also every polynomial ring $A[x_1, \ldots, x_n]$. Note that in the product $a\xi$ we have chosen to multiply the elements ξ of M from the left by the elements of A and, therefore, M is also called a *left A-module*. We can also have products ξa with the elements of A to the right and, instead of (x), $\xi(a + b) = \xi a + \xi b$, and so on. Then M is called a *right A-module*. A ring A and its polynomial rings $A[x_1, \ldots, x_n]$ are both right and left A-modules.

Noncommutative rings, quaternions

We shall now exhibit a noncommutative ring. Its elements $a, b, \ldots$ have four components ordered into quadratic schemes called *matrices*,

$$a = \begin{pmatrix} a_{11} & a_{12} \\ a_{21} & a_{22} \end{pmatrix}, \quad b = \begin{pmatrix} b_{11} & b_{12} \\ b_{21} & b_{22} \end{pmatrix}, \ldots .$$

Sums, differences, and products are defined as follows

$$a \pm b = \begin{pmatrix} a_{11} \pm b_{11}, & a_{12} \pm b_{12} \\ a_{21} \pm b_{21}, & a_{22} \pm b_{22} \end{pmatrix},$$

$$ab = \begin{pmatrix} a_{11}b_{11} + a_{12}b_{21}, & a_{11}b_{12} + a_{12}b_{22} \\ a_{21}b_{11} + a_{22}b_{21}, & a_{21}b_{12} + a_{22}b_{22} \end{pmatrix}.$$

Matrices will also appear in the chapter on geometry and linear algebra and there the rule of multiplication will be better explained. Anyway, our matrices $a, b, \ldots$ with elements in a ring A form a new ring that we shall call $M(A)$ where M stands for matrix. Null and one of $M(A)$ are the matrices

$$\begin{pmatrix} 00 \\ 00 \end{pmatrix} \quad \text{and} \quad \begin{pmatrix} 10 \\ 01 \end{pmatrix}$$

where 0 and 1 are the null and one of A. The multiplication rule gives, e.g.,

$$\begin{pmatrix} 01 \\ 00 \end{pmatrix}\begin{pmatrix} 00 \\ 10 \end{pmatrix} = \begin{pmatrix} 10 \\ 00 \end{pmatrix} \quad \text{and} \quad \begin{pmatrix} 00 \\ 10 \end{pmatrix}\begin{pmatrix} 01 \\ 00 \end{pmatrix} = \begin{pmatrix} 00 \\ 01 \end{pmatrix}.$$

Hence $M(A)$ is not commutative when $A \neq 0$, i.e., $1 \neq 0$ in A. We make the ring $M(A)$ a left A-module by putting

$$ca = \begin{pmatrix} ca_{11} & ca_{12} \\ ca_{21} & ca_{22} \end{pmatrix}$$

for all c in A and a in $M(A)$. If we want to have a division ring we can put $A = C$ and let K be the part of $M(A)$ consisting of all matrices $a = a_0 e_0 + a_1 e_1 + a_2 e_2 + a_3 e_3$ where a_0, a_1, a_2, a_3 are real numbers and

$$e_0 = \begin{pmatrix} 1 & 0 \\ 0 & 1 \end{pmatrix}, \quad e_1 = \begin{pmatrix} i & 0 \\ 0 & -i \end{pmatrix}, \quad e_2 = \begin{pmatrix} 0 & 1 \\ -1 & 0 \end{pmatrix}, \quad e_3 = \begin{pmatrix} 0 & i \\ i & 0 \end{pmatrix}.$$

These four matrices have the following multiplication table

$$e_0^2 = -e_1^2 = -e_2^2 = -e_3^2 = e_0$$
$$e_0 e_1 = e_1 e_0 = e_1, \qquad e_0 e_2 = e_2 e_0 = e_2, \qquad e_0 e_3 = e_3 e_0 = e_3$$
$$e_1 e_2 + e_2 e_1 = 0, \qquad e_2 e_3 + e_3 e_2 = 0, \qquad e_3 e_1 + e_1 e_3 = 0.$$

This shows that K is a ring with unit e_0 and that

$$(a_0 e_0 + a_1 e_1 + a_2 e_2 + a_3 e_3)(a_0 e_0 - a_1 e_1 - a_2 e_2 - a_3 e_3)$$
$$= (a_0^2 + a_1^2 + a_2^2 + a_3^2) e_0.$$

Hence a has the inverse

$$a^{-1} = (a_0^2 + a_1^2 + a_2^2 + a_3^2)^{-1}(a_0 e_0 - a_1 e_1 - a_2 e_2 - a_3 e_3)$$

when $a \neq 0$. The division ring K, whose elements are called *quaternions*, was discovered by Hamilton in 1843. He thought it would be as important as the complex numbers but in this he was mistaken.

Square matrices of type $n \times n$,

$$a = \begin{pmatrix} a_{11}, & \dots, & a_{1n} \\ & \dots & \\ a_{n1}, & \dots, & a_{nn} \end{pmatrix}$$

where the components $a_{11}, \dots$ belong to a ring A can be added and multiplied as in the case $n = 2$, and constitute a ring called *the matrix ring of order n over A*. When $n > 1$ and $A \neq 0$ this ring is not commutative.

Finite fields

In order to put our algebraic machinery to work we shall now look into the properties of fields with a finite number of elements. Let $F = (\xi, \eta, \dots)$ be such a field and suppose that it has q elements. Then we must have $\xi^q = \xi$ for every ξ in F. In fact, this holds when $\xi = 0$, and when $\xi \neq 0$ and $\xi_1, \dots, \xi_{q-1}$ are the nonzero elements of F, the products $\xi \xi_1, \dots, \xi \xi_{q-1}$ are different and nonzero and hence they must be equal to $\xi_1, \dots, \xi_{q-1}$ in some order. Hence $\xi^{q-1} \xi_1 \xi_2 \dots \xi_{q-1} = \xi_1 \xi_2 \dots \xi_{q-1}$ so that, dividing by the right side, $\xi^{q-1} = e$, the unit of F. Hence $\xi^q = \xi$ in all cases. (Note the similarity to one of our proofs of Fermat's theorem.)

Next, consider the integral multiples $0, e, 2e, \dots$ of e. There are at most q of them and since $me = ne \Leftrightarrow (m - n)e = 0$, there is a least integer $p > 1$ such that $pe = 0$. Then $0, e, 2e, \dots, (p-1)e$ are all different and p must be a prime for if $a, b > 0$ are integers and $ab = p$, then $aebe = abe^2 = pe = 0$. Since F is a field this implies that $ae = 0$ or $be = 0$ and hence that a

or b is equal to p. Note that $p\xi = pe\xi = 0$ for all ξ in F and that the elements $0, e, 2e, \ldots, (p-1)e$ constitute a subfield F_0 of F. In fact, $F_0 = \mathbf{Z}e$ is a ring and, by elementary number theory, if $0 < a < p$ there is a $0 < b < p$ such that $ab - 1$ is a multiple of p and hence $aebe = e$, i.e., ae has the inverse be. To make some use of F_0 we let n be the least integer > 0 with the property that there exist n elements $\eta_1, \ldots, \eta_n$ in F such that every ξ in F is a sum $a_1\eta_1 + \cdots + a_n\eta_n$ with $a_1, \ldots, a_n$ in F_0. Since there are at most p^n of these sums we have $q \leqslant p^n$ and we shall see that equality holds. In fact, if two of the sums were equal, $a_1\eta_1 + \cdots + a_n\eta_n = a_1'\eta_1 + \cdots + a_n'\eta_n$, but, e.g., $a_1 \neq a_1'$, then η_1 and hence any ξ in F is a sum $a_2''\eta_2 + \cdots + a_n''\eta_n$ with $a_2'', \ldots, a_n''$ in F_0, contradicting the definition of n.

To sum up, there is a prime p and an integer $n > 0$ such that F has $q = p^n$ elements. Further, $p\xi = 0$ and $\xi^q = \xi$ for every ξ in F. If $0, \xi_1, \ldots, \xi_{q-1}$ are the elements of F, arguing as in the proof of (6), this means that

$$x^q - x = x(x - \xi_1) \ldots (x - \xi_{q-1})$$

is a polynomial identity. Using this fact one can prove that for every prime p and integer $n > 0$ there exists a field F with the properties above, and that p and n determine all its algebraic properties. This field, usually denoted by $\mathrm{GF}(p^n)$, is called a *Galois field* after its discoverer (1830). A simple example is $\mathrm{GF}(2^2)$ which has four elements $0, e, \xi, \xi^{-1}$ where e is the unit. From the polynomial identity it follows that $e + \xi + \xi^{-1} = 0$. This equality and the condition that $2e = 0$ gives us all the sums and products of the four elements. Since Wedderburn proved in 1905 that every division ring with a finite number of elements is commutative, the Galois fields constitute all finite fields.

Integral domains and quotient fields

Everyone who passed elementary school knows that the rational numbers $\mathbf{Q}$ are obtained from the integers $\mathbf{Z}$ by taking quotients. The rational numbers are quotients a/b of integers a and b. Such a quotient is defined when $b \neq 0$, two quotients a/b and c/d are considered to be equal when $ad - bc = 0$, and addition and multiplication are performed according to the formulas

$$\frac{a}{b} + \frac{c}{d} = \frac{ad + bc}{bd}, \qquad \frac{a}{b}\frac{c}{d} = \frac{ac}{bd}.$$

It is of course to be expected that this will work and also result in a field when we replace the integers by a commutative ring A whose elements we now denote by $a, b, \ldots$. This is indeed the case, but under the necessary assumption that

$$a \neq 0 \quad \text{and} \quad b \neq 0 \quad \Rightarrow \quad ab \neq 0 \tag{14}$$

for all a and b in A. We leave out the simple proof. That (14) is necessary for the construction is due to the fact that if $a \neq 0$ and $b \neq 0$ but $ab = 0$, then the quotients a/b and b/a may perhaps be defined but not their product ab/ba.

The field K of quotients a/b, with $b \neq 0$, of elements of a commutative ring A which are identified, added, and multiplied as above is called the *quotient field* of A. A commutative ring $\neq 0$ with the property (14) is called an *integral domain*. The notions of divisibility all make good sense in such rings. For instance, a unit is defined as a divisor of 1, units and unit multiples of an element are its trivial divisors and the prime elements or, simply, primes are those having only trivial divisors. The property (14) is also exactly what is needed for the ring $A[x]$ of all polynomials in x with coefficients in A to have the property (11), and then $A[x]$ is also an integral domain. Let us also mention the fact proved by Gauss around 1800 that if A has a unique factorization into primes, then $A[x]$ also has this property. Repeated application shows that these two statements also apply to the polynomial rings $A[x_1, \ldots, x_n]$ in more than one variable.

Algebraic elements

Let B be an integral domain with a unit and $A \subset B$ a subring, i.e., a part of B containing the unit and such that $a - b$ and ab are in A whenever a and b are in A. We shall consider rings $A[\xi_1, \ldots, \xi_n]$ of polynomials in elements $\xi_1, \ldots, \xi_n$ of B with coefficients in A. They constitute another subring of B but it is important to note that depending on the choice of the ξ's it may happen that two such polynomials are equal although all corresponding coefficients are not the same. Let us say that an element ξ of B is *algebraic over* A if, for some $m > 0$, ξ satisfies an equation

$$\xi^m + a_{m-1}\xi^{m-1} + \cdots + a_0 = 0$$

with coefficients $a_{m-1}, \ldots, a_0$ in A. In case B is the complex numbers and A the integers, this is just the definition of an algebraic integer (see 2.3). Precisely as in this special case one proves that A-multiples, sums, and products of algebraic elements are algebraic. Provided that A has the unique factorization property we can also assert that the only algebraic elements in its quotient field are the elements of A itself. The proof is precisely as when A is the integers and applies, for instance, when A is the ring of polynomials in one variable over a field. It follows from the equation above that if ξ is algebraic over the quotient field of A, then some A-multiple $c\xi$ of ξ with $c \neq 0$ is algebraic over A. In fact, multiplying by c^m, we only have to choose $c \neq 0$ such that $ca_{m-1}, ca_{m-2}, \ldots, ca_0$ all lie in A.

Now let $\xi_1, \ldots, \xi_n$ be in B, and consider the ring $R = A[\xi_1, \ldots, \xi_n]$ of polynomials in $\xi_1, \ldots, \xi_n$ with coefficients in A. Since these polynomials are obtained from $\xi_1, \ldots, \xi_n$ by repeatedly taking A-multiples, sums, and products, all elements of this ring are algebraic over A if and only if its so-called *generators* $\xi_1, \ldots, \xi_n$ have this property. As an exercise in the handling of our algebraic machinery we shall now prove a result due to Zariski (1947) which will be decisive later on.

Lemma. *If A and $R = A[\xi_1, \ldots, \xi_n]$ are fields, all the elements of R are algebraic over A.*

PROOF. Let $C = A[\xi_1]$ be the ring of polynomials in ξ_1 with coefficients in A. When ξ_1 is algebraic over A so are all elements of C. When ξ_1 is not algebraic over A, a polynomial $c(\xi_1)$ in C vanishes only if all its coefficients vanish. This means that C is nothing but the ring of polynomials in one variable with coefficients in A. In particular, C is not a field. This proves the lemma when $n = 1$.

Next, we shall proceed by induction on the number $n > 1$ of generators. Since R is a field, any element of the quotient field K of $C = A[\xi_1]$ is in R and hence any polynomial in $\xi_2, \ldots, \xi_n$ with coefficients in K can be written as a polynomial in $\xi_1, \ldots, \xi_n$ with coefficients in A. In other words, $R = K[\xi_2, \ldots, \xi_n]$. Hence, by induction, the generators $\xi_2, \ldots, \xi_n$ are algebraic over K. But then there is an element $d(\xi_1) \neq 0$ of C such that $d(\xi_1)\xi_2, \ldots, d(\xi_1)\xi_n$ are algebraic over C. More generally, writing a given η in R as a polynomial in $\xi_2, \ldots, \xi_n$ with coefficients in C, it follows that $d(\xi_1)^m\eta$ is algebraic over C for some integer $m > 0$. Taking η in K, we shall see that this leads to a contradiction when ξ_1 is not algebraic over A. Then, as explained above, C is the ring of polynomials in one variable with coefficients in the field A. In particular, C is a unique factorization domain so that any element of K which is algebraic over C must be in C. Hence, for any η in K there is an integer $m > 0$ and a polynomial $a(\xi_1)$ in C such that $d(\xi_1)^m\eta = a(\xi_1)$. Taking $\eta = 1/b(\xi_1)$, this gives $d(\xi_1)^m = a(\xi_1)b(\xi_1)$, a contradiction when $b(\xi_1)$ is a prime polynomial but not a factor of $d(\xi_1)$. Since C has infinitely many prime polynomials, such a choice is always possible. Hence ξ_1 is algebraic over A and, since this applies to the other generators as well, the proof is finished.

We shall use this lemma to prove a famous result, Hilbert's *Nullstellensatz*. This requires the introduction of an extremely important algebraic concept, the ideal.

Ideals

A part J of a ring A which is a left (right) A-module under the ring addition, subtraction, and multiplication is called a *left (right) ideal*. An equivalent property is that $au + bv$ ($ua + vb$) belongs to J when u and v belong to J and a and b are any elements of A. When J has both these properties, J is said to be a *two-sided ideal*. In a commutative ring every left ideal is also a right ideal and conversely, and the words right and left are dropped.

Ideals appear in many ways. If, for instance, N is a part of a left A-module, all u in A that annihilate N, i.e., are such that $u\xi = 0$ for all ξ in N, form a left ideal J, for $(au + bv)\xi = a(u\xi) + b(v\xi)$ vanishes when $u\xi = 0$ and $v\xi = 0$. Conversely, if J is a left ideal in a ring A with a unit, we can construct a left A-module M such that J is the part of A that annihilates a certain element of M. As the elements of M we choose all so-called *residue classes* mod J, i.e., sets $\xi = a + J$ consisting of all sums $a + u$ of a fixed a in A and all u in J. When $a - b$ is in J, two such residue classes $\xi = a + J$ and $\eta = b + J$ are identical, and if $a - b$ does not lie in J they have no element in common. Sums and differences are given by

$$\xi \pm \eta = a \pm b + J.$$

The right side is actually a residue class determined by ξ and η for it does not depend on the choice of a in ξ and b in η. This definition makes $J = 0 + J$ the null element of M since $\xi + J = J$ for all ξ. The product of b in A and $\xi = a + J$ in M is defined by

$$b\xi = ba + J.$$

Here again, the right side is independent of the choice of a in ξ. This makes M a left A-module and if we put $\xi_0 = 1 + J$, then a is in J if and only if $a\xi_0 = 0$ where 0 now is the null element J of M.

The module we have just constructed is denoted by A mod J or, more frequently, by the symbolic quotient A/J. When J is a two-sided ideal, it is easy to see that the formula $\xi\eta = ab + J$ defines a product that makes A/J into a ring, the *residue* ring of A mod J. The simplest example is the ring $Z_m = Z/Zm$ where $m > 0$ is an integer. The unit of this ring is the class $1 + Zm$, which for clarity we shall denote by e. The ring then has m different elements $0, e, 2e, \ldots, (m - 1)e$ which are added, subtracted, and multiplied as usual with the further condition that $e^2 = e$, $me = 0$. This is, of course, just another way of doing the calculus mod m presented in Chapter 2. When m is a product ab of integers > 1 and hence $< m$, then $ae \neq 0$, $be \neq 0$ but $aebe = me = 0$, so that Z_m is not a field. But if p is a prime we have seen above that Z_p is a field, the simplest of the Galois fields.

Let us note a property of the residue ring $B = R/J$ of an ideal J in a polynomial ring $R = A[x_1, \ldots, x_n]$ with coefficients in a commutative field A, namely that, putting $e = 1 + J$, $\xi_1 = x_1 + J$, $\ldots$, $\xi_n = x_n + J$, then a polynomial P in R belongs to J if and only if $P(\xi_1, \ldots, \xi_n) = 0$ in B. To see this, note that, by construction, the residue map $P \rightarrow f(P) = P + J$ from R to R/J has the property that $f(P + Q) = f(P) + f(Q)$ and $f(PQ) = f(P)f(Q)$. Further, if $a \neq 0$ is in A then $aJ = J$. In fact, $aJ \subset J$ and $a^{-1}J \subset J$ so that, multiplying by a, $J \subset aJ$. Hence also $f(aP) = af(P)$ for all a in A, so that if $P = \sum a_{k_1 \ldots k_n} x_1^{k_1} \ldots x_n^{k_n}$ with coefficients in A, then $f(P) = \sum a_{k_1 \ldots k_n} \xi_1^{k_1} \ldots \xi_n^{k_n}$ where $\xi_k^0 = e$ for all k. Since, by definition, P is in J if and only if $f(P) = 0$ in B, the statement follows.

Noetherian rings, primary decomposition,
maximal ideals

Continuing our presentation of ideals we now consider ideals in a commutative ring A with a unit. Let B be a part of A. All sums $a_1 u_1 + \cdots + a_m u_m$ where $u_1, \ldots, u_m$ are in B and $a_1, \ldots, a_m$ are arbitrary elements of A form an ideal J. The elements of B are called *generators* of J, and if B has a finite number of elements, J is said to be *finitely generated*. A ring in which every ideal is finitely generated is said to be *noetherian* after a German mathematician, Emmy Noether. We have seen that the ring $\mathbf{Z}$ has this property and also $A[x]$ when A is a field. In fact, every ideal in those rings is generated by a single element. Having one noetherian ring, we may construct many others. A basic result in the theory of rings, proved by Hilbert in 1888, can be formulated as follows. The proof is not difficult but is too long to be given here.

Theorem 3. *When A is a noetherian ring, the polynomial rings $A[x_1, \ldots, x_n]$ are also noetherian.*

An ideal J in A different from A is said to be *prime* if $ab \in J \Rightarrow a \in J$ or $b \in J$, and *primary* if $ab \in J \Rightarrow a \in J$ or $b^n \in J$ for some integer n. When $A = \mathbf{Z}$, these are, respectively, the ideals of the form $\mathbf{Z}p$ and $\mathbf{Z}p^m$ where p is a prime number or 0 and $m > 0$ an integer. The decomposition of a natural number into powers of primes has a remarkable analogue in noetherian rings A: every ideal J not equal to A is the intersection of finitely many primary ideals. But this aspect of ideal theory shall not concern us here. Instead we turn to a special kind of prime ideals, the maximal ideals.

That an ideal J in A is prime means that its residue ring $B = A/J$ is an integral domain, for the definition above says in fact that $B \neq 0$ and that if ξ and η are in B and $\xi\eta = 0$, then either $\xi = 0$ or $\eta = 0$. That $B \neq 0$ is a field means that the ideal J is *maximal*, i.e., the only ideal $I \neq A$ containing J is J itself. In fact, let a be in A and let $\xi = a + J$ be the corresponding residue class of B. That ξ has an inverse $\eta = b + J$ means that ab can be written as $1 + c$ with c in J, and hence a and J together generate the whole ring. Hence, that every $\xi \neq 0$ in B has an inverse means precisely that every a outside of J, together with J, generates the whole ring, i.e., J is maximal.

When K is a field, all polynomials P in $A = K[x_1, \ldots, x_n]$ such that $P(c_1, \ldots, c_n) = 0$ where $c_1, \ldots, c_n$ are fixed in K, form a maximal ideal J. In fact, J is not zero and does not contain 1 and if Q is in A but outside J, then $c = Q(c_1, \ldots, c_n) \neq 0$ is in K and $Q - c$ is in J. Hence $1 = c^{-1}Q - c^{-1}(Q - c)$ is in the ideal generated by J and Q which then must be all of A. We shall now use our lemma to prove that if $K = \mathbf{C}$, ideals of this kind are the only maximal ones. In fact, let J be a maximal ideal in A and let $e = 1 + J, \xi_1 = x_1 + J, \ldots, \xi_n = x_n + J$ be residue classes. Then every element of the field A/J is a polynomial in $\xi_1, \ldots, \xi_n$ with coefficients in the field Ce. Hence, by the lemma, every ξ in A/J satisfies an algebraic equation $S(\xi e) = 0$ where $S = x^m + a_{m-1}x^{m-1} + \cdots + a_0$ is a polynomial with complex coefficients. Since we are working with complex numbers we can write S as a product $(x - b_1) \ldots (x - b_m)$ where $b_1, \ldots, b_m$ are the zeros of S. Hence $S(\xi e) = (\xi - eb_1) \ldots (\xi - eb_m) = 0$ so that, since A/J is a field, $\xi = b_k e$ for some k. In particular, there are complex numbers $c_1, \ldots, c_n$ such that $\xi_1 = c_1 e, \ldots, \xi_n = c_n e$, and hence $P(\xi_1, \ldots, \xi_n) = eP(c_1, \ldots, c_n)$ for any polynomial P in A. Hence, by a previous remark, the statement follows.

We are now about ready for the *Nullstellensatz* but we still have to note a remarkable property of maximal ideals.

Theorem 4. *In a commutative ring with a unit, every ideal not equal to the whole ring is contained in some maximal ideal.*

PROOF. We shall prove this for noetherian rings only. Let J be the given ideal and suppose that it is not contained in a maximal ideal. Then we can find an infinite strictly increasing chain of ideals $J \subset J_1 \subset J_2 \subset \ldots$ none of them equal to the whole ring. Now the union of these ideals is again an ideal and must be finitely generated. Hence all the generators are in some J_n. But then $J_m = J_n$ for all $m > n$, a contradiction.

Algebraic manifolds, Hilbert's Nullstellensatz

Let K be a commutative field and let P be a polynomial in $K[x_1, \ldots, x_n]$. A zero of P is an n-tuple $c_1, \ldots, c_n$ of elements of K such that $P(c_1, \ldots, c_n) = 0$. Sets consisting of all zeros common to one or several polynomials are called *algebraic manifolds*. It is clear that all zeros common to a collection of polynomials are also zeros of all polynomials in the ideal that they generate. We can therefore limit ourselves to zeros of ideals, meaning zeros common to all polynomials of the ideal. We have seen above that every maximal ideal of $C[x_1, \ldots, x_n]$ has a zero. Combining this with Theorem 4 we get

Hilbert's Nullstellensatz (1888). *A nonempty ideal in the ring $C[x_1, \ldots, x_n]$ without zeros must be the whole ring.*

In a less technical language: if $P_1, \ldots, P_m$ are polynomials in n variables with complex coefficients without a common zero, then there are polynomials $Q_1, \ldots, Q_m$ such that $1 = Q_1 P_1 + \cdots + Q_m P_m$. In particular, taking $m = n = 1$, this means that a complex polynomial in one variable without complex zeros must be a constant $\neq 0$. Hence our theorem generalizes the Fundamental theorem of algebra. Finally, we shall deduce the following striking statement, also called Hilbert's *Nullstellensatz*: If $P, P_1, \ldots, P_m$ are polynomials as above and P vanishes at all the common zeros of $P_1, \ldots, P_m$, then there are polynomials $Q_1, \ldots, Q_m$ and an integer $k > 0$ such that

$$P^k = Q_1 P_1 + \cdots + Q_m P_m.$$

To see this, note that the polynomials $P_0 = 1 - x_0 P, P_1, \ldots, P_m$ in the ring $A = \mathbf{C}[x_0, \ldots, x_n]$ do not have any common zero and hence there are polynomials $Q_0', \ldots, Q_m'$ in the same ring such that $1 = Q_0' P_0 + \cdots + Q_m' P_m$. But this equality is also true in the quotient field of A. Hence, if $P \neq 0$ (the case $P = 0$ is trivial), putting $x_0 = 1/P$ and multiplying by a high enough power of P gives the desired result.

Literature

Commutative algebra, touched upon here, is an important part of mathematics closely connected with algebraic geometry. The book *Commutative Algebra*, by Atiyah and MacDonald (Addison-Wesley, 1969) provides a close-up view of the subject. It is a bit tough but goes along at a brisk pace, contains plenty of exercises, and is also rather comprehensive.

3.3 Groups

Before the creation of quantum mechanics in the twenties, group theory was an exclusive mathematical specialty. In this new branch of physics, however, group theoretical arguments led to important discoveries about the structure of atoms and molecules. Group theory is now a routine part of quantum physics and quantum chemistry but was met with wide-eyed amazement by the physicists and chemists of the time whose only mathematical equipment was a course in basic analysis.

Bijections

Like rings and fields, groups are defined in an abstract way, but the definition is much simpler and the available collection of mathematical and other examples is enormous. It is instructive to start with groups of bijections.

Let us remind ourselves of the concept of a function. A function f from a set M with elements $\xi, \eta, \ldots$ to the same set M attaches to every ξ in M another element in M denoted by $f(\xi)$. When f and g are two such functions, their product or composition $f \circ g$ is defined by the formula

$$(f \circ g)(\xi) = f(g(\xi)).$$

Composition is an associative operation. If f, g, h are functions from M to M then $(f \circ g) \circ h = f \circ (g \circ h)$. But very simple examples show that in general $f \circ g$ and $g \circ f$ are different functions. A function f from M to M is said to be a *bijection* if $f(\xi)$ runs through M exactly once when ξ runs through M. In other words, for every η in M, the equation $f(\xi) = \eta$ has precisely one solution ξ. The simplest bijection is the identical map e defined by putting $e(\xi) = \xi$ for all ξ in M. It is clear that the composition $f \circ g$ of two bijections is a bijection and that $f \circ e = e \circ f = f$ for every bijection f. The equation $g(f(\xi)) = \xi$ defines a bijection g called the *inverse* of f with the property that $g \circ f = f \circ g = e$. The inverse of f will be denoted by f^{-1}. Many geometrical mappings are easily visualized bijections. We give some well-known examples when M is a plane: parallel translation, rotation around a point, reflection in a point, reflection in a line. We get their inverses by, simply speaking, translating, rotating, and reflecting back (Figure 3.1) Composition of these geometrical bijections are obtained by first performing one bijection and then the other. We illustrate this in two cases (Figure 3.2).

Symmetries

A symmetry of a plane figure or a body is a bijection of it that preserves distances. A symmetry of, e.g., a circle is either a rotation around, or a reflection in, its center, or a reflection in a line through the center. A symmetry of a square is either a rotation with multiples of a quarter turn, or a reflection in its center, or a diagonal or a line parallel to one side. The reader should illustrate these facts for himself by some figures.

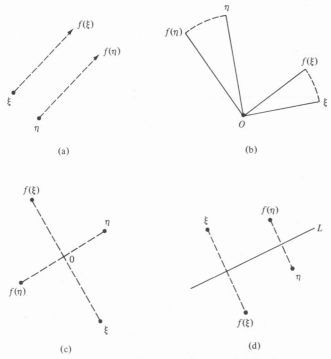

Figure 3.1 Translations, rotations, and reflections.

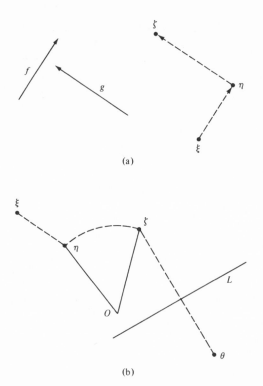

(a)

(b)

Figure 3.2 Compositions of translations, rotations, and reflections. In the figure (a), f and g are translations, $\eta = f(\xi)$ the image of ξ under f and $\zeta = g(\eta) = (g \circ f)(\xi)$ the image of η under g. In the figure (b), η is the image of ξ under a translation f, ζ is the image of η under a rotation g and θ the image of ζ under a reflection h in the line L. Hence $\theta = h(\zeta) = (h \circ g)(\eta) = (h \circ g \circ f)(\xi)$.

Permutations

Bijections of a set with a finite number of elements are called *permutations*. If f is a permutation of a set M with elements $\xi_1, \ldots, \xi_n$ then the elements $f(\xi_1), \ldots, f(\xi_n)$ are equal to $\xi_1, \ldots, \xi_n$ apart from the order. The number of different permutations is $n! = 1 \cdot 2 \ldots n$, for $f(\xi_1)$ can be chosen in n different ways and if $f(\xi_1)$ is fixed, $f(\xi_2)$ can be chosen in $n - 1$ ways and so on. There are two kinds of permutations, even and odd. Let us rearrange the sequence $f(\xi_1), \ldots, f(\xi_n)$ till it becomes $\xi_1, \ldots, \xi_n$ by steps where two neighbors change places. This can be done in many ways but one can show that the number of steps is always even or always odd. The permutation f is accordingly said to be *even* or *odd*. It is an important fact that there are as many even as odd permutations and that the composition $f \circ g$ is even if f and g are both even or both odd and odd in the remaining cases. The proofs are simple but will not be given here.

Groups of bijections

After these preparations we write down a

Definition. A nonempty set G of bijections $a, b, \ldots$ of a set is a *group* if

$$a, b \in G \quad \Rightarrow \quad a \circ b^{-1} \in G, \tag{15}$$

i.e., $a \circ b^{-1}$ belongs to G whenever a and b do.

Taking $b = a$ in (15) we see that every group contains the identity e and, taking $b = e$, that if a is in the group so is a^{-1}. The number of elements in a group is called its *order*. When the order is finite, the group is said to be *finite*. Otherwise it is *infinite*.

It is obvious that the set of *all* bijections of a set constitutes a group. When the set has n elements, this group is called the permutation group of n objects. Its order is $n!$. The even permutations also form a group now of order $n!/2$. The group of all bijections has many interesting subgroups, i.e., subsets G of this group with the property (15). Evidently, such groups can be constructed as follows. Start with an arbitrary collection A of bijections $c, d, \ldots$ and form all products $a = a_1 \circ \ldots \circ a_n$ where every a_k or a_k^{-1} is in A. Since, obviously, $b^{-1} = b_m^{-1} \circ \cdots \circ b_1^{-1}$ when $b = b_1 \circ \cdots \circ b_m$, all these products then form a group. It is the least group containing A and the elements of A are called *generators* of the group. Starting with a single element a we get a group that is called *cyclic* consisting of all powers of a, i.e., $\ldots a^{-2}, a^{-1}, a^0 = e, a^1 = a, a^2 = a \circ a, \ldots$ where $a^{-2} = a^{-1} \circ a^{-1}$ and so on. If all the powers are different, the group has infinite order and there is not much more to be said about it. When two powers are equal there are integers p and $q \neq p$ such that $a^p = a^q$, i.e., $a^{p-q} = e$. Letting n be the smallest natural number m such that $a^m = e$, it then follows that the group consists of n different elements $e, a, \ldots, a^{n-1}$ and hence has order n. A cyclic group is essentially determined by its order. Examples: a parallel translation $a \neq e$ generates an infinite cyclic group, and a reflection $a \neq e$ a cyclic group of order 2 since, then, $a \circ a = e$.

Another way of selecting subgroups of the group of all bijections of a set is to require some kind of invariance property. We give some examples. 1. All bijections a of a set M leaving a part N of M invariant in the sense that $a(\xi)$ and $a^{-1}(\xi)$ are in N whenever ξ is, constitute a group. Same conclusion under the sharper condition that $a(\xi) = \xi$ for all ξ in N. 2. An *isometry* of the geometrical plane is a bijection f that does not change distances, i.e., the distance from $f(\xi)$ to $f(\eta)$ is the same as the distance from ξ to η for all points ξ and η. It is clear that the set of all isometries is a group. Isometries are also called congruence transformations and can be visualised as follows: the image $f(A)$ of a figure A in the plane (e.g., a triangle) is obtained by lifting the figure out of the plane and putting it somewhere else in the plane after, perhaps, turning it upside down. Congruence transformations play a leading part in the geometry of Euclid's *Elements*, but the word is of course not mentioned and the concept is there only implicitly.

Affine groups

Let $\mathbf{R}^2$ be the set of pairs $x = (x_1, x_2)$ of real numbers. A function $x \rightarrow f(x) = (f_1(x), f_2(x))$ from $\mathbf{R}^2$ to $\mathbf{R}^2$ is said to be *affine* if f_1 and f_2 are polynomials in x of degree at most one, i.e., if

$$f_1(x) = a_{11}x_1 + a_{12}x_2 + h_1$$
$$f_2(x) = a_{21}x_1 + a_{22}x_2 + h_2$$

(16)

for all x. Here $a_{11}, \ldots, h_2$ are fixed real numbers called the *parameters* of f. When h_1 and h_2 vanish we say that f is *linear*. A small computation shows that if g is another affine function with parameters $b_{11}, \ldots, k_2$, then

$$(f \circ g)_1(x) = (a_{11}b_{11} + a_{12}b_{21})x_1 + (a_{11}b_{12} + a_{12}b_{22})x_2 + a_{11}k_1 + a_{12}k_2 + h_1$$
$$(f \circ g)_2(x) = (a_{21}b_{11} + a_{22}b_{21})x_1 + (a_{21}b_{12} + a_{22}b_{22})x_2 + a_{21}k_1 + a_{22}k_2 + h_2$$

so that $f \circ g$ is also an affine function. (A comparison with the section on matrix rings in 3.2 shows that matrix multiplication reflects the composition of linear functions. This explains its associativity.) To see when an affine function f is a bijection we try to solve the equations $f_1(x) = y_1$, $f_2(x) = y_2$ for x. As is well-known, this can be done for all y if and only if the determinant $d = a_{11}a_{22} - a_{12}a_{21}$ does not vanish, and then the solution is given by

$$dx_1 = a_{22}y_1 - a_{12}y_2 - a_{22}h_1 + a_{12}h_2$$
$$dx_2 = -a_{21}y_1 + a_{11}y_2 + a_{21}h_1 - a_{11}h_2$$

which means that the inverse of f is also affine. Hence all affine bijections of $\mathbf{R}^2$ form a group. We get a subgroup by, for instance, requiring that the bijections f shall transform $(0, 0)$ to $(0, 0)$. This means that $h_1 = 0$ and $h_2 = 0$, i.e., that f is linear. Another invariance condition producing a subgroup is

$$(f_1(x) - f_1(y))^2 + (f_2(x) - f_2(y))^2 = (x_1 - y_1)^2 + (x_2 - y_2)^2$$

for all x and y. When (x_1, x_2) are orthonormal coordinates in a geometric plane this is the analytic condition for f to be an isometry. It is easy to show that an arbitrary bijection f has this property if and only if it is affine and its parameters $a_{11}, \ldots, a_{22}$ are such that, for some real θ,

$$\begin{pmatrix} a_{11} & a_{12} \\ a_{21} & a_{22} \end{pmatrix} = \begin{pmatrix} \cos\theta & -\sin\theta \\ \sin\theta & \cos\theta \end{pmatrix} \quad \text{or} \quad \begin{pmatrix} -\cos\theta & \sin\theta \\ \sin\theta & \cos\theta \end{pmatrix}.$$

The geometrical interpretation is that $f = v \circ g$ where g is a rotation with the angle θ around the origin, possibly combined with a reflection in a line through the origin, and v is a parallel translation $v_1(y) = y_1 + h_1$, $v_2(y) = y_2 + h_2$.

A third invariance condition, namely

$$(f_1(x) - f_1(y))^2 - c^2(f_2(x) - f_2(y))^2 = (x_1 - y_1)^2 - c^2(x_2 - y_2)^2$$

where $c > 0$ is fixed and x and y arbitrary comes from relativity theory. It also implies that f is affine, and all affine f with this property constitute a group, as do, also, the linear ones. This last group is called the *Lorentz group* with one space dimension, after a physicist who wrote about relativity theory in the beginning of this century. The physical interpretation is that x_2 is time t, x_1 a space coordinate which we now shall call x, and that c is the velocity of light. An element of the Lorentz group that does not change the direction of time maps t, x into

$$t' = \frac{1}{2}(a + a^{-1})t + \frac{1}{2c}(a - a^{-1})x,$$

$$x' = \frac{c}{2}(a - a^{-1})t + \frac{1}{2}(a + a^{-1})x$$

where $a > 0$ is a parameter.

Abstract groups

Definition. (Cayley 1854). A *group* G is a nonempty set with elements $a, b, c, \ldots$ and a binary operation $a \circ b$ called the *group composition* or *group multiplication* such that

(i) $a, b \in G \ \Rightarrow \ a \circ b \in G$
(ii) $(a \circ b) \circ c = a \circ (b \circ c)$ for all a, b, c in G (the associative law)
(iii) G has a neutral element e with the property that $e \circ a = a \circ e = a$ for all a in G
(iv) to every a in G there is a b in G called the *inverse* of a such that $b \circ a = a \circ b = e$.

Note. There is only one neutral element, for if e, e' are two of them then $e = e \circ e' = e'$. A similar argument shows that every a has precisely one inverse.

It is clear that what we have called a group of bijections is also a group in the sense above. For the composition of bijections is associative and we have seen that a group of bijections must contain the identity map and, together with any bijection, also its inverse.

The concepts of ring and field give us some standard examples of groups. 1. G is a ring A, composition $a \circ b = a + b$, neutral element equal to the zero of A, the inverse of a equal to $-a$. This group, called the *additive group* of the ring, is commutative, i.e., $a \circ b = b \circ a$ for all a and b. A commutative group is also said to be *abelian*, after Abel, who studied such groups around 1820. 2. G is a division ring K except for its zero, composition $a \circ b = ab$, neutral element equal to the unit of K, the inverse of a equal to a^{-1}. This is called the *multiplicative group* of K. It is commutative

45

if and only if K is commutative. This example can be generalized a little bit. An element a of a ring A with a unit e is said to be *invertible* if there is a b in A such that $ba = ab = e$. All invertible elements of a ring A constitute a group under multiplication.

Homomorphisms and isomorphisms

Let us for a moment consider the additive group $\mathbf{Z}$ of integers n and the multiplicative group $2^{\mathbf{Z}}$ of integral powers 2^n of 2. Considered as groups they are both cyclic of infinite order and differ only in notation. We can also express this so that $n \to \varphi(n) = 2^n$ is a bijection from $\mathbf{Z}$ to $2^{\mathbf{Z}}$ such that $\varphi(n + m) = \varphi(n)\varphi(m)$, i.e., φ maps addition into multiplication. This bijection is the foundation of the theory of logarithms but shall here be the point of departure for the concept of an isomorphism. A function φ from a group $G = (a, b, c, \ldots)$ with the composition $a \circ b$ to another group $G' = (a', b', c', \ldots)$ with the composition $a' \circ' b'$ is said to be a *homomorphism* if

$$\varphi(a \circ b) = \varphi(a) \circ' \varphi(b)$$

for all a and b in G. A bijective homomorphism is also called an *isomorphism*, and two groups G and G' are said to be *isomorphic* if there is an isomorphism from G to G' (its inverse is then an isomorphism from G' to G). When considered as groups, isomorphic groups differ only in notation. The map $\varphi(a) = a + \mathbf{Z}m$ from $\mathbf{Z}$ to $\mathbf{Z}_m$ is an example of a surjective homomorphism. When φ is a homomorphism from G to G' and e is the neutral element of G, then $\varphi(e)$ is the neutral element of G' and the image $\varphi(G)$ of G is a subgroup of G'. The verification is left to the reader.

Group actions

We can now start proving some theorems about groups, but we shall first introduce the extremely useful concept of a *group action*. An action of a group $G = (a, b, \ldots)$ on a set $M = (\xi, \eta, \ldots)$ is simply a law of composition $a \cdot \xi$ such that

$$a \in G, \quad \xi \in M \;\Rightarrow\; a \cdot \xi \in M$$
$$a, b \in G, \quad \xi \in M \;\Rightarrow\; (a \circ b) \cdot \xi = a \cdot (b \cdot \xi) \quad \text{(the associative law)}$$
$$e \cdot \xi = \xi \quad \text{for all } \xi \text{ in } M.$$

We also say that G *operates* on M. To simplify we shall sometimes not use the signs $\circ$ and $\cdot$ and write ab and $a\xi$. There are plenty of examples of group actions. When G consists of bijections a of a set M, the composition $a \cdot \xi$ defined by $a \cdot \xi = a(\xi)$ is a group action. A group can operate on itself in various ways, e.g.,

$$a \cdot c = a \circ c \quad \text{or} \quad c \circ a^{-1} \quad \text{or} \quad a \circ c \circ a^{-1}.$$

In these formulas we can also let c belong to a group containing G, and then G acts on this bigger group.

A group action of G on M gives rise to a homomorphism from G to the group of bijections of M. This comes from the formula

$$\varphi_a(\xi) = a \cdot \xi$$

mapping elements a of G to functions φ_a from M to M. Since the equation $a \cdot \xi = \eta$ with η given in M has the unique solution $\xi = a^{-1} \cdot \eta$, all these functions are bijections. Finally, the formula

$$\varphi_a(\varphi_b(\xi)) = \varphi_a(b \cdot \xi) = a \cdot (b \cdot \xi) = (a \circ b) \cdot \xi = \varphi_{a \circ b}(\xi),$$

true for all a, b in G and ξ in M, shows that $a \to \varphi_a$, considered as a function from G to bijections of M, has the property that $\varphi_a \circ \varphi_b = \varphi_{a \circ b}$. Hence the map $a \to \varphi_a$ is a homomorphism. That $\varphi_a = \varphi_b$ means that $a \cdot \xi = b \cdot \xi$ or, equivalently, $(b^{-1} \circ a) \cdot \xi = \xi$ for all ξ. Hence, if the group action has the property that

$$a \cdot \xi \quad \text{for all} \quad \xi \quad \Rightarrow \quad a = e,$$

then the map $a \to \varphi_a$ is injective and G isomorphic to a subgroup of the group of all bijections of M. When $M = G$ and $a \cdot c = a \circ c$, the group action has indeed this property and hence we have proved

Theorem 5 (Cayley 1854). *Every group G is isomorphic to a group of bijections. When G has finite order n it is isomorphic to a subgroup of the permutation group of n objects.*

Orbits

That a group operates on a set M is illustrated by Figure 3.3. The group element a moves the "point" ξ to the "point" $a \cdot \xi$. An *orbit* of G in M is the set of points $a \cdot \xi$ coming from ξ when a varies in G. We denote it by

$$G \cdot \xi = \text{the set of } a \cdot \xi \text{ such that } a \text{ belongs to } G.$$

Examples: when M is a geometrical plane and G all rotations around a point, the orbits are circles centered at the point and the point itself. What are the orbits when G is generated by a reflection in a point or a line? The following simple lemma is the source of many theorems of group theory.

Lemma. *Let a group G operate on a set M. Two orbits are then identical or have an empty intersection. Hence the set M is a union of disjoint orbits.*

PROOF. If $a \cdot \xi = b \cdot \eta$ where a, b are in G and ξ, η are in M then $\xi = a^{-1} \circ b \cdot \eta$ so that $c \cdot \xi = c \circ a^{-1} \circ b \cdot \eta$ for all c in G. Hence $G \cdot \xi \subset G \cdot \eta$. But the opposite inclusion is proved in the same way and hence $G \cdot \xi = G \cdot \eta$.

From this lemma follows a well-known theorem about finite groups which we shall now state and prove. It is usually ascribed to Lagrange

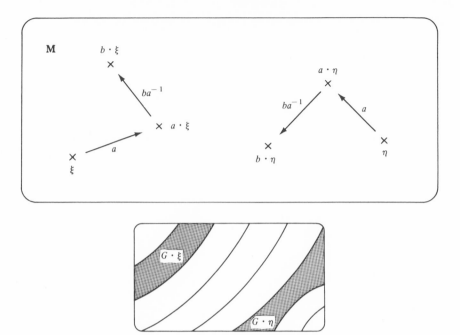

Figure 3.3 A group $G = (a, b, \dots)$ operates on a set $M = (\xi, \eta, \dots)$. The elements $\xi, a \cdot \xi, b \cdot \xi, \dots$ belong to the orbit $G \cdot \xi$ and the points $\eta, a \cdot \eta, b \cdot \eta, \dots$ to the orbit $G \cdot \eta$. The arrows represent actions of the corresponding group elements. The lower part of the figure illustrates in a very schematic way that M is the union of disjoint orbits. The orbits $G \cdot \xi$ and $G \cdot \eta$ are shaded.

(~ 1790). Fermat's theorem and Euler's generalization of it follow from the second part of the theorem when applied to, respectively, the multiplicative groups of the residue classes modulo a prime, and the residue classes modulo an integer m that are relatively prime to m.

Theorem 6. *The order of a subgroup divides the order of the group. In particular, $a^n = e$ where a is any element in the group, n its order, and e the neutral element.*

PROOF. Let G be the group and H the subgroup, write the group composition simply as ab, and let H operate on G via the product ac where a is in H and c is in G. The orbits are then the sets Hc. They have all the same number of elements as H for if $ac = bc$ then $a = b$. If G has n elements and H has m elements and the number of orbits is k, then, by the lemma, $n = mk$ and this proves the first part of the theorem. If, in particular, $H = e, a, \dots, a^{m-1}$ is generated by a single element a, then $a^m = e$ so that also $a^n = a^{mk} = e$.

Invariant subgroups

Let G and Γ be groups and φ a homomorphism from G to Γ. The set of elements a in G mapped by φ to the neutral element of Γ is called the *kernel* of the homomorphism φ. Let G have elements $a, b, \ldots$ and composition ab, and Γ elements $\xi, \eta, \ldots$, composition $\xi\eta$ and neutral element ε. Since $\varphi(ab^{-1}) = \varphi(a)\varphi(b)^{-1}$, if a and b are in the kernel of φ so is ab^{-1}. Hence this kernel is a subgroup H of G. Also, since $\varphi(aba^{-1}) = \varphi(a)\varphi(b)\varphi(a)^{-1} = \varepsilon$ when $\varphi(b) = \varepsilon$, we have $aHa^{-1} \subset H$. Changing a to a^{-1} gives the opposite inclusion $H = aa^{-1}Haa^{-1} \subset aHa^{-1}$ and hence

$$aHa^{-1} = H \tag{17}$$

for all a in G. Subgroups with this property are said to be *invariant*. Examples: when G is abelian, all subgroups are invariant. The translations $x \to x + b$ form an invariant subgroup of the group of all affine bijections $x \to f(x) = ax + b$, $a \neq 0$, of the real line. In fact, since $f \circ g(x) = acx + ad + b$ when $g(x) = cx + d$, $f \to a$ is a homomorphism into the real numbers $\neq 0$. Similarly, with f given by (16),

$$f \to \begin{pmatrix} a_{11} & a_{12} \\ a_{21} & a_{22} \end{pmatrix}$$

is a homomorphism into square nonsingular matrices of order 2. Hence the translations form an invariant subgroup of the group of affine bijections of the plane and hence also of the group of isometries. This can also be seen by geometric arguments.

When we have an invariant subgroup H of a group G we can also construct a group Γ and a homomorphism φ from G to Γ whose kernel is H. In fact, take as elements of Γ the orbits $\xi = Ha$, $\eta = Hb$ of G when H acts from the left on G. Then $\xi\eta = Hab$ defines a composition in Γ. In fact, (17) shows that $Hab = aHb = abH$ and hence replacing a by ac or ca with c in H and similarly for b does not change the orbit Hab. It is then immediate to verify that Γ is a group with neutral element $\varepsilon = H$ and that the map $a \to \varphi(a) = Ha$ is a homomorphism. Since $Ha = H$ if and only if a is in H, its kernel is precisely H. The group Γ is called the *quotient group* of G by H and is denoted by G/H. Examples: when G is the set of linear bijections $f(x) = ax + b$, $a \neq 0$, of the real line and H is the group of translations, every orbit Hf contains a linear homogeneous bijection $f_0(x) = ax$, and the quotient group is in fact isomorphic to the group of these. Similarly, the quotient of the group of isometries of the plane by the subgroup of translations is isomorphic to the group of rotations, i.e., isometries leaving a point invariant.

We are now so well-prepared that we could start proving a couple of important theorems about finite groups but, lacking space, we must limit ourselves to a look at some especially interesting ones.

Finitely generated abelian groups

An abelian group is usually written in additive form so that the composition is $a + b$, the neutral element is 0 and $-a$ the inverse of a. Every cyclic group is abelian. We say that an abelian group is *finitely generated* (over the integers) if there are finitely many elements $a_1, \ldots, a_s$ in G such that every a in G has the form $n_1a_1 + \cdots + n_sa_s$ for some choice of integers $n_1, \ldots, n_s$. The following structure theorem for such groups is proved in many textbooks. It is anonymous in the sense that it cannot be tied to any one person.

Theorem 7. *Every finitely generated abelian group is a direct sum of cyclic subgroups $G_1, \ldots, G_n$. This means that every a in G is a unique sum $a_1 + \cdots + a_n$ where $a_1 \in G_1, \ldots, a_n \in G_n$. The subgroups $G_1, \ldots, G_n$ that are of finite order can be chosen so that their orders are powers of primes and then they are unique apart from isomorphisms.*

Finite groups

Part of the mathematical literature has been devoted to making a catalogue of all nonisomorphic groups of a given order n. When this order is a prime, the situation is simple for, according to Theorem 6, such a group has no proper subgroups and must be cyclic. When the order is not a prime the situation gets livelier. There are, for instance, 14 different groups of order 16. On the other hand, there is just one of order 15. All groups of order p, p^2, pq, and p^3, where p and q are primes, are known, and they are constructed in many elementary texts.

Crystal groups and patterns

Mineral crystals and other crystals have interesting symmetry properties that can be sorted out by group theory. The crystals of a crystalline mineral are arranged in a lattice that we may think of as infinite. The lattice can then be translated in three independent directions so that it covers itself afterwards. In other words, its symmetry group contains three independent translations. It may also contain other symmetries like rotations and reflections. It is a mathematical fact that if isomorphic groups are considered to be identical there are 232 different such groups, and an experimental fact that there are crystal lattices representing all of them. The isomorphism classes of the quotients of these groups by their translation subgroups give a coarser classification, the 32 crystal classes. The same thing can be done for two-dimensional patterns repeating themselves in two independent directions. Their symmetry groups fall into 17 isomorphism classes and, taking quotients by translations, this number reduces to 12. Patterns representing all 17 classes occur, for instance, in the art of ancient Egypt. This shows that man has a very good intuitive eye for

symmetries. Some writers want to see the early discovery of all these
pattern classes as the birth of group theory. The crystal groups make tough
reading, but the pattern groups are easy to understand.

Literature

A. Speiser, *Gruppentheorie* (Springer-Verlag, 1937), and Coxeter, *Introduction to Geometry* (Wiley, 1963). There is also a less technical book, *Symmetry*, by H. Weyl (Princeton, 1952).

Solvable groups

A group G is said to be *solvable* if there is a finite chain of successive
subgroups $G = G_0 \supset G_1 \supset \cdots \supset G_n$ where every group is invariant in the
preceding one, all factor groups G_{k+1}/G_k are cyclic, and G_n is the identity.
In the sixties the American mathematicians Feit and Thompson proved a
very difficult result, namely that all groups of odd order are solvable.

Galois theory

We now have all the mathematical tools that are necessary to explain—naturally without proofs—how Galois used group theory to see why
certain equations can be solved by radicals, i.e., by extraction of nth roots
and rational operations, and others cannot. Let $P(x) = x^n + a_{n-1}x^{n-1}
+ \cdots + a_0$ be a prime polynomial with rational coefficients and let
$\xi_1, \ldots, \xi_n$ be all its zeros in the complex plane. Construct the ring of all
polynomials

$$b_0 + b_1\xi_1 + \cdots + b_n\xi_n + b_{11}\xi_1^2 + \cdots$$

in these zeros with rational coefficients $b_0, b_1, \ldots$ and let K be its quotient
field. This is then an algebraic number field containing the rational
numbers $\mathbf{Q}$. We denote its elements by $\xi, \eta, \ldots$. Consider bijections φ of
K that are also field isomorphisms, i.e., such that

$$\varphi(\xi + \eta) = \varphi(\xi) + \varphi(\eta), \qquad \varphi(\xi\eta) = \varphi(\xi)\varphi(\eta) \tag{18}$$

for all ξ and η in K and at the same time leave the elements of Q invariant,

$$a \in \mathbf{Q} \implies \varphi(a) = a. \tag{19}$$

Applying these rules a number of times we get

$$\varphi(P(\xi)) = \varphi(\xi)^n + a_{n-1}\varphi(\xi)^{n-1} + \cdots + a_0 = P(\varphi(\xi))$$

for all ξ. In particular, if $P(\xi)$ vanishes so does $P(\varphi(\xi))$. Hence the
numbers $\varphi(\xi_1), \ldots, \varphi(\xi_n)$ form a permutation of the numbers $\xi_1, \ldots, \xi_n$.
It is also clear that $\varphi(\xi)$ is determined for all ξ in K when we know
$\varphi(\xi_1), \ldots, \varphi(\xi_n)$. Now it is obvious that all bijections φ of K with the
properties (18) and (19) form a group. It is called the *Galois group* of K,
and we have just seen that it can be considered as a subgroup of the

permutation group π_n of n objects. The connection between solvability by radicals and the Galois group is given by the following theorem by Galois (1830).

Theorem 8. *When P is a rational prime polynomial the equation $P(x) = 0$ is solvable by radicals if and only if the corresponding Galois group is solvable.*

It is rather easy to see that the Galois group is equal to π_n except for special choices of the coefficients. That it is impossible to solve a general equation of degree more than four is then a consequence of the following property of the group π_n which is not difficult to prove. A group is said to be *simple* when it has no invariant subgroups except the neutral element and the group itself. These are called the *trivial subgroups*.

Theorem 9. *When $n > 4$, the group of even permutations of n objects is simple not cyclic, and is the only nontrivial invariant subgroup of the permutation group of n objects.*

Our last two theorems are prime examples of powerful mathematics bringing simplicity and clarity to what at the outset seems to be a hopeless mess.

Constructions by ruler and compass

Galois theory also provides the solution of two problems put by the Greeks. In the *Elements* geometric constructions are performed with the aid of a ruler and a compass. It was then easy to construct a square whose area is the double of the area of a given square, or to bisect a given angle into equal parts. But all attempts to construct a cube whose volume is the double of the volume of a given cube, or to trisect a given angle into equal parts, failed. Galois theory shows that it is impossible to do these two things exactly in a finite number of steps. The general problem is the following one: what are the lengths that can be constructed with the aid of a ruler and a compass and a line interval of length 1? Or, equivalently, what points in the complex plane can we construct starting from 0 and 1? For simplicity we refer to these points as complex numbers $a, b, \ldots$. Simple arguments show that if a and b are constructible so are $a \pm b$, ab, ab^{-1} and $\sqrt{a}$. In particular, all constructible numbers form a field and every constructible number turns out to be algebraic. Hence, to every such number ξ there is a rational prime polynomial P such that $P(\xi) = 0$. Form the corresponding field generated by all zeros of P and let G be its Galois group.

Theorem 10. *The number ξ is constructible if and only if the order of G is a power of 2.*

It follows that zeros of prime polynomials of degree three are not constructible. In fact, the corresponding Galois group has order 6 or 3. The doubling of the cube is equivalent to the construction of $\xi = \sqrt[3]{2}$. The corresponding polynomial $x^3 - 2$ is a prime polynomial and hence ξ is not constructible. The trisection of the angle turns out to correspond to the construction of the zeros of polynomials $4x^3 - 3x + a$ with rational a. With only rare exceptions they are rational prime polynomials and then their zeros are not constructible. In *What is Mathematics?* by Courant and Robbins (Oxford) there are elementary proofs of these assertions which do not use Galois theory.

Group representations

A representation of a group G is, by definition, a homomorphism of G into the group of invertible elements of a ring A, i.e., a map φ from G to A such that $\varphi(ab^{-1}) = \varphi(a)\varphi(b)^{-1}$ for all a, b in G. When A is a ring of square matrices with complex elements we say that φ is a *complex matrix representation* of G. Since the beginning of this century, all such representations have been known in principle when G is finite, and in detail when G is, for instance, the group of rotations around a point in three-dimensional space. These later ones play an important part in nonrelativistic quantum mechanics. They are fully explained in H. Weyl's book, *The Theory of Groups and Quantum Mechanics* (1931; Dover, 1950).

History

The history of group theory is rather intricate and it is difficult to sort out what really happened before the terminology was fixed. This was done at the end of the nineteenth century, in particular through the book by C. Jordan, *Traité des Substitutions* (1870), which gave a full treatment of Galois theory. (See G. A. Miller, *History of the Theory of Groups up to 1900*, in *Collected Works*, vol. I (1935), pp. 427–467).

3.4 Documents

Ars Magna

Proofs in Ars Magna are written in the style of Euclid. Cardano's formula also appears in modern notation but then only with numerical coefficients. At this time, negative numbers were a novelty, complex numbers were not known, and algebraic calculations were not trusted to provide proofs. Here is Cardano's description of the formula (3*), the equation being $x^3 + px = q$.

"Cube one third of the coefficient of x; add to it the square of one-half of the constant of the equation; and take the square root of the whole. You will duplicate this and to one of the two you add one-half of the number you have already squared and from the other you subtract one-half of the same. You will then have a binomium and its apotome. Then subtracting the cube root of the apotome from the cube root of the binomium, the remainder or that which is left is the value of x."

Hieronimo Cardano 1501–1576

Galois on permutation groups

Galois died in a duel at the age of 21. He had published an article on what is now known as the Galois fields but his main work remained in unfinished manuscripts. In these he states among other things the following three theorems, here given in verbatim translation.

Evariste Galois 1811–1822

"Theorem I. The permutations common to two groups form a group. Theorem II. When a group is contained in another one, the latter is the union of a certain number of groups similar to the first one, which is called a divisor. Theorem III. If the number of permutations of a group is divisible by p (a prime), this group contains a substitution whose period has p terms."

Theorem I requires no commentary. In Theorem II a subgroup is acting on a group with all the orbits also referred to as groups. Theorem III is an impressive result rediscovered by Cauchy in 1844. In modern terms it says that if p divides the order of a group, then the group has a cyclic subgroup of order p. There is no very easy proof.

Cayley on matrices

(*In Philosophical Transactions of the Royal Society of London*, 1858, pp. 17–37).

Arthur Cayley   1821–1896

"The term matrix may be used in a more general sense but in the present memoir I consider only square and rectangular matrices and the term matrix used without qualification is understood as meaning a square matrix; in this restricted sense, a set of quantities arranged in the form of a square.... It will be seen that matrices (attending only those of the same order) comport themselves as single quantities; they may be added, multiplied or compounded together etc.: the law of addition of matrices is precisely similar to that for addition of ordinary algebraic quantities; as regards their multiplication (or composition) there is the peculiarity that matrices are not in general convertible; it is nevertheless possible to form the powers (positive or negative, integral or fractional) of a matrix.... I obtain the remarkable theorem that any matrix whatever satisfies an algebraical equation of its own order, the coefficient of the highest power being unity, and those of the other powers functions of the terms of the matrix, the last coefficient being in fact the determinant...."

When he says "convertible," Cayley means that multiplication of matrices is in general not commutative. In the last passage Cayley refers to the by now famous Hamilton–Cayley theorem which can be explained as follows. Let $A = (a_{jk})$ be an $n \times n$ matrix with elements in a commutative ring R with a unit and let

$$D(x) = \det(a_{jk} - x\delta_{jk})$$
$$= (-x)^n + (a_{11} + a_{22} + \cdots + a_{nn})(-x)^{n-1} + \cdots + \det(a_{jk})$$

be its characteristic polynomial (see the next chapter, section 4.3). Then $D(A) = 0$ in the ring of $n \times n$ matrices with elements in R. This spectacular result is easy to prove but has not turned out to be very important compared to the general usefulness of matrices.

Literature

A Survey of Modern Algebra, by Birkhoff and MacLane (Macmillan, 1963), and *The Theory of Groups*, by M. Hall (Macmillan, 1959) are both easy to read and contain a lot of material. *Algebra*, by S. Lang (Addison-Wesley, 1965) is a rather tough, encyclopedic book. Van der Waerden's classic, *Moderne Algebra* (from 1932) is now available in paperback under the title of *Algebra* (Springer-Verlag, 1968).

4
GEOMETRY AND LINEAR ALGEBRA

4.1 *Euclidean geometry*. History. Non-Euclidean geometry. Conic sections. 4.2 *Analytical geometry*. Coordinate systems. Equations of curves and surfaces. Vectors. Distance, angle, inner product. Reconstruction of Euclidean geometry. The origin and use of vectors. 4.3 *Systems of linear equations and matrices*. Systems of linear equations. Linear functions. Matrix algebra. Square matrices and their inverses. Determinants and characteristic polynomial. The solvability of systems of linear equations. The origin and use of matrices. 4.4 *Linear spaces*. Definitions and examples. Linear functions. Invariant subspaces, eigenvectors, eigenvalues, spectrum. Complement, codimension, projection. Operations on linear functions. Pathology of bases. 4.5 *Linear spaces with a norm*. Distance and Banach spaces. The contraction theorem. 4.6 *Boundedness, continuity, and compactness*. Bounded linear functions. Dual spaces and linear functionals. Compactness and compact operators. 4.7 *Hilbert spaces*. Inner products and Euclidean spaces and Hilbert spaces. Orthogonal complements and orthogonal projection. Dirichlet's problem. 4.8 *Adjoints and the spectral theorem*. Adjoints. The spectral theorem. The spectral theorem and quadratic forms. The spectral theorem and vibrating systems. 4.9 *Documents*. Euclid on parallel lines. Felix Klein on geometry. Hilbert on functional analysis and the spectral theorem.

No beginner's course in mathematics can do without linear algebra. According to current international standards it is presented axiomatically. It is a second generation mathematical model with its roots in Euclidean geometry, analytical geometry, and the theory of systems of linear equations. This brings pedagogical difficulties. Beginners with a shaky background in geometry and algebraic computation who also have difficulties with abstractions are really not ripe for the study of linear algebra. On the other hand, there is no need to exaggerate the difficulties. The theory is very simple, has few theorems and is free from complicated proofs. It is also a must. Not being familiar with the concepts of linear algebra such as linearity, vector, linear space, matrix, etc., nowadays amounts almost to being illiterate in the natural sciences and perhaps in the social sciences as well.

We shall devote the first part of this chapter to the three sources of linear algebra: Euclidean geometry, analytical geometry, and systems of linear equations. After a section on matrices we then pass to linear algebra proper and its objects, linear spaces and linear maps between them.

The rest of the chapter deals with linear analysis, an extremely useful hybrid of algebra and analysis obtained by introducing the notion of length into the algebraic machinery. Linear analysis in linear spaces of infinite dimension, usually called functional analysis, is a successful twentieth century invention. We shall review some of its basic concepts and results including the spectral theorem for self-adjoint linear operators. It will be applied to the analysis of small vibrations of mechanical systems.

The reader should realize that this chapter covers a lot of ground and that it is not meant for quick consumption. In many places some previous experience of the subject is required.

4.1 Euclidean geometry

History

Euclid's *Elements* was written in Alexandria around 300 B.C. It was preserved in handwritten copies until printing started, around 1500. The oldest copies now in existence are from about 1000 A.D. (T. L. Heath's English translation from 1908, with Commentary, is available as a Dover publication.) Until the beginning of this century, the *Elements* was *the* textbook of mathematics in secondary school. The strength of this position and also Heath's conservative leanings are abundantly clear from his preface: "It is of course not surprising that, in these days of short cuts, there should have arisen a movement to get rid of Euclid . . . a rush of competitors anxious to be first in the field with a new text book on the more 'practical' lines which now find so much favour." The Swedish poet C. M. Bellman wrote

> When I think of Euclid, even now
> I have to wipe my sweaty brow,

an echo of the despair of many generations of schoolchildren.

Thirteen parts of the *Elements* have been preserved. The first four parts deal with triangles, parallelograms, and circles. They explain and prove well-known geometric theorems, for instance that the sum of the three angles of a triangle is two right angles, the theorem of Pythagoras, the theorem that a circular arc is seen under a constant angle from the remaining part of the circle. The proofs rely on propositions that are not proved—in our terminology, *axioms* or *postulates*. They are a mixture of general logical principles like "those which are equal to the same are mutually equal," and geometric propositions such as the famous axiom of the parallels. The latter can be formulated as follows. Through a point outside a straight line there passes precisely one straight line parallel to the first one, i.e., the two never meet however far out they are prolonged.

Parallel lines.

The cord is one of the oldest measuring tools of civilization. Euclidean geometry can be thought of as a mathematical model for measuring with a cord. The straight line corresponds to the stretched cord, and its points to the least parts of the cord. A plane is given by two intersecting lines and the sphere corresponds to the surface generated by one end of the cord when the other end is fixed. That Euclidean geometry deals with abstractions is clear already from the first definition of the *Elements*: a *point* is that which has no part. The fit between model and reality is phenomenally close. The sum of the angles of a triangle is in fact always 180° within the limits of precision of the measuring equipment. Navigational failures or geometric mishaps in carpentry are never caused by any defects in the geometrical theorems employed. The only essential contribution to Euclidean geometry made since classical times are tables of the trigonometric functions giving various ratios between the sides of a right triangle as functions of one of the angles.

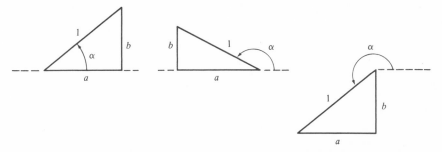

Figure 4.1 *The trigonometric functions.* The sine, cosine, tangent, and cotangent of an angle α are defined by $\sin \alpha = b$, $\cos \alpha = a$, $\tan \alpha = b/a$, and $\cot \alpha = a/b$ where a and b are given by the figure. The length a is considered negative to the left, the length b negative downwards.

Non-Euclidean geometry

With the rise of modern science after 1500 Euclidean geometry, too, became an object of scientific curiosity. The axiom of parallels was obviously very special and there were many efforts to deduce it from the other axioms. All of them failed and around 1830, Bolyai and Lobachevski, independently of each other, proved that further efforts were futile. They constructed a so-called non-Euclidean plane E^* that differs from the Euclidean plane E in one respect only. The axiom of parallels does not hold in E^* but all the other axioms of E do hold. Fifty years later Poincaré illustrated this discovery by drawing a map of E^* in the form of a big circular disk C in the Euclidean plane (Figure 4.2). He chose the length scale at the point P on the map to be $1 : (1 - d^2/R^2)^{-1}$ where R is the radius of the disk and d the distance on the map from P to the center of the disk. For instance, the two arcs α and β on the map are then equally

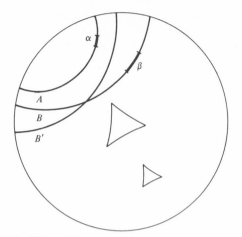

Figure 4.2   Poincaré's map of the non-Euclidean plane E^*.

long in E^*. Another consequence of the choice of scale is that infinite straight lines in E^* are circular arcs on the map meeting the boundary of C at right angles. There are three such arcs on the figure: A, B, B'. In particular, the arcs α and β of A and B correspond to two equally long line segments in E^*. The two curved triangles on the map correspond to two straight triangles in E^* of equal size. A path on the map going out to the boundary of C is infinitely long in E^* and the boundary of C has no counterpart in E^*. But close to the center of the disk where d/R is small, the scale is very close to 1 : 1 and distances in E^* and on the map are very similar. In E^* there is also a counterpart to the group of congruence transformations of the Euclidean plane. There is, for instance, a bijection of E^* that preserves distances and maps a given triangle onto another given triangle with sides of the same size as the first one. The two curved triangles of the figure illustrate this situation. It turns out that all of Euclid's axioms hold in E^* except the axiom of parallels. This is illustrated by the figure. In E^*, B and B' are two straight lines with just one common point, but neither of them meets the straight line A. The non-Euclidean plane also has other remarkable features. The sum of the angles of a triangle, for instance, is always *less* than two right angles and decreases as the size of the triangle increases. In fact, this is so on the map and hence also in E^* since the map preserves angles. It is only the length scale that changes from point to point.

The discovery of non-Euclidean geometry was a psychological breakthough. Euclidean geometry was no longer identical with Reality; there were other models. To Bolyai, the non-Euclidean plane was a new world. The beginning of the nineteenth century also saw the birth of projective geometry. It was inspired by the theory of perspectives, and the projective plane is obtained from the Euclidean plane by adding to it the so-called line at infinity. This, and the discovery of non-Euclidean geometry, broke the monopoly of Euclidean geometry. In 1872 Felix Klein pointed out that every geometry has a group of bijections corresponding to the congruence transformations of Euclidean geometry; and around the turn of the century Hilbert made a famous logical analysis of geometrical axioms. In Hilbert's analysis the words "point" and "line" are completely neutral, everything is expressed in the cold logical language of sets and relations, and the sole purpose of pictures is to illustrate the logic. This way of dealing with axioms is now completely accepted. On the other hand, pictures are and will always be indispensible to our intuition. Used without prejudice they can also represent very abstract concepts. A picture of two overlapping disks, for instance, is a very efficient and commonly used way of conveying the idea of the union and intersection of two completely arbitrary sets.

Conic sections

After this flight in time, let us return to Greek mathematics, about 300–200 B.C. The circle was not the only curve studied. Four lost parts of the *Elements* dealt with ellipses, hyperbolas, and parabolas or, using a

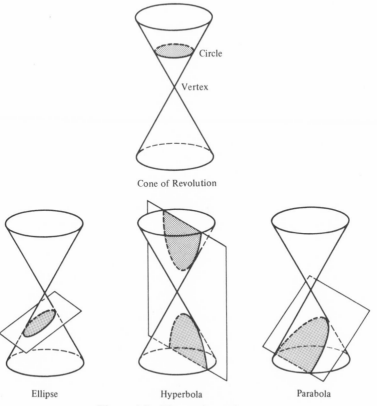

Figure 4.3   The conic sections.

common name, conic sections. These are the curves one gets when a plane intersects a cone of revolution (Figure 4.3). There is a complete theory of conics in a treatise by Apollonius (200 B.C.). He shows, for instance, that an ellipse (hyperbola) is the locus of a point moving in such a way that the sum (difference) of its distances to two given points, the *foci*, is constant. Further details are given in the Figures 4.4–6.

Since the time of Apollonius, the conic sections have had a fantastic career in physics. Kepler discovered around 1610 that the planets move in ellipses with the sun in one of the foci, and Newton proved in his book

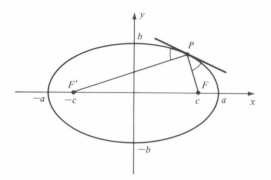

Figure 4.4   *Ellipse*. The sum of the distances from a point P on the ellipse to the points F and F' is constant $= 2a$. At P there is a tangent drawn to the ellipse. The two angles marked are equal. Rays of light from F are reflected by the ellipse to F' and therefore these two points are called *foci*. The equation of the ellipse in the coordinate system of the figure is $(x/a)^2 + (y/b)^2 = 1$. The coordinates of the two foci are $(\pm c, 0)$ where $c^2 = a^2 - b^2$.

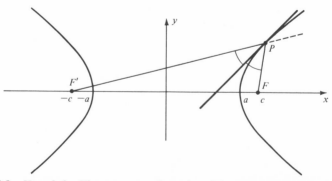

Figure 4.5   *Hyperbola*. There are two branches. The difference between the distances from a point P on the hyperbola to F and F' equals $\pm 2a$. At P there is a tangent drawn to the hyperbola. The two angles marked are equal. Rays of light coming from F and reflected in the hyperbola seem to come from F'. The equation of the hyperbola in the coordinate system of the figure is $(x/a)^2 - (y/b)^2 = 1$. The coordinates of the two foci are $(\pm c, 0)$ where $c^2 = a^2 + b^2$.

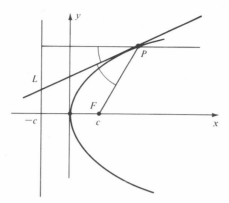

Figure 4.6 *Parabola*. One branch. A point P on the parabola has the same distance to its directrix, the straight line L, and to its focus F. At P there is a tangent drawn to the parabola. The two angles marked are equal. Rays of light coming from F and reflected in the parabola become parallel. The equation of the parabola in the coordinate system of the figure is $y^2 = 4cx$.

Principia Mathematica . . . (1686) that this follows from the law of gravitation and the laws of mechanics. The cornerstone of quantum mechanics is the spectral theorem for self-adjoint linear transformations, descendants of the conic sections.

Later we shall deduce Newton's results about planetary motion and then we shall need the equation of conics in polar coordinates. It is given below in the text of Figure 4.7.

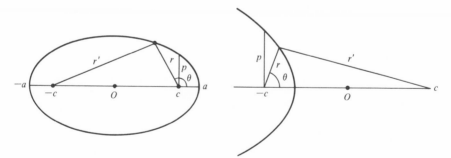

Figure 4.7 *The equation of conics in polar coordinates*. With notations as in the left part of the figure we get $r + r' = 2a$, $r'^2 = r^2 \sin^2\theta + (2c + r \cos \theta)^2$ so that $(2a - r)^2 = r^2 + 4cr \cos \theta + 4c^2$. Putting $c = ea$ so that $0 \leqslant e < 1$ this gives

$$r = p(1 + e \cos \theta)^{-1}, \qquad p = a(1 - e^2). \tag{*}$$

The number e, which vanishes when the ellipse is a circle, is called the *eccentricity* and p the *parameter*. For a hyperbola we use the right part of the figure. The branch marked there also has the equation (*) but with $p = a(e^2 - 1)$ where now $e = c/a > 1$. When $e = 1$ we get the equation of a parabola.

63

4.2 Analytical geometry

Coordinate systems

Analytical geometry is the systematic use of the fact that there is a natural correspondence between the real numbers and the points on a straight line, between pairs of real numbers and the points in a plane, and between triples of real numbers and points in space. Computations with numbers can then be interpreted geometrically and geometric problems can be reformulated as algebraic problems. A book by Descartes, *Géometrie*, from 1637, is generally considered to have initiated analytical geometry. We shall restrict ourselves to parallel coordinates on the line, in the plane, and in space, presenting them in the form of three figures with text. Our constructions use everyday geometry, i.e., Euclidean geometry.

Note that parallel coordinates by no means constitute the only possibility. There are plenty of ways of attaching numbers to points, many of them in current use, e.g., polar coordinates and the longitude-latitude system on the earth's surface.

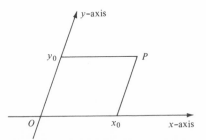

Figure 4.8 *Coordinates on the line.* The point P has the *coordinate* x_0 defined as the distance from P to a fixed point O, measured with respect to some scale. The point O is called the *origin*. The distance shall be positive on one side of O, marked by the direction of the arrow, and negative on the other side.

Figure 4.9 *Parallel coordinates in the plane.* Two *axes*, the *x-axis* and the *y-axis*, meet in a point O, the origin. Lines through P parallel to the axes meet them in two points. Taken together, their coordinates, x_0 and y_0, are the coordinates of P. When the axes are at right angles and the scale of length is the same on the two axes, the coordinate system is said to be *rectangular* or *orthonormal*. The square of the distance between two points P and Q with coordinates (x_0, y_0) and (x_1, y_1) is then given by the theorem of Pythagoras

$$\text{dist}(P, Q)^2 = (x_1 - x_0)^2 + (y_1 - y_0)^2.$$

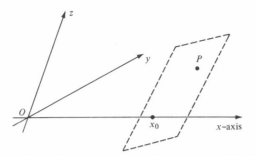

Figure 4.10   *Parallel coordinates in space.* Three axes, the *x*-axis, the *y*-axis, the *z*-axis meet in a point *O*, the origin. Pairwise they determine planes through the origin called the *xy-plane, the yz-plane, the xz-plane.* A plane through *P* parallel to the *yz*-plane meets the *x*-axis in a point. Its coordinate x_0 is called the *x-coordinate* of *P*. Its *y-coordinate* and *z-coordinate* are defined analogously. The numbers (x_0, y_0, z_0) are the *coordinates* of *P*. When the axes are at right angles and the scale of length is the same on the three axes, the coordinate system is said to be *rectangular* or *orthonormal*. The square of the distance between two points *P* and *Q* with coordinates (x_0, y_0, z_0) and (x_1, y_1, z_1) is then given by the theorem of Pythagoras

$$\text{dist}(P, Q)^2 = (x_1 - x_0)^2 + (y_1 - y_0)^2 + (z_1 - z_0)^2.$$

Equations of curves and surfaces

We have already shown that the equations of the ellipse, hyperbola, and parabola are, respectively,

$$\left(\frac{x}{a}\right)^2 + \left(\frac{y}{b}\right)^2 - 1 = 0, \qquad \left(\frac{x}{a}\right)^2 - \left(\frac{y}{b}\right)^2 - 1 = 0, \qquad y^2 - 4cx = 0.$$

This means that a point *P* with coordinates (x, y) lies on one of these curves if and only if its coordinates satisfy the corresponding equation. In this connection one might ask the following question: Has every curve an equation $f(x, y) = 0$, and does every such equation correspond to a curve? Here $f(x, y)$ is a real function of *x* and *y*. The answer to both questions is yes, but under certain natural restrictions. To go into this would carry us too far afield just now, but the answer is easy for linear functions *f*, i.e., those of the form $f(x, y) = ax + by + c$ where *a*, *b*, *c* are real numbers. The simplest theorems of Euclidean geometry show that if $a^2 + b^2 > 0$ then $f(x, y) = 0$ is the equation of a straight line; that every straight line has an equation of this form; and that two equations, $f(x, y) = 0$ and $f'(x, y) = 0$ where $f'(x, y) = a'x + b'y + c'$ with $a'^2 + b'^2 > 0$, are equations of the same straight line if and only if *f* and *f'* are multiples of each other, i.e., there is a real number $h \neq 0$ such that $f'(x, y) = hf(x, y)$ or, which is the same thing, $a' = ha$, $b' = hb$, $c' = hc$. For planes in space there is an analogous situation. Their equations are of the form $ax + by + cz + d = 0$ where $a^2 + b^2 + c^2 > 0$. A straight line in space can be considered as the

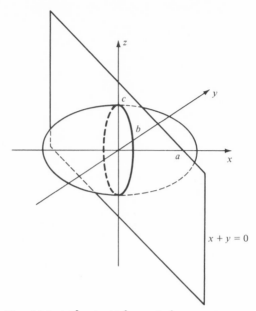

Figure 4.11   The ellipsoid $(x/a)^2 + (y/b)^2 + (z/c)^2 = 1$ and its intersection (thick line) with the plane $x + y = 0$.

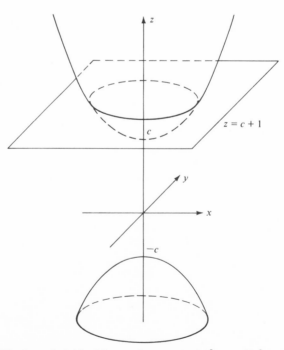

Figure 4.12   The hyperboloid with two sheets $-(x/a)^2 - (y/b)^2 + (z/c)^2 = 1$ and its intersection (thick line) with the plane $z = c + 1$.

66

intersection of two planes. To express that a point P in space belongs to a given line we need two equations $ax + by + cz + d = 0$ and $a'x + b'y + c'z + d' = 0$ for its coordinates x, y, z. The general rule is that one equation $f(x, y, z) = 0$ defines a surface and two such equations define a curve. We illustrate this with two figures, one of the ellipsoid, and one of a hyperboloid with two sheets.

Vectors

Introducing coordinates is not the only way to combine algebra and geometry. We can also make a little algebraic machinery out of ordered line segments. An ordered line segment is a straight line PQ from a point P to a point Q taken in this order. They are added according to the rule $PQ + QR = PR$, illustrated by the left part of Figure 4.13.

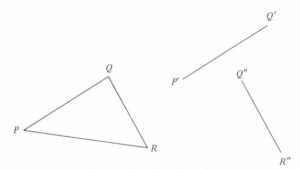

Figure 4.13   Equality of vectors.

This definition has the disadvantage that we can only add two segments when the second one starts where the first one ends. If we want to add arbitrary segments, something more has to be done. The solution is to prescribe that two ordered line segments are *equal* when one is a parallel translation of the other, as are PQ and $P'Q'$, and QR and $Q''R''$, of Figure 4.13. To mark this change of scene we introduce a new term. Ordered line segments PQ considered in this way are called *vectors* and will be denoted by $\overrightarrow{PQ}$ or by single letters $u, v, \ldots$. Simply expressed, a vector is an arrow that can be moved parallel to itself in the plane or in space without losing its identity. Vectors are added according to Figure 4.14. A careful analysis starting from the axioms of Euclidean geometry shows that

$$u + v = v + u \quad \text{(the commutative law)} \tag{A1}$$

$$u + (v + w) = (u + v) + w \quad \text{(the associative law)} \tag{A2}$$

for all u, v, w. The null vector $\overrightarrow{PP}$, denoted by $\underline{0}$, is such that

$$u + \underline{0} = u \quad \text{(existence of a null vector)} \tag{A3}$$

for all u, and the vector v given by $\overrightarrow{QP}$ when u is given by $\overrightarrow{PQ}$ is such that

$$u + v = \underline{0} \quad \text{(existence of an opposite vector)}$$

67

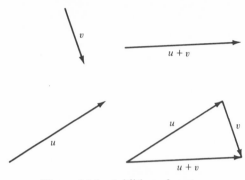

Figure 4.14   Addition of vectors.

We can also multiply numbers $a, b, \ldots$ and vectors $u, v, \ldots$. Geometrically, the vector au is parallel to u and $|a|$ times as long as u with the same (opposite) direction when $a > 0$ $(a < 0)$. When $a = 0$, $au = \underline{0}$. Another analysis, starting again with the axioms of Euclid, shows that the following rules hold:

$$1u = u \tag{A5}$$

$$(ab)u = a(bu) \tag{A6}$$

$$(a + b)u = au + bu \tag{A7}$$

$$a(u + v) = au + av \tag{A8}$$

for all numbers a, b and vectors u, v. These are the familiar rules of arithmetic, the only difference being that one has to remember that the sum of two vectors is a vector and that the product of a number and a vector is a vector.

Every vector of the form $a_1 u_1 + \cdots + a_n u_n$ where $a_1, \ldots, a_n$ are numbers is said to be a linear combination of the vectors $u_1, \ldots, u_n$. If such a linear combination is equal to $\underline{0}$ but, e.g., $a_1 \neq 0$, multiplying by $1/a_1$ and moving the multiples of $u_2, \ldots, u_n$ to the right-hand side shows u_1 to be a certain linear combination of $u_2, \ldots, u_n$. We say that $u_1, \ldots, u_n$ are *linearly independent* if

$$x_1 u_1 + \cdots + x_n u_n = \underline{0} \quad \Rightarrow \quad x_1 = 0, \ldots, x_n = 0$$

when $x_1, \ldots, x_n$ are numbers. In other words, only the trivial linear combination $0u_1 + \cdots + 0u_n$ can be $\underline{0}$. The geometrical interpretation is that one vector is linearly independent if it is not zero, two if they are not parallel, and three vectors are linearly independent if they do not lie in one plane when issuing from the same point. Hence, according as our construction of vectors has taken place on a line, in a plane, or in space, they have the following property, called the dimension axiom:

there are one (two, three) linearly

independent vectors but not more.   (A9)

Our vector algebra gives us at once an algebraic definition of parallel coordinates. For if e_1, e_2, e_3 are three linearly independent vectors in space, O a fixed and P a variable point, according to (A9), the vectors $\overrightarrow{OP}, e_1, e_2, e_3$ are linearly dependent, and this is possible only if $\overrightarrow{OP}$ is a linear combination of the others,

$$\overrightarrow{OP} = x_0 e_1 + y_0 e_2 + z_0 e_3.$$

Here the numbers x_0, y_0, z_0 are uniquely determined by P. In fact, if x_0', y_0', z_0' give the same right-hand side we get $(x_0 - x_0')e_1 + (y_0 - y_0')e_2 + (z_0 - z_0')e_3 = \underline{0}$ and hence $x_0' = x_0, y_0' = y_0, z_0' = z_0$. When e_1, e_2, e_3 are unit vectors, i.e., have length 1, the numbers x_0, y_0, z_0 are precisely the parallel coordinates introduced above using figures. The general case just amounts to a change of scale, separately on each axis, and our coordinates are still referred to as parallel coordinates.

Using the algebraic definition of parallel coordinates it is very easy to see what becomes of the coordinates when we replace O by another point O' and e_1, e_2, e_3 by three other linearly independent vectors e_1', e_2', e_3'. We then have the earlier formula for $\overrightarrow{O'P}$,

$$\overrightarrow{O'P} = x_0' e_1' + y_0' e_2' + z_0' e_3'$$

and $\overrightarrow{O'P} = \overrightarrow{O'O} + \overrightarrow{OP}$ and hence

$$x_0' e_1' + y_0' e_2' + z_0' e_3' = w + x_0 e_1 + y_0 e_2 + z_0 e_3$$

where $w = \overrightarrow{O'O}$. Writing all vectors on the right-hand side as linear combinations of e_1', e_2', e_3' we see that each new coordinate x_0', y_0', z_0' is of the form $a + bx_0 + cy_0 + dz_0$.

Distance, angle, inner product

One advantage of vector algebra is that it has an algebraic formalism, free of coordinates, for the Euclidean concepts of distance and angle. It is based on the inner product (u, v) of two vectors u and v defined by

$$(u, v) = |u|\,|v|\cos \alpha$$

where $|u|$ is the length of u, $|v|$ the length of v and α the angle between them when they start from the same point. Sometimes the word *scalar product* is used because (u, v) is a real number and real numbers are called scalars when appearing together with vectors. The inner product has the following remarkable properties, proved by Euclidean geometry and true for all numbers a, b and vectors u, v, w:

$$(u, u) \geqslant 0, \qquad (u, v) = (v, u) \tag{B1}$$

$$(u, u) = 0 \quad \Rightarrow \quad u = \underline{0} \tag{B2}$$

$$(au + bv, w) = a(u, w) + b(v, w) \tag{B3}$$

$$(u, av + bw) = a(u, v) + b(u, w). \tag{B4}$$

Note that $(u, u) = |u|^2$ is the square of the length of u. A vector of length 1 is said to be a *unit vector* and two vectors u and v are said to be *orthogonal* when $(u, v) = 0$. The formulas (B3) and (B4) say that the functions $u \to (u, v)$ and $v \to (u, v)$ are linear. That a function $u \to f(u)$ from vectors to numbers or from vectors to vectors is *linear* means that $f(au + bv) = af(u) + bf(v)$ for all vectors u, v and numbers a, b. When $u = x_1 e_1 + x_2 e_2 + x_3 e_3$ and $v = y_1 e_1 + y_2 e_2 + y_3 e_3$ are two linear combinations of e_1, e_2, e_3 we can use (B3) and (B4) several times, writing the inner product (u, v) as

$$(u, v) = x_1 y_1 (e_1, e_1) + x_1 y_2 (e_1, e_2) + x_1 y_3 (e_1, e_3)$$
$$+ x_2 y_1 (e_2, e_1) + x_2 y_2 (e_2, e_2) + x_2 y_3 (e_2, e_3)$$
$$+ x_3 y_1 (e_3, e_1) + x_3 y_2 (e_3, e_2) + x_3 y_3 (e_3, e_3).$$

When e_1, e_2, e_3 are an *orthonormal* system, i.e., mutually orthogonal unit vectors, this gives $(u, v) = x_1 y_1 + x_2 y_2 + x_3 y_3$, which, when $u = v$, is the theorem of Pythagoras. The formula also shows that $(u, e_1) = x_1$, $(u, e_2) = x_2$, $(u, e_3) = x_3$ and hence

$$u = (u, e_1)e_1 + (u, e_2)e_2 + (u, e_3)e_3.$$

The coordinates of u in an orthonormal system are simply the products of u with the axis vectors.

Reconstruction of Euclidean geometry

Without revealing precisely how, we have deduced the algebraic rules A1–A9 for vectors and B1–B4 for the inner product from Euclidean geometry. It is also possible to follow a reversed path resulting in an axiomatic definition of Euclidean geometry equivalent to that of Euclid. We shall then use the term *real linear space* for every set M of objects $u, v, \ldots$ where sums and products with numbers have been defined with the properties A1–A8. If, in addition, A9 holds we say that M has *dimension* 1, 2, *or* 3, and if there is a function (u, v) from $M \times M$ to numbers having the properties B1–B4 we say that M is a real linear space of the corresponding dimension equipped with an inner product.

We can now give a strict, abstract definition of what is meant by a *Euclidean plane* or *space* E. It runs as follows: E is a set with elements $P, Q, R, \ldots$ equipped with certain bijections from E to a real linear inner product space V whose dimension is 2 when E is a plane and 3 when E is a space. To every P in E there is precisely one bijection f_P and it is required that $f_P(R) = f_P(Q) + f_Q(R)$ for all P, Q, R in E. (When V already comes from a Euclidean plane or space, $f_P(Q)$ is the vector $\overrightarrow{PQ}$.) The elements of E are then called *points*, and three points P, Q, R are said to be on the same line when any two of the vectors $f_P(Q), f_Q(R), f_R(P)$ are linearly dependent. With these definitions and a similar one of planes it is possible to deduce all the axioms and theorems of Euclidean geometry, including the axiom of parallels. The length of a line segment from P to Q is then, of course, given by $(u, u)^{1/2}$ where $u = f_P(Q)$. Working without inner product we get the so-called *affine geometry*. There all the axioms and theorems of the *Elements* hold that do not use distance, e.g., the axiom of parallels and the theorem that the medians of a triangle meet in one point. (Much more about the interplay of algebra and geometry can be found in E. Artin's book, *Geometric Algebra* (Interscience, 1957).)

The origin and use of vectors

The vector algebra we have just described has appeared in many forms in many mathematical papers since the seventeenth century and has no principal inventor. For over 100 years vector algebra has been used in physics, where various forces have been illustrated by arrows and denoted by symbols that have been added and multiplied by real numbers. The physical counterpart of vector addition is the parallelogram of composition of forces implicit in the work of Archimedes (200 B.C.). Vectors were introduced in the schools in the 1950s as part of the modernization of mathematics.

4.3   Systems of linear equations and matrices

Systems of linear equations

A system of p linear equations for n unknown numbers $x_1, \ldots, x_n$ looks like this:

$$
\begin{aligned}
a_{11}x_1 + a_{12}x_2 + \cdots + a_{1n}x_n &= y_1 \\
a_{21}x_1 + a_{22}x_2 + \cdots + a_{2n}x_n &= y_2 \\
&\cdots \\
a_{p1}x_1 + a_{p2}x_2 + \cdots + a_{pn}x_n &= y_p.
\end{aligned}
\tag{1}
$$

The left sides are all of the form $a_1x_1 + \cdots + a_nx_n$ where $a_1, \ldots, a_n$ are given numbers. Such expressions are called *linear combinations* of $x_1, \ldots, x_n$ with coefficients $a_1, \ldots, a_n$. In (1) there are p linear combinations with doubly indexed coefficients so that a_{jk} is the coefficient of x_k in the jth equation. The system is said to be *underdetermined, overdetermined,* or *quadratic* according as $p < n$, $p > n$, or $p = n$, i.e., the number of equations is less than, greater than, or equal to the number of unknowns. When the right sides $y_1, \ldots, y_p$ all vanish, the system is said to be *homogeneous*.

A great number of more or less practical problems lead to two linear equations with two unknowns, and some of them occur in cuneiform texts, 4000 years old. General systems (1), especially quadratic ones, appear in all kinds of applied mathematics and play an important part in numerical analysis. They are also of theoretical importance.

Our task is now to say something about the solutions of (1) without unnecessary assumptions about the coefficients and the right sides. If, e.g., $n = p = 2$ we find that (1) has the unique solution

$$
x_1 = \frac{a_{22}y_1 - a_{12}y_2}{D}, \qquad x_2 = \frac{-a_{21}y_1 + a_{11}y_2}{D}
$$

when the determinant

$$D = a_{11}a_{22} - a_{12}a_{21}$$

does not vanish. But if D vanishes we seem to face chaos. The system may have no solutions (as for instance $x_1 + x_2 = 1$, $x_1 + x_2 = 0$), or it may have an infinity of solutions (as for instance when all coefficients and right sides vanish). We shall see next that this chaos can be analyzed in a simple way even when p and n are arbitrary. This depends on the following simple fact: adding a multiple of one equation to another one changes (1) to a system having exactly the same solutions. In fact, writing $x_1, \ldots, x_n$ simply as x, let us note that if x is a solution of (1), adding b times the first equations to the second one shows that x also solves a system obtained from (1) by replacing the second equation by

$$(a_{21} + ba_{11})x_1 + (a_{22} + ba_{12})x_2 + \cdots + (a_{2n} + ba_{1n})x_n = y_2 + by_1.$$

Conversely, if x is a solution of this new system, subtracting b times its first equation from its second one shows that x solves (1).

Armed with the principle just proved we shall now have a closer look at the system (1). To start with there is the embarrassing possibility that the system may contain null equations, i.e., those of the type $0x_1 + \cdots + 0x_n = t$. If a single null equation with $t \neq 0$ appears, the system has no solution. On the other hand, if, e.g., $a_{11} \neq 0$, we may eliminate x_1 from all equations except the first one by multiplying the first equation by $b_k = a_{k1}/a_{11}$ and subtract the result from the kth equation. We then get a new system with the same solutions as (1) where x_1 appears only in the first equation and the right sides $y_1, \ldots, y_p$ are replaced by certain linear combinations of $y_1, \ldots, y_p$, namely in order $y_1, y_2 - b_2 y_1, \ldots, y_p - b_p y_1$. If instead $a_{21} \neq 0$, the same method allows us to eliminate x_1 from all equations except the second one, etc. If we allow ourselves to relabel the unknowns and the equations and repeat our procedure of elimination sufficiently many times, we see that (1) has the same solutions as at least one system of the form

$$
\begin{aligned}
c_{11}z_1 + \cdots \qquad\qquad\quad + c_{1n}z_n &= h_1(y) \\
c_{22}z_2 + \cdots \quad + c_{2n}z_n &= h_2(y) \\
&\ \ \vdots \\
c_{rr}z_r + \cdots + c_{rn}z_n &= h_r(y) \\
0 &= h_{r+1}(y) \\
&\ \ \vdots \\
0 &= h_p(y)
\end{aligned}
\tag{1'}
$$

where $h_1(y), \ldots, h_p(y)$ are certain linear combinations of $y_1, \ldots, y_p$ and $z_1, \ldots, z_n$ are the unknowns $x_1, \ldots, x_n$ taken in some order, and none of the numbers $c_{11}, c_{22}, \ldots, c_{rr}$ is equal to zero. If, for instance, (1) consists of null equations only, we put $r = 0$ and $h_1(y) = y_1, \ldots, h_p(y) = y_p$. This is of course an extreme case but our reasoning covers it.

The system $(1')$ is easy to analyze. It has solutions if and only if $h_{r+1}(y) = 0, \ldots, h_p(y) = 0$, and when these conditions are satisfied one gets every solution by fixing $z_{r+1}, \ldots, z_n$ at will and after that successively computing $z_r, z_{r-1}, \ldots, z_1$ from the first r equations.

We can now at once draw a number of important conclusions from the fact that (1) and $(1')$ have the same solutions. Since the systems are both homogeneous or both inhomogeneous, since $r \leqslant p$ and since z_n can be chosen arbitrarily in the homogeneous system $(1')$ when $r < n$, we get

The theorem of underdetermined systems. *Every underdetermined homogeneous system of linear equations has at least one nontrivial solution, i.e., one where not all the unknowns are equal to zero.*

The system $(1')$ is said to have *rank r*. This is also called the rank of the system (1), for in spite of the fact that all the coefficients and the right sides of $(1')$ depend upon how the process of elimination is performed, the number r is the same for all systems $(1')$ coming from (1). To see this, make (1) homogeneous and imagine together with $(1')$ a second system, $(1'')$, arrived at in the same way, but of rank s, and add to both systems the equations $z_{r+1} = 0, \ldots, z_n = 0$. The resulting systems will then have the same solutions and in the first case only the trivial one $z_1 = 0, \ldots, z_n = 0$. But if $s < r$, the second system has less than n equations not counting the null equations and this contradicts the preceding theorem. Hence $s \geqslant r$ so that, by symmetry, $s = r$.

It is clear that the rank r of (1) cannot exceed p or n. A comparison of (1) and $(1')$ shows that $r = n$ means that the homogeneous system (1) has only the trivial solution and that $r = p$ means that (1) has solutions whatever the right sides are. In fact, a look at the elimination process shows that this last property is shared by (1) and $(1')$. When $n = p$, these observations give

The theorem of quadratic systems. *The following properties of a quadratic system of linear equations are equivalent*

  (i) *the rank equals the number of unknowns*
  (ii) *the system has solutions for all right sides*
  (iii) *the homogeneous system has only the trivial solution.*

A quadratic system (1) with $p = n$ also has a determinant D defined by

$$D = a_{11}a_{22} - a_{12}a_{21}$$

when $n = 2$ and by the impressive formula

$$D = \sum_{k_1=1}^{n} \sum_{k_2=1}^{n} \cdots \sum_{k_n=1}^{n} \varepsilon(k_1, \ldots, k_n)a_{1k_1} \cdots a_{nk_n} \tag{2}$$

in the general case. Here $\varepsilon(k_1, \ldots, k_n)$ equals 0, 1, -1 according as two

73

numbers of the sequence $k_1, \ldots, k_n$ are equal or the sequence is an even or odd permutation of $1, \ldots, n$. One can show that

$$D \neq 0 \qquad \qquad \text{(iv)}$$

is equivalent to each of the three conditions above. Compared to what we have done so far, the proof of this is cumbersome and we shall not give it. But, for completeness, let us sketch how to analyze whether or not (1) has solutions. A sequence c of numbers $c_1, \ldots, c_p$ is said to be a *linear relation* for (1) if $c_1 f_1(x) + \cdots + c_p f_p(x) = 0$ for all $x_1, \ldots, x_n$. Here $f_1(x), \ldots, f_p(x)$ are the left sides of (1) and the condition amounts to a homogeneous linear system of n equations for $c_1, \ldots, c_p$ obtained by putting the coefficients of $x_1, \ldots, x_n$ in the expression $c_1 f_1(x) + \cdots + c_p f_p(x)$ equal to zero. The fact is that (1) has solutions if and only if $c_1 y_1 + \cdots + c_p y_p = 0$ for all linear relations c of (1). To prove this, note that replacing in this condition, e.g., c_1 by $c_1 + bc_2$ (a mere change of notation) amounts to stating it for the system obtained from (1) by adding b times the first equation to the second one. Hence, by a repetition of this procedure, the condition actually says the same thing when stated for (1) and the equivalent system (1'). But in the latter case it is obviously true, the linear relations c' of (1') being those with $c'_1 = 0, \ldots, c'_r = 0$.

So far we have used the word *number* without saying what kind of numbers appear in (1). They may be rational, real, or complex and, in fact, numbers in any field. For, passing between (1) and (1'),, we have only used the rules of computation of a field, e.g., the rule that any element $\neq 0$ of a field has an inverse. Hence what we have done holds when all the numbers of (1), coefficients, unknowns, and right-hand sides, belong to a field.

Our two theorems have appeared more or less explicitly in the mathematical literature since the seventeenth century. In its modern formulation, the first one is due to Jacobi (~ 1840). The second one was first stated and proved with the aid of determinants (Cramer, ~ 1750). The concept of rank was introduced by Kronecker in 1864.

After very little work and with a minimum of terminology we now have mastered what will turn out to be the essentials of linear algebra. Next we proceed to express the results in two other languages, one using the concept of a linear function and the other the calculus of matrices.

Linear functions

It is convenient to denote the set of unknowns $x_1, \ldots, x_n$ of (1) by a single symbol. We put $x = (x_1, \ldots, x_n)$ and call x an *n-tuple* with *components* $x_1, \ldots, x_n$. The set of all such n-tuples will be denoted by K^n where K is the field in which we are doing the arithmetic. The solutions of (1) can be considered as elements of K^n. The reader who does not feel at ease with K can think of $\mathbf{R}$ all the time.

Two n-tuples are added by adding corresponding components and one n-tuple is multiplied by a number of multiplying all its components by that number. This means that

$$x + y = (x_1 + y_1, \ldots, x_n + y_n), \qquad ax = (ax_1, \ldots, ax_n). \qquad (3)$$

Using these definitions we can, for instance, make the following clean statement: if x and y are solutions of a homogeneous system (1) so are $x + y$ and ax whatever the number a. The proof is left to the reader. Putting $\underline{0} = (0, \ldots, 0)$ we immediately verify that the operations (3) obey the computation rules A1 to A8 of geometric vectors.

Returning to (1), the left sides can be thought of as functions

$$x \rightarrow f(x) = c_1 x_1 + \cdots + c_n x_n \qquad (4)$$

from K^n to K. Here $c_1, \ldots, c_n$ are fixed numbers that determine the function and are themselves determined by it, for if

$$e_k = (0, \ldots, 0, 1, 0, \ldots, 0) \qquad (5)$$

with 1 at the kth place, then $c_k = f(e_k)$. Functions of this kind are said to be *linear*. In order to handle all the left sides of (1) at the same time we introduce functions

$$x \rightarrow f(x) = (f_1(x), \ldots, f_p(x)) \qquad (6)$$

from K^n to K^p whose components are functions from K^n to K. Such a function is said to be linear when its components are linear. Putting in particular

$$f_j(x) = \sum_{k=1}^{n} a_{jk} x_k = a_{j1} x_1 + \cdots + a_{jn} x_n \qquad (7)$$

the left sides of (1) give us a linear function from K^n to K^p.

Assume now for a moment that f and g are just arbitrary functions from K^n to K^p and let a and b be two numbers. The sum $af + bg$ is then a function from K^n to K^p defined by

$$(af + bg)(x) = af(x) + bg(x) \qquad (8)$$

where x is in K^n and the right side denotes addition in K^p. Also, let f be a function from K^n to K^p and g a function from K^n. The composed function $f \circ g$ from K^q to K^p is then defined by

$$(f \circ g)(x) = f(g(x)) \qquad (9)$$

when x is in K^q. That there is a branch of mathematics called linear algebra depends, in the last analysis, upon

The theorem of linear functions. *Linear combinations and compositions of linear functions are linear.*

Here, a linear combination of a number of functions $f_1, \ldots, f_r$ from K^n to K^p is any function $a_1 f_1 + \cdots + a_r f_r$ where $a_1, \ldots, a_r$ are numbers. The function $af + bg$ of (8) is an example of this and we have earlier described the right side of (4) as a linear combination of the variables $x_1, \ldots, x_n$.

The proof of the theorem is only a matter of bookkeeping. We have to verify that the components of the right sides of (8) and (9) are linear combinations of $x_1, \ldots, x_n$ and $x_1, \ldots, x_q$ respectively when f and g are linear functions. If the components

$$f_j(x) = \sum_{k=1}^{n} a_{jk} x_k \quad \text{and} \quad g_j(x) = \sum_{k=1}^{n} b_{jk} x_k \tag{10}$$

of f and g are linear combinations of $x_1, \ldots, x_n$, then, according to (8), the components

$$(af + bg)_j(x) = af_j(x) + bg_j(x) = \sum_{k=1}^{n} (aa_{jk} + bb_{jk}) x_k \tag{11}$$

of $af + bg$ have the same property. If the components

$$f_i(y) = \sum_{k=1}^{n} a_{ik} y_k \quad \text{and} \quad g_k(x) = \sum_{j=1}^{q} b_{kj} x_j \tag{12}$$

of f and g are linear combinations of, respectively, $y_1, \ldots, y_n$ and $x_1, \ldots, x_q$ then, according to (9), the components

$$(f \circ g)_i(x) = f_i\big(g_1(x), \ldots, g_q(x)\big) = a_{i1} g_1(x) + \cdots + a_{in} g_n(x)$$

of $f \circ g$ are linear combinations of $x_1, \ldots, x_q$. More precisely, we get

$$(f \circ g)_i(x) = \sum_{j=1}^{q} c_{ij} x_j \quad \text{where} \quad c_{ij} = \sum_{k=1}^{n} a_{ik} b_{kj}. \tag{13}$$

(We suppose here that the reader knows how to handle the sign of summation.) With this the theorem is proved but before we finish with the linear functions for this time we shall prove a result which we could have started with.

Theorem. *A function f from K^n to K^p is linear if and only if*

$$f(ax + by) = af(x) + bf(y) \tag{14}$$

for all x and y in K^n and all numbers a and b.

Since a function is linear if and only if its components are linear and f has the property (14) if and only if its components have that property, it suffices to consider the case $p = 1$. If f has the property (14) and $e_1, \ldots, e_n$ are defined by (5), then $x = x_1 e_1 + \cdots + x_n e_n$ and hence

$$f(x) = x_1 f(e_1) + \cdots + x_n f(e_n)$$

is linear. Conversely,, if $f(x) = c_1 x_1 + \cdots + c_n x_n$ is linear, (14) follows from the fact that

$$c_1(ax_1 + by_1) + \cdots + c_n(ax_n + by_n)$$
$$= a(c_1x_1 + \cdots + c_nx_n) + b(c_1y_1 + \cdots + c_ny_n).$$

Returning once more to the system (1) we can now write it as a single equation

$$f(x) = y$$

where f is a linear function from K^n to K^p and y is an element of K^p. This puts the system into a new perspective. That it is solvable, i.e., has solutions, means that y belongs to the set of values of f. Further, each one of the three conditions of the theorem of quadratic systems mean that f is a bijection from K^n to K^n. In fact, according to (14), the equality $f(x) = f(x')$ is equivalent to $f(x - x') = \underline{0}$ and hence f is injective if and only if the homogeneous system has only the trivial solution $x = \underline{0}$. We shall return to this point of view in the section on linear spaces.

Matrix algebra

Strip the linear functions of every piece of clothing and there remain the matrices and the calculus of matrices. Let f be a linear function given by (6) and (7). The method is to disregard everything about the function except the scheme of its coefficients

$$A = \begin{bmatrix} a_{11} & a_{12} & \cdots & a_{1n} \\ a_{21} & a_{22} & \cdots & a_{2n} \\ \cdots & & & \\ a_{p1} & a_{p2} & \cdots & a_{pn} \end{bmatrix} \tag{15}$$

We describe this as a rectangular scheme of numbers having p rows and n columns and call it a *matrix of type $p \times n$*. Linear combinations $aA + bB$ of matrices of the same type, and products AB of matrices A of type $p \times n$ and B of type $n \times q$, are defined so that they correspond to the functions $af + bg$ and $f \circ g$, where the pair f, g is given by the formula (10) in the first case and by (12) in the second case. Writing for simplicity (15) as $A = (a_{jk})$, (11) shows that

$$A = (a_{jk}), \quad B = (b_{jk}) \quad \Rightarrow \quad aA + bB = (aa_{jk} + bb_{jk})$$

when A and B are of the same type, and (13) shows that

$$A = (a_{ik}), \quad B = (b_{kj}) \quad \Rightarrow \quad AB = (c_{ij}) \quad \text{where} \quad c_{ij} = \sum_k a_{ik}b_{kj}$$

when A has as many columns as B has rows. Otherwise the product is not defined. To memorize the rule of multiplication note that the element at the place i, j of AB is the product of the ith row of A by the jth column of

B and that the product

$$(a_1, a_2, \ldots, a_n) \begin{pmatrix} b_1 \\ n_2 \\ \vdots \\ b_n \end{pmatrix} = a_1 b_1 + \cdots + a_n b_n$$

of a row matrix and a column matrix is a number. Linear combinations of matrices of the same type obey the rules A1 to A8 for geometric vectors, $\underline{0}$ denoting the matrix of the proper type all of whose elements vanish. Since composition of functions is associative, so is matrix multiplication, that is,

$$(AB)C = A(BC)$$

whenever the two sides are defined. In fact, the identities $(f_1 + f_2) \circ g(x) = (f_1 + f_2)(g(x)) = f_1(g(x)) + f_2(g(x)) = f_1 \circ g(x) + f_2 \circ g(x)$ and the analogous ones for $g \circ (f_1 + f_2)$ show matrix multiplication to be doubly distributive.

$$(A_1 + A_2)B = A_1 B + A_2 B, \qquad A(B_1 + B_2) = AB_1 + AB_2.$$

The matrices AB and BA are of the same type if and only if A and B are square of the same type $n \times n$. They are then said to have *order n*. When the order exceeds one, there are simple examples to show that $AB \neq BA$. Matrix multiplication of square matrices is not in general commutative. Finally, let us note that numbers go right through products. We have $aAB = AaB$ and we read Aa as aA when A, B are matrices and a is a number.

Interchanging the rows and columns of a matrix is called *transposition*. When $A = (a_{jk})$ is of type $p \times n$, the transposed matrix $A^t = (a^t_{jk})$ of type $n \times p$ is defined by the equation

$$a^t_{jk} = a_{kj}.$$

For transposition we have the rules of computation

$$(aA + bB)^t = aA^t + bB^t, \qquad (AB)^t = B^t A^t$$

when A, B are matrices and a, b numbers. For matrices with complex elements there is also the operation of *complex conjugation* $A \to \bar{A}$ where all the elements are conjugated, and the operation $A \to A^* = \bar{A}^t$, called *taking the adjoint*. Here we get $(aA + bB)^* = \bar{a}A^* + \bar{b}B^*$ and $(AB)^* = B^* A^*$.

Returning to the system (1) we can now write it in the form

$$AX = Y \qquad (1'')$$

where $X = (x_1 \ldots x_n)^t$ and $Y = (y_1 \ldots y_p)^t$ are column matrices with n and p elements respectively. The formula $A(aX_1 + bX_2) = aAX_1 + bAX_2$ exhibits in a simple way the fact that linear combinations of solutions X of the homogeneous system $AX = \underline{0}$ are also solutions of that system.

Matrix algebra as now presented is a simple and efficient piece of machinery for handling most of linear algebra. Here are a few examples.

Square matrices and their inverses

Square matrices of order n correspond to linear functions from K^n to K^n. Adding or multiplying them we get matrices of the same kind so that all square matrices of order n form a ring, noncommutative when $n > 1$. This ring has an identity element, the unit matrix of order n,

$$E = E_n = \begin{bmatrix} 1 & 0 & 0 & \cdots & 0 \\ 0 & 1 & 0 & \cdots & 0 \\ \cdots & & & & \\ 0 & & \cdots & & 1 \end{bmatrix}$$

corresponding to the function $x \to x$. That A corresponds to a bijection of K^n means that A has an *inverse* A^{-1}, necessarily unique, such that $A^{-1}A = AA^{-1} = E$. When A is *invertible*, i.e., has an inverse, the system (1″) has the unique solution $Y = A^{-1}X$. Products of invertible elements are invertible and we have $(AB)^{-1} = B^{-1}A^{-1}$. A matrix A such that $A^{-1} = A^t$ is said to be *orthogonal*. All such matrices constitute a group under multiplication, the orthogonal group, most important when $K = \mathbf{R}$. Unitary matrices and the unitary group are defined in the same way from the condition that $A^{-1} = A^*$ when $K = \mathbf{C}$.

Determinants and characteristic polynomial

The *determinant* det A of a square matrix A is defined by the formula (2). One can show that det $AB = $ det A det B, and that A is invertible if and only if det $A \neq 0$. The definition of a determinant and a little computation show the function $z \to \det(A - zE)$ to be a polynomial of degree n. It is called the *characteristic polynomial* of A. More precisely,

$$\det(A - zE) = \det A + \cdots + (a_{11} + a_{22} + \cdots + a_{nn})(-z)^{n-1} + (-z)^n.$$

In particular, by the fundamental theorem of algebra, for every square matrix with complex elements, there is at least one complex number z such that $\det(A - zE) = 0$. This means that $(A - zE)X = 0$, i.e., $AX = zX$, for at least one complex column matrix $X \neq 0$, a result that will be used later on.

The solvability of systems of linear equations

Let A, X, Y be matrices of types $p \times n$, $n \times 1$, and $p \times 1$ respectively. Then $AX = Y$ is a system of p linear equations for n unknowns. When $A = (a_{jk})$ we have the system (1). What we have said about linear relations and the solvability of (1) can also be expressed so that the system $AX = Y$ is solvable if and only if $Z'A = 0 \Rightarrow Z'Y = 0$ for all Z of type $p \times 1$. For square systems with det $A \neq 0$ the solution is unique and given by Cramer's formula (~ 1750), namely

$$x_k = \det(A_1, \ldots, Y, \ldots, A_n)/\det A$$

where $k = 1, \ldots, n$ and $A_1, \ldots, A_n$ are the columns of A, and Y appears in the place number k in the parenthesis on the right.

The origin and use of matrices

Matrix algebra has existed in its present form since the middle of the nineteenth century. It was invented by Hamilton, Cayley, and Sylvester and for a long time remained an algebraic specialty until it became a tool of quantum mechanics in the 1920s. Nowadays matrices are part of a general education in mathematics. They are widely used in numerical analysis and in all the other branches of applied mathematics.

4.4 Linear spaces

Definitions and examples

A lot of mathematical objects, e.g., geometric vectors, matrices of the same type, real functions, etc., can be added and multiplied by numbers so that the usual rules for computation hold. Such objects are usually called *vectors*. A set V of vectors such that addition and multiplication by numbers does not lead outside V is said to be a *linear space*, and a function F from one linear space V to another is said to be *linear* if $F(au + bv) = aF(u) + bF(v)$ for all vectors u, v in V and numbers a, b. These concepts are the ultimate abstractions of linear algebra, a branch of mathematics that could be described as the theory of linear spaces and linear maps between them. Using a softer language, we can say that linear algebra is a mathematical model for computations frequently performed with geometric vectors or systems of linear equations or functions. A theorem of linear algebra can be interpreted in many ways, depending on the nature of the linear space. Abstraction also pays in another way. The theorems of linear algebra are easy to grasp and prove for anyone who has made himself at home in the abstract landscape that we now proceed to describe.

A linear space or a vector space V is a set of elements $u, v, \ldots$ where sums $u + v$ and products au by numbers a are defined in such a way that the rules of computation A1 to A8 of 4.2 hold. The elements of V are called *vectors* and the numbers are called *scalars*. When V has only one element, necessarily equal to $\underline{0}$, we say that V is *trivial*. According as the scalars are rational, real, or complex numbers we say that V is a rational, real, or complex linear space or a linear space over $\mathbf{Q}$, $\mathbf{R}$, or $\mathbf{C}$. In the sequel, we let K denote one of these fields. It could also be any field.

There are plenty of examples of linear spaces. All geometric vectors in a plane or in space constitute linear spaces over the real numbers. All matrices of the same type, or, more interestingly, all solutions X of a linear homogeneous system $AX = \underline{0}$ with elements and coefficients in a field K are linear spaces over K. An important example is the set of all functions $t \to u(t)$ from an arbitrary set with elements $t, \ldots$ to numbers. Addition and multiplication are then of course defined so that $(u + v)(t) = u(t) + v(t)$ and $(au)(t) = au(t)$ for all t.

A sum $a_1v_1 + \cdots + a_pv_p$ where $v_1, \ldots, v_p$ are vectors and $a_1, \ldots, a_p$ are numbers is called a *linear combination* of $v_1, \ldots, v_p$ with coefficients $a_1, \ldots, a_p$. When $a_1 = 0, \ldots, a_p = 0$, the linear combination is called *trivial*. A set of vectors $v_1, \ldots, v_p$ is said to be *linearly independent* when none of them is a linear combination of the others. In the opposite case, when at least one of the vectors is a linear combination of the others, the set is said to be *linearly dependent*. Now it is obvious that, e.g., v_1 is a linear combination of $v_2, \ldots, v_p$ if and only if $a_1v_1 + \cdots + a_pv_p = 0$ for at least one choice of $a_1, \ldots, a_p$ such that $a_1 \neq 0$ (multiply by $1/a_1$). Hence the set is linearly dependent if and only if some nontrivial linear combination $a_1v_1 + \cdots + a_pv_p$ vanishes and linearly independent when only the trivial one vanishes. It is an important fact that more than q linear combinations of q vectors are *always* linearly dependent. For, if $v_1, \ldots, v_p$ are linear combinations of $u_1, \ldots, u_q$ (e.g., $v_1 = c_1u_1 + \ldots, \quad \ldots, \quad v_p = c_pu_p + \ldots$), then we can write $a_1v_1 + \cdots + a_pv_p$ as $b_1u_1 + \cdots + b_qu_q$ where the coefficients $b_1, \ldots, b_q$ are linear combinations of $a_1, \ldots, a_p$ (e.g., $b_1 = c_1a_1 + \cdots + c_pa_p$). By the theorem of underdetermined systems they all vanish for some nontrivial choice of $a_1, \ldots, a_p$ when $p > q$.

A part U of a linear space V is said to be a *linear subspace* when $u, v \in U \Rightarrow au + bv \in U$ for all numbers a and b. When U consists of all linear combinations of a finite number of fixed vectors, we say that U is *finitely generated* with these vectors as *generators*. When one of them is a linear combination of the others it is of course superfluous. Taking away the superfluous generators one at a time and assuming that $U = \underline{0}$, we see that U is also generated by a collection of linearly independent vectors $v_1, \ldots, v_n$. Such a collection is said to be a *basis* of U. Every u in U is then a linear combination $a_1v_1 + \cdots + a_nv_n$ which is unique, since any two such linear combinations are equal only when their coefficients coincide. From what we have proved above it follows that all bases have the same number of elements and that every set of linearly independent vectors in U can be completed to a basis (just add one vector at a time keeping the collection linearly independent). The number of elements of a basis of U is said to be its *dimension* and we denote it by $\dim U$. When U is not finitely generated, we say that its dimension is infinite and write $\dim U = \infty$. The trivial space has no basis and is said to have dimension 0.

The vector spaces coming from a Euclidean line, plane, and space have, respectively, dimensions 1, 2, and 3. All n-tuples $x = (x_1, \ldots, x_n)$ with components in K constitute a linear space called K^n with addition and multiplication by scalars according to (3). Its dimension is n and a basis is, for instance, $v_1 = (1, 0, \ldots, 0)$, $v_2 = (0, 1, 0, \ldots, 0)$, $\ldots$, $v_n = (0, \ldots, 0, 1)$. When $a_1, \ldots, a_n$ are numbers $\neq 0$, the vectors $(a_1, \ldots), (0, a_2, \ldots), \quad \ldots, \quad (0, \ldots, 0, a_n)$ also form a basis whatever numbers occur in the dotted places after a_1, after a_2, etc. The space K^n

differs only in notation from the linear space of all functions from a set with n elements to K. More examples: the complex numbers $\mathbf{C}$ form a linear space of dimension 2 over the real numbers a basis being, for instance, 1 and i. Any field F is a vector space over the field k generated by the unit of F. Hence, if F is finitely generated (over k) it has a basis $v_1, \ldots, v_n$ and consists of all $v = a_1 v_1 + \cdots + a_n v_n$ with $a_1, \ldots, a_n$ in k. When F has finitely many elements, $k = \mathbf{Z}_p$ for some prime p, and then F has p^n elements. (See the section of the preceding chapter on finite fields.)

Finally, we shall give some examples of vector spaces of infinite dimension. The real numbers $\mathbf{R}$ and hence also the complex numbers $\mathbf{C}$ are vector spaces of infinite dimension over the rational numbers $\mathbf{Q}$. In fact, if every real number were a linear combination with rational coefficients of finitely many fixed real numbers, the real numbers would be countable and this is not the case. Another example is the linear space of all functions from an interval to K. In fact, dividing the interval into a collection of subintervals and choosing for each one a function $\neq 0$ vanishing outside it, we get a set of linearly independent functions. Not even the subspaces of continuous or differentiable functions (when $K = \mathbf{R}$ or $\mathbf{C}$) have finite dimension. Restricting ourselves further to the linear space of all polynomials $v(x) = a_0 + a_1 x + \cdots + a_n x^n$ in one real variable with real coefficients, we also get a linear space of infinite dimension. In fact, such a polynomial vanishes in an interval if and only if all its coefficients vanish, and hence the polynomials $1, x, x^2, \ldots$ are linearly independent. On the other hand, all polynomials of degree at most n constitute a linear subspace of dimension $n + 1$ with basis $1, x, \ldots, x^n$.

The elements of a linear space can, of course, also be thought of as points. A straight line between two points u and v would then be the set of points $tu + (1 - t)v$ where t is any real number. When $0 < t < 1$ these points are said to lie *between* u and v and constitute a *line segment* with end points u and v. A subset of a linear space containing, together with any two points u and v, also the line segment between them, is said to be *convex*. All this is just as in Euclidean space. In the last example above, all polynomials $a_0 + a_1 x + \cdots + a_n x^n$ with positive coefficients or with coefficients adding up to a fixed number constitute convex sets.

Linear functions

A function F from one linear space U to another one, V, is said to be linear if

$$F(au + bv) = aF(u) + bF(v) \tag{16}$$

for all vectors u, v in U and numbers a, b. We also use the words linear map, transformation, or operator and, when $V = K$ consists of scalars, linear form. When $e_1, \ldots, e_n$ is a basis of U, (16) shows that $F(x_1 e_1 + \cdots + x_n e_n) = x_1 F(e_1) + \cdots + x_n F(e_n)$ for all numbers $x_1, \ldots, x_n$. Conversely, taking this formula with $F(e_1), \ldots, F(e_n)$ arbitrary in V as a definition we get a linear function F from U to V. (The reader should check this statement as an exercise.) Briefly: a linear function from a linear space of finite dimension is uniquely determined by its values on a basis, and these values can be chosen arbitrarily. Note that if $f_1, \ldots, f_p$ is a basis of the space V, the equations

$$F(e_j) = \sum_{k=1}^{p} a_{kj} f_k, \qquad (j = 1, \ldots, n),$$

define a $p \times n$ matrix $A = (a_{kj})$. (Actually, the map $F \to A$ is a bijection from the set of linear functions from U to V to the set of matrices of this type.)

It follows from (16) that the *image*

$$\mathrm{Im}\, F = \{ F(u);\ u \in U \}$$

and the *kernel*

$$\ker F = \{ u \in U;\ F(u) = \underline{0} \}$$

are linear subspaces of V and U respectively. When $\dim U < \infty$ and $e_1, \ldots, e_n$ is a basis of U extended from a basis $e_1, \ldots, e_p$ of $\ker F$, then $F(x_1 e_1 + \cdots + x_n e_n) = x_{p+1} F(e_{p+1}) + \cdots + x_n F(e_n)$, i.e., $\mathrm{Im}\, F$ is generated by the vectors $F(e_{p+1}), \ldots, F(e_n)$. These vectors are also linearly independent since $x_1 e_1 + \cdots + x_n e_n$ belongs to $\ker F$ if and only if $x_{p+1} = 0, \ldots, x_n = 0$. Hence we have proved

The theorem of the image and the kernel. *If U and V are linear spaces and F is a linear function from U to V and $\dim U < \infty$, then*

$$\dim \ker F + \dim \mathrm{Im}\, F = \dim U.$$

This remarkably concise result contains almost everything we know about systems of linear equations. In fact, if $A = (a_{kj})$ is the matrix defined above, the equation $F(x_1 e_1 + \cdots + x_n e_n) = y_1 f_1 + \cdots + y_p f_p$ is equivalent to the matrix equation $AX = Y$ where $X = (x_1, \ldots, x_n)^t$ and $Y = (y_1, \ldots, y_p)^t$. The case $Y = \underline{0}, p < n$ gives the theorem of underdetermined systems. In fact, $\dim \ker F \geqslant n - p$ is then positive. When $p = n$, the theorem shows that $\dim \ker F = 0 \Leftrightarrow \dim \mathrm{Im}\, F = \dim V$ where the last equality means that $\mathrm{Im}\, F = V$. Hence the map $X \to AX$ is injective and surjective at the same time. This is the theorem of quadratic systems. We also see that the rank of A equals $\dim \mathrm{Im}\, F$, called, in the sequel, the *rank* of F.

Invariant subspaces, eigenvectors, eigenvalues, spectrum

Let U be a linear space, let F be a linear map from U to U and, for simplicity, write $F(u)$ as Fu. A subspace V of U is said to be invariant if $FV \subset V$. When V has dimension 1 and v is a basis of V, this means that $Fv = \lambda v$ where λ is a number. This is also expressed by saying that v is an *eigenvector* of F and λ the corresponding *eigenvalue*. All solutions u of the equation $Fu = \lambda u$ constitute a linear space, the *eigenspace* of λ, and its dimension is called the *multiplicity* of λ. In other words, that λ is an eigenvalue of F means precisely that $\ker(F - \lambda I) \neq 0$ where $I : u \to u$ is the identity map from U to U. When U has a finite basis $e_1, \ldots, e_n$ and $Fe_j = \Sigma a_{jk}e_k$, it is clear that λ is an eigenvalue of F if and only if $\det(A - \lambda E) = 0$ where $A = (a_{jk})$ and E is the unit matrix of order n. Hence, by the fundamental theorem of algebra, F has at least one eigenvalue provided U is a complex space of finite dimension.

When the dimension of U is finite, the set of eigenvalues of F is called the *spectrum* of F. In view of the theorem of the image and the kernel, the spectrum then consists of all numbers λ for which $F - \lambda I$ is *not* a bijection. When the dimension of U is infinite, this property will serve to define the spectrum. We give an example. Let U be the real linear space of all real polynomials $u(t)$ in one real variable t and let F be multiplication by t so that $(Fu)(t) = tu(t)$. If $(F - \lambda I)u = 0$, then $(t - \lambda)u(t) = 0$ for all t so that $u(t) = 0$ for all t. Hence $\ker(F - \lambda I) = 0$ for all λ. On the other hand, the equation $(F - \lambda I)u = v$, i.e., $(t - \lambda)u(t) = v(t)$ for all t, has a solution u which is a polynomial only if $v(\lambda) = 0$ and hence $\mathrm{Im}(F - \lambda I) \neq U$ for all λ. Hence $F - \lambda I$ is never a bijection, the spectrum of F is the whole real axis, but F fails to have eigenvalues. (It follows from the properties of continuous functions (see Chapter 5) that if U is the space of real continuous functions from an interval J on the real axis, F as defined above has the spectrum J but no eigenvalues.)

Complement, codimension, projection

A linear space U is said to be the *sum* of two of its subspaces V and W if $U = V + W$ in the sense that every u in U can be written in at least one way as $v + w$ where $v \in V$ and $w \in W$. When this can be done in precisely one way, i.e., when $u = \underline{0} \Rightarrow v = \underline{0}$ and $w = \underline{0}$, the sum is said to be *direct* and we write $U = V \dotplus W$ with a dot over the plus sign. An equivalent condition is that $U = V + W$ and $V \cap W = 0$. Under these circumstances we also say that V and W are complementary subspaces of U, each the complement of the other. When W and W' are two complements of V in U, we can write every w in W as $v + w'$ where $v \in V$ and $w' \in W'$ are uniquely determined by w. Hence, by the symmetry of W and W', $F(w) = w'$ defines a bijection F between the two complements W and W'. This bijection is linear, for if $w_1 = v_1 + w_1'$ and $w_2 = v_2 + w_2'$ then $aw_1 + bw_2 = av_1 + bv_2 + aw_1' + bw_2'$ for all numbers a and b, i.e., $F(aw_1 + bw_2) = aF(w_1) + bF(w_2)$. Hence F maps linearly independent vectors into linearly independent ones, and from this it follows that all complements of a given linear subspace V have the same dimension. This is also called the *codimension* of V and we denote it by codim V. When dim U is finite we may extend a basis of V to a basis of U and from this it follows that dim V + codim V = dim U. But it may happen that dim $U = \infty$, dim $V = \infty$, but codim $V < \infty$. Example: U consists of all polynomials $u(x) = a_0 +$

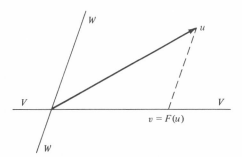

Figure 4.15 Projection on a subspace along a complement.

$a_1 x + \ldots$ and V of all such polynomials with $a_0 = 0, \ldots, a_p = 0$. Then V has the codimension $p + 1$ and a complement of V is generated by, e.g., the polynomials $1, x, \ldots, x^p$.

When $U = V + W$ is the direct sum of V and $u = v + w$ is the corresponding decomposition of an element u of U, the formula $F(u) = v$ defines a linear map from U to V such that ker $F = W$, Im $F = V$. It is called the *projection* of U onto V along W and is illustrated by Figure 4.15 where dim $U = 2$.

Operations on linear functions

Starting from two linear functions we may construct others. This will be clear from the following theorem which shows why we call linear algebra the theory of linear spaces and linear maps between them.

Theorem. *Linear combinations, compositions, and inverses of linear functions are linear.*

The proof is just a collection of straightforward verifications. If F and G are linear functions from U to V so is the linear combination $H = cF + dG$, for

$$H(au + bv) = cF(au + bv) + dG(au + bv)$$

$$= acF(u) + bcF(v) + adG(u) + bdG(v) = aH(u) + bH(v).$$

If F is linear from U to V and G linear from V to W and $H = G \circ F$, then

$$H(au + bv) = G(F(au + bv)) = G(aF(u) + bF(v))$$

$$= aG(F(u)) + bG(F(v)) = aH(u) + bH(v).$$

If F is linear from U to V with inverse G then $G(aF(u) + bF(v)) = au + bv = aG(F(u)) + bG(F(v))$, where $F(u)$ and $F(v)$ are arbitrary in V.

The linear space of all linear functions F from U to V will be denoted by $L(U, V)$. When $e_1, \ldots, e_n$ is a basis of U and $f_1, \ldots, f_p$ a basis of V, the formula $F(e_j) = \Sigma a_{kj} f_k$ gives a linear bijection $F \to A = (a_{kj})$ from $L(U, V)$ to the linear space of all matrices of type $p \times n$. Since its dimension is pn we know that dim $L(U, V) = $ dim $U \cdot$ dim V when both

85

dimensions of the right side are finite. (Putting $0 \cdot \infty = 0$, the formula holds generally).

Pathology of bases

Using a transfinite version of mathematical induction it is possible to show that every linear space has a basis. A basis of $\mathbf{R}$ considered as a linear space over the rational numbers $\mathbf{Q}$ consists of an uncountable collection of real numbers e_α such that every real number x is a unique finite sum $\Sigma x_\alpha e_\alpha$ with rational coefficients x_α. The functions $x \to x_\alpha$ are then linear over $\mathbf{Q}$ but with some pathological properties. They are, for instance, not bounded in any interval. Their bad behavior is one of many reasons to introduce the notion of size into linear algebra. This will be done next.

4.5 Linear spaces with a norm

Distance and Banach spaces

Let U be a real or complex linear space of finite dimension. How are we to measure the length $|u|$ of a vector u and the distance $|u - v|$ between two vectors u and v? One way would be to introduce a basis $e_1, \ldots, e_n$ of U and put, e.g.,

$$|u| = \left(|x_1|^2 + \cdots + |x_n|^2\right)^{1/2}$$

where $x_1, \ldots, x_n$ are the coordinates of $u = x_1 e_1 + \cdots + x_n e_n$. But this length and other lengths using coordinates, e.g., $|u| = |x_1| + \cdots + |x_n|$, depend on the choice of a basis. The accepted way out is simply to suppose that there is measure of length in U, more precisely a function $u \to |u|$ from U to real numbers $\geqslant 0$, such that

$$|u| = 0 \quad \Rightarrow \quad u = \underline{0},$$

$$|au| = |a|\,|u|,$$

$$|u + v| \leqslant |u| + |v| \qquad \text{(the triangle inequality)}$$

for all vectors u, v in U and all numbers a. Such a function is also called a *norm*. The two functions $|u|$ proposed above have these properties and are indeed norms. When U consists of bounded scalar functions $t \to u(t)$ from some set I, we get a natural norm by putting

$$|u| = \sup|u(t)| \quad \text{when } t \text{ belongs to } I. \tag{17}$$

(See Chapter 5, Limits, continuity, and topology. Most of what now follows presupposes some knowledge of that chapter.) When every function $t \to |u(t)|$ has a largest value, we may replace sup in (17) by max. This is the case, for instance, when I is finite or when I is a closed bounded interval and all the u's are continuous functions.

Just as in the Euclidean case, the inequality $|u - u_0| \leqslant r$ where u_0 and r are fixed, represents a ball with center u_0 and radius r. By the triangle inequality it is convex. Here is a picture of it:

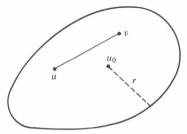

Figure 4.16 A ball with center u_0 and radius r in a normed linear space.

Let us now consider an infinite sequence $u_1, u_2, \ldots$ of points of U having a limit v (the elements of U, the vectors, are now thought of as points). This means that the distances $|u_k - v|$ tend to zero as $k \to \infty$. When U consists of functions from a set I with the norm (17), this means that the functions u_k tend to v uniformly. In the general case, the distances $|u_k - u_j| \leqslant |u_k - v| + |u_j - v|$ (use the triangle inequality) tend to zero as the smaller of the numbers j and k tends to ∞. A sequence with this last property is said to be a *Cauchy sequence*. When U has finite dimension, every Cauchy sequence has a unique limit. This is an easy consequence of the corresponding property of numbers, real or complex. But if dim $U = \infty$, things are different. Example: according to Weierstrass's approximation theorem (see Chapter 9) every continuous function from a closed bounded interval can be approximated uniformly by polynomials. But not every such function is a polynomial. To avoid such complications, one usually assumes about U that every Cauchy sequence has a limit point (necessarily unique). Since the 1920s, linear spaces with this property have been called *complete normed spaces* or *Banach spaces*, after a celebrated Polish mathematician. The most telling example is perhaps the space of all continuous (real or complex) functions from a closed bounded interval.

The norm turns out to be a powerful tool and one can do lots of things with it. We shall now state and prove a theorem with important applications in analysis (there are two of them in Chapter 7). The theorem has to do with a contraction, i.e., a function T from a space to itself that does not increase distances. This means that

$$|T(u) - T(v)| \leqslant |u - v|$$

when $T(u)$ and $T(v)$ are defined. Note that T does not have to be linear, nor defined everywhere. The reader is asked to illustrate the inequality above, the theorem itself, and its proof with some figures. The theorem is not supposed to make a deep impression before one has seen the applications.

The contraction theorem

Theorem (Contraction theorem). *Let U be a Banach space and let $u \to T(u)$ be a contraction such that*

$$|u - u_0| \leqslant r, \quad |v - u_0| \leqslant r \quad \Rightarrow \quad |T(u) - T(v)| \leqslant c|u - v|$$

where $c < 1$ and $r > 0$ are given numbers and u_0 a given point of U. Then, if v is so close to u_0 that

$$|v - u_0| \leqslant (1 - c)r - |T(u_0) - u_0|, \tag{18}$$

the equation

$$u - T(u) = v - u_0 \tag{19}$$

has a unique solution u such that $|u - u_0| \leqslant r$.

Note that when the right side of (18) is negative, the theorem says nothing, but if $|T(u_0) - u_0|$ is small enough, there are vectors satisfying (18). The theorem can be used in two ways. First, if we put $v = u_0$ and $T(u_0)$ is so close to u_0 that $|T(u_0) - u_0| \leqslant (1 - c)r$, the theorem says that there is a u such that $T(u) = u$, in other words, u is a fixed point under T. Second, assuming that u_0 is a fixed point, the theorem says that the equation $T(u) = v$ has a unique solution u close to u_0 when v is sufficiently close to u_0. Finally, when T is linear and such that $|T(u)| \leqslant c|u|$ for all u and some $c < 1$, the theorem simplifies enormously. Taking $u_0 = 0$ and r arbitrary, it says that $u \to u - T(u)$ is a bijection of U.

After these remarks we shall now prove the theorem. If there were two solutions, u' and u'', we get $u' - u'' = T(u') - T(u'')$ so that $|u' - u''| \leqslant c|u' - u''|$ and hence $u' = u''$ since $c < 1$. To construct a solution of (19) we use successive approximations according to the following scheme:

$$u_1 - T(u_0) = v - u_0$$
$$u_2 - T(u_1) = v - u_0$$
$$u_3 - T(u_2) = v - u_0$$
$$. \; . \; . \; .$$

The first approximation to (19) is then $u = u_0$, the second one is u_1 as defined above, and so on. The success depends on the fact that $T(u)$ changes less than u itself when a small vector is added to u. If, for instance, $T(u) = u_0$ for all u, our scheme gives the solution $u = v$ already in the first step. The first equation above and the triangle inequality give

$$|u_1 - u_0| \leqslant |v - u_0| + |T(u_0) - u_0|$$

so that, by (18),

$$|u_1 - u_0| = \leqslant (1 - c)r.$$

Hence the distance from u_1 to u_0 is less than r, $T(u_1)$ is defined and so is

u_2. Subtracting the first equation from the second one we have

$$|u_2 - u_1| = |T(u_1) - T(u_0)| \leqslant c|u_1 - u_0| \leqslant c(1 - c)r$$

so that

$$|u_2 - u_0| \leqslant |u_2 - u_1| + |u_1 - u_0| \leqslant (1 + c)(1 - c)r.$$

Here again, the right side is less than r, $T(u_2)$ is defined and so is u_3. After k such steps we find that $|u_{k+1} - u_k| \leqslant c^k(1 - c)r$ and

$$|u_{k+1} - u_0| \leqslant (1 + c + c^2 + \cdots + c^k)(1 - c)r < r$$

so that our procedure can be continued indefinitely. We also get

$$k \geqslant j \quad \Rightarrow \quad |u_{k+1} - u_j| \leqslant |u_{k+1} - u_k| + \cdots + |u_{j+1} - u_j|$$
$$\leqslant (c^k + \cdots + c^j)(1 - c)r$$

where the right side does not exceed $c^j r$ and hence tends to zero as j tends to infinity. Hence, since the space U is complete, the sequence $u_0, u_1, \ldots$ has a limit u such that $|u - u_0| \leqslant r$. Since $u_{k+1} - T(u_k) = v - u_0$ for all k, and the contraction property entails that $T(u_k)$ tends to $T(u)$ as u_k tends to u, the element u is the solution of (19).

Schemes of approximation are an important part of mathematics. They were used by the Greeks to approximate areas, by Newton to find zeros of real functions, by Cauchy and Picard in the nineteenth century to construct solutions of differential equations, and they are used every day in numerical mathematics. The contraction theorem is just an adaptation of older results to Banach spaces. It has a companion, also with important applications to analysis, namely the Schauder fixed point theorem (1930). It says that a continuous mapping of a compact convex set of a Banach space into itself has at least one fixed point (the word compact is explained below). Both these results are examples of analysis in infinite-dimensional spaces, a branch of mathematics usually called *functional analysis*. The word functional was originally applied to functions from spaces of infinite dimension, themselves consisting of scalar functions from intervals.

4.6 Boundedness, continuity, and compactness

Bounded linear functions

Let U and V be Banach spaces and F a linear function from U to V. If dim U is finite and $e_1, \ldots, e_n$ is a basis of U, then $F(x_1 e_1 + \cdots + x_n e_n) = x_1 F(e_1) + \cdots + x_n F(e_n)$ so that, taking norms in V on both sides,

$$|F(x_1 e_1 + \cdots + x_n e_n)| \leqslant c_0(|x_1| + \cdots + |x_n|)$$

where c_0 is the largest of the numbers $|F(e_1)|, \ldots, |F(e_n)|$. On the other hand, we have $|x_1| + \cdots + |x_n| \leqslant c_1|x_1 e_1 + \cdots + x_n e_n|$ for some number c_1 where the bars on the right side now denote the norm in U. In fact, $|x_1| + \cdots + |x_n| = 1$ defines a compact part K of $\mathbf{R}^n$ (or $\mathbf{C}^n$), and the properties of the norm show $h(x) = |x_1e_1 + \cdots + x_ne_n|$ to be a positive continuous function from K, and hence it has a positive minimum m there. If $|x| = |x_1| + \cdots + |x_n|$ we then have $h(x/|x|) = h(x)/|x| \geqslant m$ for all $x \neq 0$ so that $h(x) \geqslant m|x|$ for all x and our assertion holds with $c_1 = 1/m$. Finally, putting $c = c_0c_1$, we get

$$|F(u)| \leqslant c|u| \tag{20}$$

for all u in U. A linear function F is said to be bounded when this inequality holds for some $c < \infty$ depending on F but independent of u. We have just shown that every linear function from U to V is bounded when U has finite dimension. Applying this to the case when U and V are the same linear space of finite dimension but with different norms $|u|_1$ and $|u|_2$ and $F(u) = u$ for all u, we see that the two norms are equivalent in the sense that there are two positive numbers c_1 and c_2 such that, for all u,

$$c_1|u|_1 \leqslant |u|_2 \leqslant c_2|u|_1.$$

When dim U is infinite, linear functions F from U to V are no longer automatically bounded so we put down (20) as a requirement and let $L(U, V)$ be the set of linear bounded functions from U to V. When F is such a function, $|F(u) - F(v)| = |F(u - v)| \leqslant c|u - v|$. Hence, if $u_1, u_2, \ldots$ is a sequence in U converging to u, $F(u_k)$ converges to $F(u)$ in V. In other words, the function F is continuous. In particular, if all $F(u_k)$ vanish, $F(u)$ vanishes. Hence the kernel of F is a closed linear space, i.e., it contains its limit points. Let us also remark that a linear function is bounded and continuous at the same time. In fact, we already know that bounded $\Rightarrow$ continuous. When F is not bounded, there is a sequence of vectors $u_k \neq 0$ such that $|F(u_k)|/|u_k| \to \infty$ when $k \to \infty$ and then $v_k = u_k/|F(u_k)| \to 0$ while $|F(v_k)| = 1$, contradicting the continuity. Hence continuous $\Rightarrow$ bounded.

Let $|F|$ be the best constant in (20), i.e., the greatest lower bound of all c for which (20) holds. It is immediately seen that the map $F \to |F|$ is a norm on the linear space $L = L(U, V)$. This space is also complete for if $|F_k - F_j| \to 0$ as $\min(k, j) \to \infty$, $F(u) = \lim F_k(u)$ exists for every u in U and F is clearly linear and bounded with $|F| = \lim|F_k|$. Hence the space L is in fact a Banach space. In addition, the norm $|F|$ behaves very decently under composition. The computation $|(F \circ G)(u)| = |F(G(u))| \leqslant |F| \, |G(u)| \leqslant |F| \, |G| \, |u|$, true for all u, shows that $|(F \circ G)| \leqslant |F| \, |G|$. It follows, for instance, that $L(U, U)$ is a normed ring. Since it is also a Banach space it is called a *Banach algebra*. Its elements are usually referred to as *bounded linear operators*.

Let us now consider just one bounded linear operator $F : U \to U$. When $|F| < 1$, $I - F$ is a bijection by the contraction theorem. Hence, when a is a number and $|a| > |F|$, i.e., $|F/a| < 1$, then $F - aI = -a(I - F/a)$ is a bijection. It follows that the spectrum of F is contained in the interval $|a| \leqslant |F|$ (a circle when U is a complex space). It is an important fact, announced by Mazur in 1938 and proved by Gelfand in 1941, that the spectrum of F is never empty. We have seen earlier that this does not mean that F has eigenspaces. Actually, F need not even have closed invariant subspaces different from 0 and U. This was proved by Per Enflo in 1976. We should perhaps remark here that well-behaved operators, e.g., the self-adjoint ones to be treated in the last section of this chapter, do in general have plenty of closed invariant subspaces.

Dual spaces and linear functionals

Let U be a Banach space. The space $L(U, K)$, where $K = R$ or C is the space of scalars of U, is called the *dual space* of U and is usually denoted by U'. Its elements are the bounded linear functions from U to K. They are traditionally called *linear functionals*. When U has finite dimension, linear form and linear functional are the same thing and we have dim U' = dim U. When $U = C(I)$ is the Banach space of all continuous scalar functions from a compact interval I and the norm is taken to be $|u| = \max|u(t)|$, every function

$$f(u) = c_1 u(t_1) + c_2 u(t_2) + \cdots + c_n u(t_n)$$

where $c_1, \ldots, c_n$ are fixed numbers and $t_1, \ldots, t_n$ are fixed points in I and n is arbitrary, is a linear functional on U. In fact, f is linear and $|f(u)| \leq (|c_1| + \cdots + |c_n|)|u|$. It is an important result in integration theory that every linear functional on $C(I)$ can be written as a Riemann-Stieltjes integral $\int_I u(t)dg(t)$, where g is the difference between two increasing bounded functions. (In the complex case, Re g and Im g have this property. See Chapter 8 on integration.) This remark shows that we are now in deep water. When the dimension of U is infinite, even to show that linear functionals exist is a problem. It was solved in the twenties by Hahn and Banach. They proved that if V is a closed linear subspace of a Banach space U and u_0 is a point in U outside V there is a linear functional f in U' such that $f = 0$ on V but $f(u_0) \neq 0$. This result is still the flagship of linear functional analysis and has scores of important applications to analysis. The gist of the proof is as follows. If W is a linear subspace of a real linear space U and v is outside W and f is a linear scalar function from W then $f(w + av) = f(w) + af(v)$ is a linear function from $W + Rv$ whatever $f(v)$ is. If U has a norm and $|f(w)| \leq |w|$ on W, it is possible to choose $f(v)$ so that the same inequality holds on $W + Rv$. In fact, using the homogeneity, this amounts to the inequality $|f(w + v)| \leq |w + v|$ for all w in W, which in turn results from the two inequalities $-|w_1 + v| - f(w_1) \leq f(v) \leq |w_2 + v|$ $- f(w_2)$ for all w_1 and w_2 in W. That there is room for an $f(v)$ between the two extremes follows immediately from the triangle inequality. Hence we have unlimited possibilities of extending linear functionals so that the inequality $|f(u)| \leq c|u|$ with a fixed c is respected, and this opens the way to a proof by induction.

It follows from the Hahn–Banach theorem that dim $U = \infty \Rightarrow$ dim $U' = \infty$. It also follows easily that if V is a closed linear subspace of U of finite codimension p, then there are p linearly independent linear functionals $f_1, \ldots, f_p$ such that V is defined by the system of equations $f_1(u) = 0, \ldots, f_p(u) = 0$. Conversely, if $f_1, \ldots, f_p$ are linearly independent linear functionals, the same system of equations defines a closed linear subspace of codimension p. Another important theorem by Banach says that the inverse of a continuous bijection between Banach spaces is continuous. A consequence of this remarkable fact is that any two norms on a Banach space are equivalent. Note that our proof in the finite-dimensional case is of no use when the dimension is infinite.

Compactness and compact operators

A subset of a Banach space is said to be *bounded* when the norm is bounded on it, and *closed* when it contains its limit points. It is said to be *precompact* when every sequence of points in the set has a convergent subsequence, and *compact* when it is precompact and closed. Precompact sets have to be bounded. In $\mathbf{R}^n$ and, more generally, Banach spaces of finite dimension, bounded sets are precompact. But when the dimension is infinite, it is not difficult to construct infinite sequences $e_1, e_2, \ldots$ of unit vectors such that $|e_j - e_k| > 1 - \delta$ when $j \neq k$ and δ is any positive number given in advance. It follows that no ball with a positive radius, however small, is precompact. This basic property of infinite-dimensional Banach spaces is sometimes expressed by saying that they are not *locally compact*.

By definition, a bounded function F from a Banach space U to another one, V, sends bounded sets into bounded sets. When F sends bounded sets into precompact sets, e.g., when F has finite rank, F is said to be *compact* or a *compact operator*. It is very easy to verify that all such operators form a closed linear subset of $L(U, V)$ and that a product $F \circ G$ is compact when one of the factors is compact. An interesting property of compact operators is that the theorem of the image and the kernel, phrased as $\dim \ker A = \operatorname{codim} \operatorname{Im} A$ holds for all $A = E + F$ in $L(U, U)$ when F is compact and E is the identity map from U to U. The following theorem was proved by F. Riesz in 1918, although stated in a different language.

The theorem of compact operators. *When U is a Banach space and $F \in L(U, U)$ is compact, then $\ker(E + F)$ and $\operatorname{Im}(E + F)$ are closed and*

$$\dim \ker(E + F) = \operatorname{codim}(E + F) \tag{21}$$

is finite. In addition, when one side vanishes, $E + F$ is a continuous bijection with a continuous inverse also of the form $E + G$ where G is compact.

That $\ker(E + F)$ is closed we know already. The rest of the theorem is more difficult and we cannot go into the proof. The theorem says that, for instance, there are $p = \dim \ker(E + F)$ linearly independent vectors $u_1, \ldots, u_p$ in U and linear functionals $f_1, \ldots, f_p$ from U such that the equation $u + F(u) = v$ has a solution if and only if $f_1(v) = 0, \ldots, f_p(v) = 0$, and that two solutions differ by a linear combination of $u_1, \ldots, u_p$. The theorem also gives information about the eigenvectors and eigenvalues of F, e.g., that every eigenvalue $z \neq 0$ has finite multiplicity. In fact, since $F - zE = -z(E - z^{-1}F)$ and $z^{-1}F$ is compact, $\ker(F - zE)$ has finite dimension. Riesz also proved that 0 is the only possible limit of a convergent sequence of eigenvalues. Hence the spectrum of F consists of at most countably many eigenvalues, all of finite multiplicity and clustering only at the origin, which may or may not be an eigenvalue but which belongs to the spectrum.

A concrete version of the theorem just stated and commented on was proved in 1900 by Ivar Fredholm. He considered the case when $U = C(I)$ where I is a compact interval and F is given by (we write $F(u) = Fu$)

$$Fu(s) = \int_I K(s, t)u(t)\, dt$$

where K is a continuous function. Such an F is compact, and the linear equation $u + Fu = v$, considered by Fredholm, is then

$$u(s) + \int_I K(s, t)u(t)\, dt = v(s).$$

It is called an integral equation of the second kind. Since many important physical problems, e.g., the Dirichlet problem (treated below) can be formulated as such integral equations, Fredholm's work was sensational.

Returning to the abstract theory, it was for a long time an open question whether every compact operator is the uniform limit of a sequence of compact operators of finite rank. Many of them are, as a matter of fact. In 1972, Per Enflo gave the answer. It is negative.

4.7 Hilbert spaces

Inner products, Euclidean spaces, and Hilbert spaces

A real function $u, v \rightarrow (u, v)$ from the product $U \times U$ of a real linear space with itself is called an *inner product* when it has the properties B1–B4 of section 4.2. The fact that

$$(tu + sv, tu + sv) = t^2(u, u) + 2st(u, v) + s^2(v, v) \geq 0$$

for all t, s gives the inequality $(u, v)^2 \leq (u, u)(v, v)$, called Schwarz's inequality and usually written as $|(u, v)| \leq |u|\,|v|$ where $|u| = (u, u)^{1/2}$. Putting $s = 0$ in the equality above we get $|tu| = |t|\,|u|$, and putting $s = t = 1$, we get $|u + v|^2 \leq |u|^2 + 2|u|\,|v| + |v|^2 = (|u| + |v|)^2$, i.e., the triangle inequality $|u + v| \leq |u| + |v|$. Besides, it follows from B2 that $|u| = 0 \Rightarrow u = 0$ so that $|u|$ is a norm. A complete linear space with an inner product is called a *Hilbert space* after a famous German mathematician. The elements of the space Hilbert worked with around 1905 were infinite sequences $u = (x_1, x_2, \ldots)$ of real number such that $|u|^2 = x_1^2 + x_2^2 + \cdots < \infty$ with

93

the inner product $(u, v) = x_1 y_1 + x_2 y_2 + \ldots$ where $v = (y_1, y_2, \ldots)$. A *Euclidean linear space* is a Hilbert space of finite dimension.

Complex linear spaces can be fitted with inner products in much the same way as real ones. One then allows the inner product to have complex values, and B1 to B4 are modified as follows,

$$(u, u) \geqslant 0, \qquad (u, v) = \overline{(v, u)}, \qquad (u, u) = 0 \quad \Rightarrow \quad u = \underline{0}$$

$$(au + bv, w) = a(u, w) + b(v, w), \qquad (u, av + bw) = \bar{a}(u, v) + \bar{b}(u, w)$$

for all vectors u, v, w and complex numbers a and b. Then we have Schwarz's inequality $|(u, v)| \leqslant |u| \, |v|$, and $u \to |u|$ is still a norm. A complex linear space with an inner product is also called a Hilbert space when it is complete. Allowing the numbers $x_1, x_2, \ldots$ of our earlier example to be complex, requiring that $|u|^2 = |x_1|^2 + |x_2|^2 + \ldots < \infty$ and putting $(u, v) = x_1 \bar{y}_1 + x_2 \bar{y}_2 + \ldots$ we get such a space.

In mathematical practice most Hilbert spaces appear as spaces of functions where the inner product involves a sum or an integral. Let, for instance, M be a set and U the set of real or complex functions $t \to u(t)$ from M. Then, if M is finite and $r(t) > 0$ for all t,

$$(u, v) = \sum u(t) \overline{v(t)} \, r(t)$$

is an inner product on M. If M is countable we can use the same formula provided U is restricted by the condition that $\sum |u(t)|^2 r(t) < \infty$. When M is an interval I we can change the sum to an integral

$$(u, v) = \int u(t) \overline{v(t)} \, r(t) \, dt$$

where r is positive everywhere, provided we restrict ourselves to functions such that $\int |u(t)|^2 r(t) \, dt < \infty$. The completeness may still be problematic but can be saved with a suitably defined integral (the Lebesgue integral). Inner products may also involve derivatives and multiple integrals. But the details of this are too numerous for us and we retreat to the abstract situation.

A finite or infinite set of unit vectors $e_1, e_2, \ldots$ that are pairwise orthogonal, i.e., such that $(e_j, e_k) = \delta_{jk} = 1$ when $j = k$ and $= 0$ otherwise, is called an *orthonormal system*. Such systems are important both conceptually and practically. The coefficients $x_1, \ldots, x_p$ of a linear combination

$$u = x_1 e_1 + \cdots + x_p e_p$$

are then inner products, $x_1 = (u, e_1), \ldots, x_p = (u, e_p)$ and $|u|^2 = |x_1|^2 + \cdots + |x_p|^2$. In particular, the vectors of an orthonormal system are linearly independent. Hence, if $\dim U = p$, the vectors $e_1, \ldots, e_p$ form a basis of U, and we say that the system is *complete*. When an orthonormal system $e_1, e_2, \ldots$ has infinitely many elements and $|x_1|^2 + |x_2|^2 + \ldots < \infty$, the vectors

$$u_p = x_1 e_1 + \cdots + x_p e_p$$

form a Cauchy sequence, for

$$q > p \quad \Rightarrow \quad |u_q - u_p|^2 = |x_{p+1}|^2 + \cdots + |x_q|^2.$$

We write the limit element u as a linear combination with infinitely many terms,

$$u = x_1 e_1 + x_2 e_2 + x_3 e_3 + \cdots = \sum_1^\infty x_k e_k$$

Passing to the limit we also get $x_k = (u, e_k)$. When every u in U can be obtained in this way, we say that the orthonormal system is complete and constitutes a *basis* of U. Such bases always exist, perhaps with more than countably many elements, but to go into this would again carry us too far afield. Instead we shall do a little geometry in Hilbert space, which works regardless of dimension.

Orthogonal complement and orthogonal projection

The *orthogonal complement* $V^\perp$ of a linear subspace V of U consists of all w in U for which $(w, V) = 0$, i.e., w is orthogonal to every v in V. It is clear that $V^\perp$ is a linear subspace and that it is closed. In fact, by Schwarz's inequality, $|(w_k, v) - (w, v)| = |(w_k - w, v)| \leqslant |w_k - w| \, |v|$ so that $(w_k, v) \to (w, v)$ when $w_k \to w$. We shall now prove what amounts to the ultimate version of the theorem of Pythagoras.

The projection theorem. *Let U be a Hilbert space and V a closed linear subspace. Then every u in U is a unique sum $v + w$ where v is in V and w in $V^\perp$ and $|u|^2 = |v|^2 + |w|^2$.*

The last statement follows since $(v, w) = 0$. We say that v is the *orthogonal projection* of u onto V. Figure 4.17 illustrates the theorem. The proof depends on the so-called parallelogram identity $|u - v|^2 + |u + v|^2 = 2|u|^2 + 2|v|^2$. We write it as

$$|v_p - v_q|^2 = 2|u - v_p|^2 + 2|u - v_q|^2 - 4|u - (v_p + v_q)/2|^2$$

where u is fixed and $v_1, v_2, \ldots$ is a sequence in V. The right side is not larger than $2|u - v_p|^2 + 2|u - v_q|^2 - 4d^2$ where $d = \inf |u - v|$ for all v in V is the distance from u to V. If we choose the sequence $v_1, v_2, \ldots$ so that $|u - v_p|^2$ tends to d^2 as p tends to infinity, it follows that $|v_p - v_q|$ tends to zero as p and q tend to infinity. Hence the sequence is a Cauchy sequence

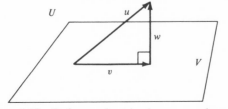

Figure 4.17 Orthogonal projection on a subspace.

95

and if v is its limit element, we have $|u - v| = d$ and hence $|u - v| \leqslant |u - v - tv'|^2$ for all v' in V and numbers t. When U is a real space it follows from this that $0 \leqslant 2t(u - v, v') + t^2(v', v')$ for all t, and this is possible only when $(u - v, v') = 0$. Hence $u - v$ is orthogonal to V and $u = v + (u - v)$ is the sum of $v \in V$ and $u - v \in V^\perp$. Since $V \cap V^\perp = \underline{0}$, the sum is unique. We leave it to the reader to prove the theorem in the complex case. The projection theorem has the following important corollary.

Theorem. *Every linear functional on a Hilbert space is an inner product.*

In other words, if $u \to f(u)$ is linear with scalar values and $|f(u)| \leqslant c|u|$ for some c and all u, then there is a v in U such that $f(u) = (u, v)$ for all u. To prove this, note that $\ker f$ is a closed linear subspace V of U not equal to U except in the uninteresting case when $f = 0$. Otherwise, f cannot vanish on $V^\perp$, so that this space must contain a w such that $f(w) = 1$. Since $u = u - f(u)w + f(u)w$ and $f(u - f(u)w) = f(u) - f(u) = 0$, it follows that $(u, w) = f(u)(w, w)$. Putting $v = w/(w, w)$ gives the desired result.

Dirichlet's problem

We are now going to apply our last theorem to a problem in mathematical physics. First, let u and v be smooth real functions from an open bounded part I of $\mathbf{R}^n$ and consider the integral

$$D(u, v) = \int_I (\partial_1 u(x)\partial_1 v(x) + \cdots + \partial_n u(x)\partial_n v(x))dx_1 \ldots dx_n$$

where $\partial_k = \partial/\partial x_k$. When $u = v$ this is called the *Dirichlet integral* of u and will be denoted by $D[u]$. Since $D(u, v) = D(v, u)$ is a linear function of u and v we have $D[u + v] = D(u + v, u + v) = D(u, u) + 2D(u, v) + D(v, v)$. If u and w are smooth functions and w vanishes on the boundary J of I, then by integration by parts,

$$D(u, w) = -\int_I w(x)\Delta u(x)dx_1 \ldots dx_n$$

where $\Delta = \partial_1^2 + \cdots + \partial_n^2$ is Laplace's operator. When u is harmonic in I, i.e., when Δu vanishes there, the integral vanishes and we have

$$D[u + w] = D[u] + D[w].$$

This formula is called *Dirichlet's principle*. Putting $v = u + w$, it can be expressed somewhat loosely as follows: among all functions v equal to a given function f on the boundary J of I, the harmonic one has the least Dirichlet integral. Here we are led to the question of existence and uniqueness in Dirichlet's problem, stated around 1840: find a harmonic function u in I taking given values on the boundary. For $n > 1$ this is a nontrivial problem but for $n = 1$ the harmonic functions are the polynomials of degree at most 1, $u(x) = cx + d$, and the problem is elementary.

For $n = 2$ and 3 Dirichlet's problem arises in the theory of elasticity, in the theory of electricity, and in hydrodynamics. We give just one example. When $n = 2$, we can think of the graph $y = v(x)$ as describing the position of a thin elastic membrane lying over the plane region I. By basic but somewhat idealized physics, its potential energy in this position has the form $aD[v] + b$, $a > 0$. Held in the position $y = f(x)$ at the boundary J, the membrane assumes a position of equilibrium $y = u(x)$ which minimizes the energy and hence should correspond to the solution of Dirichlet's problem. Figure 4.18 illustrates this.

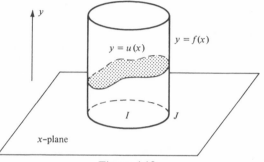

Figure 4.18

Our interpretation gives a physical existence proof of Dirichlet's problem, but we also need a mathematical one. There are at least five of them and all have been important in analysis. Riemann, who treated a similar problem in his doctoral thesis in 1851, *assumed* without discussion that there is a unique function minimizing Dirichlet's integral. Later it became clear that this has to be *proved*, which Hilbert did in 1901. We shall present a modernized version of this proof.

Let K be the space of all real, twice continuously differentiable functions from $\mathbf{R}^n$ restricted to I, and let H be the linear subspace whose elements vanish on the boundary J of I. Then the integral $(v, w) = D(v, w)$ is almost an inner product on H, the only flaw being that $(v, v) = 0$ implies that v is constant and not that v vanishes. But when we restrict (v, w) to H, it becomes an inner product. Further, assume that f can be extended from J to a function in K which we still denote by f. Then $w \to (f, w)$ is a linear functional on H and hence, by our last theorem, must be an inner product. In other words, there is g in H such that $(f, w) = (g, w)$ for all w in H. Then the function $u = f - g$ equals f on J for w vanishes there. Further, since $(u, w) = D(u, w) = 0$ for all w in H, an integration by parts (see above) shows u to be harmonic, $\Delta u = 0$ in H. It also follows that u satisfies Dirichlet's principle $D[u + w] = D[u] + D[w]$ for all w in H. If u and u' are two solutions of Dirichlet's problem, then $u - u'$ is in H and hence $(u, u - u') = 0$, $(u', u - u') = 0$ so that $(u - u', u - u') = 0$ and hence $u = u'$. This seems to clinch the matter but, unfortunately, we have repeated Riemann's mistake. It is true that the space H is linear and has an inner product, but it is not complete and hence is not a Hilbert space. On the other hand, limits of Cauchy sequences in the space can be represented by functions vanishing on J at least in a generalized sense. If we use Lebesgue integrals and show that $(u, w) = 0$ for all w in H implies that u is harmonic also when u is not a smooth function, our proof works but we have to skip the details.

4.8   Adjoints and the spectral theorem

Adjoints

When U is a Hilbert space and $F \in L = L(U, U)$ is a bounded linear operator and v an element of U, the function $u \rightarrow (Fu, v)$ is a linear functional for $|(Fu, v)| \leqslant |F| \, |u| \, |v|$. Hence, there is to every v in U a F^*v in U such that $(Fu, v) = (u, F^*v)$ for all u. The rules for computing with an inner product prove $v \rightarrow F^*v$ to be a linear function and since, with $|u| = 1$,

$$|F^*v| = \sup|(u, F^*v)| = \sup|(Fu, v)| \leqslant |F| \, |v|,$$

F^* is bounded. The operator F^* is said to be the *adjoint* of F. We have $F^{**} = F$ which, together with the inequality above, gives $|F^*| = |F|$. If $e_1, e_2, \ldots$ is an orthonormal basis of U and $Fe_j = \Sigma_k a_{jk} e_k$, $F^* e_j = \Sigma_k a_{jk}^* e_k$, we get $a_{jk}^* = (F^* e_j, e_k) = (e_j, F e_k) = \overline{(F e_j, e_k)} = \bar{a}_{kj}$. Hence, when $\dim U < \infty$, the matrices (a_{jk}) and (a_{jk}^*) are adjoint to each other. We say that F is *self-adjoint* when $F^* = F$, i.e., when $(Fu, v) = (u, Fv)$ for all u and v. An equivalent property is that (Fu, u) be real for all u. When F is self-adjoint, then $FV \subset V \Rightarrow FV^\perp \subset V^\perp$, i.e., if F maps a linear subspace into itself, then F maps the orthogonal complement into itself. The verification is immediate. This property can be used to prove a spectral theorem due to Hilbert, who proved it in 1909.

The spectral theorem

Theorem (Spectral theorem). *A compact self-adjoint linear operator on a Hilbert space $\neq 0$ has a complete orthonormal set of eigenvectors. The eigenvalues are real with the origin as their only limit point.*

The theorem is very easy to prove when U is a complex space of finite dimension. In fact, let F be the operator. By an earlier result it has an eigenvector $u \neq 0$. Hence, putting $V = Cu$ we have $FV \subset V$ and, consequently, $FV^\perp \subset V^\perp$. If $\dim V^\perp > 0$ we can find an eigenvector in $V^\perp$ and so on. Finally, we get $n = \dim U$ pairwise orthogonal eigenvectors which we can normalize so that they become unit vectors $e_1, \ldots, e_n$ constituting an orthonormal basis. Since $Fu = zu$ implies that $(Fu, u) = z(u, u)$ when z is a complex number, all eigenvalues are real. The real case reduces to the complex case via a choice of an orthonormal basis and a passage to matrices. The proof in the general infinite-dimensional case is not difficult.

The spectral theorem has a counterpart for self-adjoint bounded non-compact operators and even for unbounded operators. In its final form it was proved by von Neumann in 1928.

We shall now apply the spectral theorem to quadratic forms and vibrating systems.

The spectral theorem and quadratic forms

A *real quadratic form* in n variables is a function

$$f(x) = \sum_{j,\,k=1}^{n} a_{jk} x_j x_k$$

where the sum is a linear combination of the products $x_1 x_1, \ldots, x_j x_k, \ldots$ with real coefficients a_{jk}. Since $x_j x_k = x_k x_j$ we can assume without restriction that $a_{jk} = a_{kj}$ for all j and k. The associated bilinear form

$$f(x, y) = \sum_{j,\,k=1}^{n} a_{jk} x_j y_k$$

is then symmetric, i.e., $f(x, y) = f(y, x)$ for all x and y. Next, let

$$g(x) = \sum_{j,\,k=1}^{n} b_{jk} x_j x_k$$

be another real quadratic form which is positive definite, i.e., $g(x) > 0$ when $x \neq 0$. It is clear that the associated bilinear form $(x, y) = g(x, y)$ is then an inner product on $\mathbf{R}^n$. Further, the equation

$$f(x, y) = (Ax, y)$$

defines a self-adjoint linear map $A : \mathbf{R}^n \to \mathbf{R}^n$. In fact, the function $y \to f(x, y)$ is linear and hence equal to an inner product (u, y) where $u \in \mathbf{R}^n$ depends on x. The function $x \to u$ is linear for if $f(z, y) = (v, y)$ then $f(ax + bz, y) = af(x, y) + bf(z, y) = a(u, y) + b(v, y) = (au + bv, y)$ for all y and all numbers a and b. Finally, putting $u = Ax$ we have $(Ax, y) = f(x, y) = f(y, x) = (Ay, x) = (x, Ay)$ for all x and y. Now, by the spectral theorem, $\mathbf{R}^n$ has an orthonormal basis $e_1, \ldots, e_n$ of eigenvectors of A, $Ae_j = \lambda_j e_j$ where the eigenvalues $\lambda_1, \ldots, \lambda_n$ are real. Introducing the linear forms $u_k = (x, e_k)$ and $v_k = (y, e_k)$ we get $x = u_1 e_1 + \cdots + u_n e_n$ and $y = v_1 e_1 + \cdots + v_n e_n$ and hence

$$f(x, y) = (Ax, y) = \sum u_j v_k (Ae_j, e_k) = \sum u_j v_k \lambda_j (e_j, e_k)$$

so that

$$f(x, y) = \lambda_1 u_1 v_1 + \cdots + \lambda_n u_n v_n \tag{22}$$

and, replacing A by the identity map,

$$g(x, y) = u_1 v_1 + \cdots + u_n v_n. \tag{23}$$

In particular, we have managed to write $g(x)$ as a sum of squares of linear forms and $f(x)$ as a linear combination of the same squares. The eigenvalues $\lambda_1, \ldots, \lambda_n$ of A are now also said to be eigenvalues of f with respect to g.

Taking $g(x) = x_1^2 + \cdots + x_n^2$ and interpreting $x_1, \ldots, x_n$ as coordinates in an orthonormal system with axis vectors $(1, 0, \ldots, 0)$ etc., $u_1, \ldots, u_n$ are coordinates in another orthonormal system with axis vectors $e_1, \ldots, e_n$. With the aid of this we can find out the geometric meaning of, e.g., the equation $f(x) = 1$ when $n = 2$. It is a curve in $\mathbf{R}^2$ with the equation $\lambda_1 u_1^2 + \lambda_2 u_2^2 = 1$. If $\lambda_1 > 0, \lambda_2 > 0$ it is an ellipse with the half axes $1/\lambda_1^{1/2}$ and $1/\lambda_2^{1/2}$, if $\lambda_1\lambda_2 < 0$ it is a hyperbola, if $\lambda_1 = 0$ and $\lambda_2 > 0$ or $\lambda_1 > 0$ and $\lambda_2 = 0$, it represents two parallel straight lines. When both λ_1 and λ_2 are $\leqslant 0$, the equation $f(x) = 1$ has no solutions. A little additional computation shows that every equation $f_2(x) + f_1(x) + f_0 = 0$ where f_2 is a real quadratic form, f_1 is a linear form, and f_0 is a number, represents a conic section, perhaps degenerate, provided the equation has solutions. When $n = 3$ we get the so-called surfaces of degree two and when $n > 3$ the hypersurfaces of degree two. In all this, the formula (22) is the basic tool. The conic sections, the second degree surfaces, the quadratic forms, and the self-adjoint operators have a common core of substantial mathematics, and form a chain of concepts of increasing generality developed over a span of 2000 years.

The spectral theorem and vibrating systems

The most important applications of the spectral theorem are to classical mechanics and quantum mechanics. Let S be a mechanical system consisting of a number of rigid bodies which may be in different configurations or positions. We assume that there is a bijection between these positions and some open set in $\mathbf{R}^n$ and we shall speak of points x in this set as *positions* of the system. When the system moves in time, its positions move along a curve $t \to x(t)$ in $\mathbf{R}^n$, the velocity at $x(t)$ being $x'(t) = dx(t)/dt$. A pair $x, \dot{x}$ of a position x and an arbitrary velocity $\dot{x}$ at that position is called a *state* of the system. The mechanics of S is given by two functions of states, a potential energy $V(x)$ and a kinetic energy $T(x, \dot{x})$. Both are assumed to be smooth real functions. Putting $V_j = \partial V/\partial x_j$ and $T_j = \partial T/\partial \dot{x}_j$, motions of the system are governed by the Lagrange equations

$$\frac{d}{dt} T_j(x, x') + V_j(x) = 0,$$

where $j = 1, \ldots, n$, and $x = x(t)$. We shall consider small movements close to a position of equilibrium taken to be at $x = 0$. We shall further assume that, for small x,

$$V(x) = 2^{-1} \sum_{j, k=1}^{n} a_{jk} x_j x_k \quad \text{and} \quad T(x, \dot{x}) = T(\dot{x}) = 2^{-1} \sum_{j, k=1}^{n} b_{jk} \dot{x}_j \dot{x}_k$$

are quadratic forms in x and $\dot{x}$ respectively and that they are positive definite. These assumptions are quite general. In fact, at a point of stable equilibrium the potential energy has a minimum and hence may be approximated by a positive definite quadratic form. The kinetic energy has of course a minimum when the velocities vanish. With these assumptions we can rewrite the Lagrange equations as

$$\sum_k b_{jk} x''(t) + \sum_k a_{jk} x_k(t) = 0, \qquad j = 1, \ldots, n,$$

or, multiplying by the components of an y in $\mathbf{R}^n$ and summing,

$$g(x''(t), y) + f(x, y) = 0, \tag{24}$$

where f and g are the bilinear forms associated with, respectively, $V(x)$ and $T(\dot{x})$. Next, using the spectral theorem, rewrite (24) according to (22) and (23), i.e., in terms of a basis $e_1, \ldots, e_n$ of $\mathbf{R}^n$ orthonormal with respect to the inner product $(x, y) = g(x, y)$. Putting $x(t) = u_1(t)e_1 + \cdots + u_n(t)e_n$ and $y = v_1 e_1 + \cdots + v_n e_n$ we then get

$$\sum \left(u_j''(t)v_j + \omega_k^2 u_j(t)v_j \right) = 0$$

where $\omega_1^2, \ldots, \omega_n^2$ are the eigenvalues of f with respect to g, necessarily positive since f is positive definite. Here $v_1, \ldots, v_n$ are arbitrary and hence (24) falls apart into n scalar equations

$$u_1''(t) + \omega_1^2 u_1(t) = 0, \ldots, u_n''(t) + \omega_n^2 u_n(t) = 0. \tag{25}$$

Since every solution of the equation $w''(t) + \omega^2 w(t) = 0$ is of the form $a \cos(\omega t + \alpha)$, we can write (25) as

$$x(t) = a_1 \cos(\omega_1 t + \alpha_1)e_1 + \cdots + a_n \cos(\omega_n t + \alpha_n)e_n.$$

In other words, the movement is a linear combination of simple or harmonic oscillations $t \to a_k \cos(\omega_k t + \alpha_k)e_k$ with frequencies $\omega_k/2\pi$ and certain amplitudes a_k. They are called the *eigenvibrations* or *proper modes* of the system. When the amplitudes are small the function $x(t)$ is small for all t and this is a justification a posteriori of our approximation of potential and kinetic energies.

Since our mechanical system has been quite general, it is clear that harmonic oscillations are very important in nature. As a consequence of classical mechanics and the spectral theorem we can also say with some confidence that every vibration is a linear combination of harmonic oscillations. This principle is the basis of the analysis, in physics, of sound and light. In quantum mechanics, the spectral theorem has a still more important position. In this theory, the observable quantities correspond to self-adjoint operators on a complex Hilbert space of infinite dimension.

4.9 Documents

Euclid on parallel lines

Here is a verbatim translation of Euclid's fifth postulate about parallel lines. It is equivalent to the simpler formulation given in section 4.1.

> "That, if a straight line meeting two straight lines making interior angles on the same side less than two right angles, the two straight lines, if produced indefinitely, meet on that side on which the angles are less than two right angles."

The Heath edition of the *Elements* records seven notable attempts from Ptolemy ($\sim$150) to Legendre ($\sim$1780) to prove this postulate from the other ones given by Euclid.

Felix Klein on geometry

The following is a nontechnical passage from Klein's Erlangen program (1872).

"There are transformations of [ordinary space] leaving invariant geometric properties of space configurations. In fact, geometric properties are in themselves independent of position, absolute magnitude and orientation of the object under consideration. The properties of a space configuration do not change under movements in space, similarity transformations and reflections and all transformations generated by them. The totality of all these transformations we shall call the *capital group* of space transformations: *geometric properties are invariant under the transformations of the capital group*. This can be turned around: *geometric properties are characterized by being invariant under transformations of the capital group* ....

Felix Klein 1849–1925

"Let us now regard our spatial intuition as mathematically irrelevant and let us consider space as a manifold of several dimensions whose elements are points. In analogy with the space transformations we shall consider transformations of our manifold. They also constitute *groups* but we shall no longer limit ourselves to special groups. All groups will be considered in their own rights. The following extensive problem then presents itself as a generalization of geometry:

"*Let there be a manifold and a group of transformations of it into itself. It is required to investigate those configurations which are invariant under the transformations of the group.*"

In the technical part of this paper, Klein reviews contemporary geometry from this point of view.

Hilbert on functional analysis and the spectral theorem

The beginning of functional analysis was a theory of infinite systems of linear equations with an infinite number of unknowns. Here is how Hilbert looked at the situation in an expository article from 1909 (*Collected Works*, vol. 3, no. 6). The passages within brackets fill in some of the context and explain Hilbert's terminology.

"[In the beginning it is remarked that functions can be defined by infinite sequences of numbers $x_1, x_2, \ldots$, e.g., their Fourier coefficients (see Chapter 9), and that relations between functions amount to relations between the corresponding sequences.] Because of its general nature, the problem of determining infinitely many unknowns from infinitely many equations seems impossible at first; there is the danger of losing oneself in vague and difficult speculations with no gain of insight into deeper problems. But if we are not mislead by such considerations we shall find, with Siegfried [a Wagner hero], that the conjurers of fire retreat and a magnificent prize beckons: *the methodological unification of algebra and analysis*.

"... let us consider the procedure [of solving equations] for a *finite* number of variables. The left sides of the equations are functions of a number of variables and the difficulty of obtaining solutions depends on the nature of these functions... [linear, quadratic, continuous, differentiable].

"The first task of an analysis of an infinity of variables must be to transfer in a proper way all these notions to infinitely many variables. At first we have to note that a linear expression

$$a_1 x_1 + a_2 x_2 + \ldots$$

1862-1943

Hilbert

in an infinity of variables is a *function* of these only when the series converges and this happens only when we restrict them by inequalities. These restricting inequalities can be chosen in many ways but... [a choice is made]. The restriction for the infinitely many variables $x_1, x_2, \ldots$ is simply that the sum of their squares be finite.... In particular, it turns out that our linear expression $a_1 x_1 + a_2 x_2 + \ldots$ is a function of the infinitely many variables $x_1, x_2, \ldots$ if and only if the sum of the squares of the coefficients $a_1, a_2, \ldots$ is finite. [Here we are in the Hilbert space H of sequences $x = (x_1, x_2, \ldots)$ with the inner product $(a, x) = a_1 x_1 + a_2 x_2 + \ldots$ and norm $\|x\| = (x, x)^{1/2}$.]

We now turn to the concept of *continuity*. [Next Hilbert explains that a function $F(x)$ of $x = (x_1, x_2, \ldots)$ is said to be continuous if $F(x^{(n)}) \to F(x)$ as $n \to \infty$ and $x^{(n)} = (x_1^{(n)}, x_2^{(n)}, \ldots)$ tends to x in the sense that $x_k^{(n)} \to x_k$ for all k. Continuity under this kind of convergence, now called *weak convergence*, is a rather strong requirement.]

103

Using our definition it is easy to see that, precisely as for a finite number of variables, a continuous function of a continuous function is continuous. But above all, we have the fact that a continuous function of infinitely many variables must have a minimum which, because of its generality and precision, is a substitute for Dirichlet's principle. [Hilbert means that if F is continuous, it attains its lower bound on any closed bounded set in H. This follows very simply from the fact that a bounded sequence of elements in the space contains a (weakly) convergent subsequence. After noting that linear functions are continuous in his sense, Hilbert passes to quadratic functions

$$Q(x) = \sum a_{pq} x_p x_q, \qquad a_{qp} = a_{pq},$$

where the right side is supposed to converge. He notes that continuity at the origin $(0,0,\ldots)$ implies continuity everywhere for Q but that continuity is an extra requirement.]

The linear and quadratic functions just defined are homogeneous in the infinitely many variables $x_1, x_2, \ldots$ and are therefore called forms.... The notions of bilinear forms, linear and orthogonal transformations and invariants also extend and give us a theory of forms of infinitely many variables—a new branch of science between algebra and analysis. Its methods are algebraic but its results belong to analysis.

The most important theorems concern bilinear and quadratic forms. The most important result is the following: a continuous quadratic form becomes, after a suitable orthogonal transformation, a sum of squares of the new variables $x_1', x_2', \ldots$ so that

$$Q(x) = k_1 x_1'^2 + k_2 x_2'^2 + \ldots$$

where $k_1, k_2, \ldots$ are certain constants tending to zero. [This is the spectral theorem for compact, self-adjoint operators. In fact, putting $Q(x,y) = \sum a_{pq} x_p y_q$, the equality $Q(x,y) = (Ax, y)$ turns out to define a self-adjoint compact operator A from H to H. If $e_1, e_2, \ldots$ is a complete orthonormal set of eigenvectors of A and $k_1, k_2, \ldots$ the corresponding eigenvalues, and $x_1', x_2', \ldots$ the coordinates of x with respect to $e_1, e_2, \ldots$, we get the formula above.]

Literature

Linear algebra is treated in innumerable textbooks, most of them quite adequate. The acknowledged masterpiece is *Finite-dimensional Vector Spaces* (formerly van Nostrand, now Springer-Verlag), by Paul Halmos. The field of functional analysis has been in a state of constant growth since the 1930s, and the literature is now truly enormous. Its bible is a book by S. Banach from 1932, *Théorie des opérations linéaires* (Math. Monographs, Warsaw). This work retains its freshness but is now obsolete. It is advisable to start with a simple text like *Elements of Functional Analysis*, by Maddox (Cambridge University Press), and then go on to an encyclopedic book like *Functional Analysis*, by Reed and Simon (*Methods of Mathematical Physics*, vol. 1, Academic Press (1972).

5
LIMITS, CONTINUITY, AND TOPOLOGY

5.1 *Irrational numbers, Dedekind's cuts, and Cantor's fundamental sequences*. Cuts. Limits of sequences. Subsequences and upper and lower limits. Cauchy sequences. Cantor's fundamental sequences. Epistemology of the real numbers and the continuum hypothesis. 5.2 *Limits of functions, continuity, open and closed sets*. Limits. Convergence of sequences of functions and uniform convergence. Several variables. Open and closed subsets of $\mathbf{R}^p$. Uniform continuity and continuous functions from compact sets. Back to one variable. 5.3 *Topology*. Topological spaces. Neighborhoods, continuity and homeomorphisms. Topology and intuition. The fundamental theorem of algebra. Algebraic curves. 5.4 *Documents*. Dedekind on Dedekind cuts. Poincaré on topology.

More than 2000 years ago the Greeks discovered that the rational numbers were not enough to measure lengths in geometry, for instance not the diagonal of a square when its side is put equal to 1. Accepting the concept of length in geometry, we must also accept that there are more real numbers than rational ones. This fact is the origin of a great deal of philosophizing about the enigma of the continuum. The simplest way out is to think of the real numbers as infinite decimal fractions, but then there are difficulties extending the laws of arithmetic to the real numbers. Dedekind's cuts and Cantor's fundamental sequences are better solutions, technically speaking, and reflect two equivalent basic facts, the principle of the least upper bound and Cauchy's convergence principle. The theory of limits and continuity, which is fundamental in analysis, is based on these principles and the geometric notions of length and distance. One can also go a step further and replace length and distance by the concept of a neighborhood. Then one is no longer tied to the real numbers or parts of n-dimensional space, which can be abandoned in favor of general sets. This happens in general topology, touched upon in the last part of the chapter. It ends with a section on topology and intuition where general constructions give way to two concrete results. The entire chapter requires of the reader a certain maturity and a ready acceptance of abstract reasoning.

5.1 Irrational numbers, Dedekind's cuts, and Cantor's fundamental sequences

Cuts

It is proved in the *Elements* that the diagonal of a square is not rationally proportional to its side. In modern language, this means that $\sqrt{2}$ is not a rational number. The ingenious proof runs as follows. Suppose that $\sqrt{2}$ were a rational number p/q where p and q are integers. Squaring, we get $2q^2 = p^2$ so that p must be even, so $p = 2r$ where r is another integer. Then $2q^2 = 4r^2$ and hence $q^2 = 2r^2$ so that q also must be even. But, dividing both p and q by a power of 2, we can always assume from the beginning that either p or q is odd and this gives the desired contradiction. If we assume with the Greeks that every interval has a length, the situation is interesting but not critical: there are incommensurable lengths and that is all. But in the number model, the situation is different. Suppose for instance that we master the rational numbers and know how to compute with them. What then *is* a number like $\sqrt{2}$, and what do we mean by the sum $1 + \sqrt{2}$? Can we define sums and products where irrational numbers are involved, and do the laws of arithmetic still hold for them? These questions, long overshadowed by more tangible problems, were answered about a century ago by Dedekind and Cantor.

When Dedekind started giving lectures in mathematics to freshmen he found it convenient to begin with some simple assumption to make analysis logically coherent. His choice was

every monotone bounded sequence of real numbers has a limit.   (1)

If asked to do so, Dedekind could also have proved this principle using his definition of real numbers. His idea was to describe the position of an irrational number ξ relative to the rational numbers $\mathbf{Q}$ by considering the class S of rational numbers $< \xi$. Whatever ξ is we can assert that S has the following so-called *cut property* relative to the rational numbers $\mathbf{Q}$,

$$S \neq \varnothing, \quad S \neq \mathbf{Q} \quad \text{and} \quad a \in S, \quad b \in \mathbf{Q}, \quad b < a \quad \Rightarrow \quad b \in S.$$

Any subset S of $\mathbf{Q}$ with this property is called a *Dedekind cut* in the rational numbers. If S has a largest number ξ, necessarily rational, it follows that S consists of all rational numbers $\leqslant \xi$, and if its *complement* $\mathbf{Q} \setminus S$ has a smallest element ξ, also necessarily rational, then S consists of all rational numbers $< \xi$. When neither of these cases occurs we say that S *represents* an irrational number ξ. In all cases we shall write S as (ξ). If we think of ξ as a point on an infinite straight line, (ξ) is then the set of rational points to the left of the point ξ which may or may not be included in the set.

We can now start computing with numbers defined by Dedekind cuts. A sum $\xi + \eta$, for instance, is defined by letting the corresponding cut

$(\xi + \eta)$ consist of all sums $a + b$ where $a \in (\xi)$ and $b \in (\eta)$. Although technically more complicated, it is also possible to find a cut $(\xi\eta)$ defining a product $\xi\eta$. Straightforward verifications then show that the laws of computation with rational numbers extend to our new numbers. If we let $\xi > 0$ mean that (ξ) contains some positive rational number and $\xi < \eta$ that $\eta = \xi + \zeta$ with $\zeta > 0$, we shall find that the three possibilities $\xi < \eta$, $\eta < \xi$, $\xi = \eta$ are exhaustive and mutually exclusive and that $\xi < \eta$, $\eta < \zeta \Rightarrow \xi < \zeta$. In other words, the order properties of the rational numbers extend to numbers defined by Dedekind cuts. In particular, the set D of these numbers $\xi, \eta, \ldots$ has all the properties of the rational numbers that made it possible to make Dedekind cuts in $\mathbf{Q}$, and so we can start making Dedekind cuts in D too. But this will not give us any new numbers for it turns out that every cut T in D either has a largest element ξ in D and hence consists of all $\eta \leqslant \xi$, or $D \setminus T$ has a smallest element ξ in D so that T consists of all $\eta < \xi$. The proof borders on a triviality: if a subset T of D has the cut property relative to D and S is the set of rational numbers a belonging to some (η) where η is in T, S has the cut property relative to the rational numbers and hence defines a ξ in D. If this ξ belongs to the set T, it must be its largest element and if it does not, it must be the smallest element of the complement $D \setminus T$.

Let us now define the real numbers as Dedekind cuts in the rational numbers. We shall then have the same arithmetical laws and the same order laws for the real numbers $\mathbf{R}$ as for the rational ones, but we have gained an additional property, namely

every cut in $\mathbf{R}$ either contains a largest number

$$\text{or its complement contains a smallest number.} \quad (2)$$

From this it follows, for instance, that

$$\text{every subset of } \mathbf{R} \text{ bounded from above has a least upper bound.} \quad (3)$$

That a set M is bounded from above just means that it has an upper bound, i.e., a number $\eta \geqslant$ every number in M. To see that $(2) \Rightarrow (3)$ just note that the property

$$\eta \in S \quad \Leftrightarrow \quad \text{there is a } \zeta \text{ in } M \text{ such that } \eta \leqslant \zeta$$

defines a cut $S = (\xi)$ in the real numbers and that

$$\xi \text{ is the least upper bound of } M. \quad (4)$$

In fact, ξ is an upper bound and $\leqslant$ every other upper bound. The least upper bound may or may not belong to M. Another name for it is the *supremum* of M, and we write $\xi = \sup M$ or $\xi = \sup \eta$ when $\eta \in M$. The *greatest lower bound* is defined analogously. It is also called the *infimum* of M and it is denoted by inf M. The counterpart of (3) is the statement that every subset of R bounded from below has a greatest lower bound. The property (3) is sometimes called the axiom of the least upper bound and it is the point of departure of analysis in many textbooks.

Limits of sequences

So far we have not yet arrived at Dedekind's basic principle (1) of analysis. To do this we have to know what is meant by the limit of an infinite sequence of numbers, $a_1, a_2, \ldots$. We say that such a sequence has the *limit a* if the difference $a_n - a$ is as small as we wish if only n is large enough. The notations used for this are

$$\lim_{n \to \infty} a_n = a \quad \text{or} \quad \lim a_n = a \quad \text{or} \quad n \to \infty \;\Rightarrow\; a_n \to a$$

and the way to express this in words is to say that a_n tends to a as n tends to infinity. Formal proofs that sequences have or do not have limits use the well-known ε, ω-criterion. It says that $\lim a_n = a$ if and only if to every $\varepsilon > 0$ there is a number ω such that

$$n > \omega \;\Rightarrow\; |a_n - a| < \varepsilon. \tag{5}$$

We suppose that the reader has had some practice in the use of this criterion. It provides clear and convincing proofs that there are sequences without limits, e.g., $1, -1, 1, -1, \ldots$, that a sequence has at most one limit, and that

$$\lim(a_n \pm b_n) = \lim a_n \pm \lim b_n$$
$$\lim a_n b_n = \lim a_n \lim b_n$$
$$\lim 1/a_n = 1/\lim a_n$$

provided $\lim a_n$ and $\lim b_n$ exist and, in the last formula, $\lim a_n \neq 0$. A sequence with a limit is said to be *convergent*. Using (5) one proves that a convergent sequence is bounded from above and from below.

Finally, we shall now use (3) and (5) to prove (1). A sequence $a_1, a_2, \ldots$ is said to be *increasing* if $a_1 \leqslant a_2 \leqslant \cdots$, *decreasing* if $a_1 \geqslant a_2 \geqslant \cdots$, and *monotone* if it has at least one of these properties. Let us first consider an increasing sequence. That it is bounded means that there is a number c such that $a_n \leqslant c$ for all n. Then, according to (3), the set whose elements are $a_1, a_2, \ldots$ has a least upper bound a. We shall see that $\lim a_n = a$. In fact, since a is the least upper bound, to every $\varepsilon > 0$ there are numbers in the sequence larger than $a - \varepsilon$. Let a_ω be one of them. Then

$$n > \omega \;\Rightarrow\; |a_n - a| = a - a_n \leqslant a - a_\omega < \varepsilon,$$

proving (1) for increasing sequences. For decreasing ones the proof is the same using the greatest lower bound.

Subsequences and upper and lower limits

Deleting elements from a sequence $a_1, a_2, \ldots$ we get subsequences, e.g., $a_1, a_3, a_5, \ldots$ or $a_2, a_4, a_6, \ldots$. The general form of a subsequence is $a_{n_1}, a_{n_2}, \ldots$ where $n_1 < n_2 < \cdots$ is a strictly increasing sequence of integers. To simplify the notations we shall write the sequence as (a_n) and subsequences as (a_{n_k}). We shall see that

$$\text{every bounded sequence has a convergent subsequence.} \qquad (6)$$

To prove this put $b_n = \sup(a_n, a_{n+1}, \dots)$ and $c_n = \inf(a_n, a_{n+1}, \dots)$. If the sequence (a_n) is bounded, the sequences (b_n) and (c_n) exist and are bounded. It is also clear that (b_n) is decreasing and (c_n) is increasing. Hence, by (1), they have limits, b and c respectively. We remark in passing that these limits are called the *upper* and *lower limits* of (a_n), and that they are denoted by $\limsup a_n$ and $\liminf a_n$ respectively. Now choose for every n an integer k_n such that $a_{k_n} \geqslant b_n - 2^{-n}$. Since $a_{k_n} \leqslant b_n$ this gives us a convergent subsequence (a_{k_n}) with the limit b. We could also have chosen k_n such that $a_{k_n} \leqslant c_n + 2^{-n}$ getting a subsequence with limit c.

Cauchy sequences

If a sequence (a_n) has the limit a, then $|a_n - a_m| \leqslant |a_n - a| + |a_m - a|$ is arbitrarily small when n and m are sufficiently large. We write this also as

$$n, m \to \infty \quad \Rightarrow \quad |a_n - a_m| \to 0.$$

Sequences with this property are called Cauchy sequences and the nontrivial fact that

$$\text{every Cauchy sequence is convergent} \qquad (7)$$

is called Cauchy's principle of convergence. It is used when one wants to prove that a sequence converges but one does not know its limit. To prove (7) let (a_n) be a Cauchy sequence. We then know that to every $\varepsilon > 0$ there is an ω such that

$$n \quad \text{and} \quad m > \omega \quad \Rightarrow \quad |a_n - a_m| < \varepsilon.$$

Letting $\varepsilon = 1$, for instance, we see that the sequence is bounded. Hence there is a convergent subsequence (a_{m_k}) with a limit a. Fixing n, putting $m = m_k$, and letting k tend to infinity gives $|a_n - a| \leqslant \varepsilon$ when $n > \omega$, and hence (a_n) converges to the limit a.

Cantor's fundamental sequences

Cantor's construction of the real numbers differs from Dedekind's in that it does not use the fact that rational numbers are ordered. It starts from the set F of Cauchy sequences of rational numbers, now called *fundamental sequences*, and the subset N of F that consists of sequences tending to zero. These are called *null sequences*. If the sum and product of two sequences (a_n) and (b_n) are defined as $(a_n + b_n)$ and $(a_n b_n)$ respectively, it is easy to show that F becomes a commutative ring and N a two-sided ideal in F. Cantor's definition of the real numbers is then simply

$$\mathbf{R} = F/N.$$

In other words, a real number ξ is an equivalence class $(a_n) + N$ of fundamental sequences that differ only by null sequences. As in Dedekind's definition, the arithmetical laws and the order laws for $\mathbf{R}$ follow, although the ordering now has a more technical definition. The distance $|\xi - \eta|$ between two real numbers represented by the sequences (a_n) and (b_n) is defined as the class containing the Cauchy sequence $(|a_n - b_n|)$. A sequence $\xi_1, \xi_2, \ldots$ of real numbers is said to be a fundamental sequence if $|\xi_n - \xi_m|$ tends to 0 as n and m tend to infinity. The point of Cantor's construction is that every such sequence is convergent. For if its elements are represented by the sequences $(a_{1n}), (a_{2n}), \ldots$, a little reflection shows that there are rational numbers a_p such that $a_{pn} - a_p \to 0$ as first n and then p tends to infinity, that $\xi = (a_p)$ is a fundamental sequence, and hence that $|a_{pn} - a_n| \leqslant |a_{pn} - a_p| + |a_p - a_n| \to 0$ as first n and then p tends to infinity. Hence ξ_p tends to ξ as p tends to infinity. Cauchy's principle of convergence holds also in Cantor's construction of the real numbers.

Epistemology of the real numbers and the continuum hypothesis

So far we have treated sequences of numbers as if we knew everything about them. It was, for instance, understood that a sequence either has or does not have a limit. But, with a certain justification, we may also say that this depends on how well we know the sequence. To start with, it has infinitely many elements, it cannot be tabulated, and it cannot be constructed unless we know some definite rule for writing down its elements. Doubts of this kind have inspired *intuitionism*, a mathematical school of though with a three-valued logic where a third category, called *undecidable*, completes the classical pair, either-or. It has had a limited influence. Most mathematicians prefer their logical paradise as it is.

In connection with his work on set theory Cantor introduced the cardinal numbers. The definition is simple. Two sets are said to have the same *cardinal number* or *cardinality* if there is a bijection between them, and a set bijective with the set $(1, 2, \ldots, n)$ of the first n natural numbers is said to have the cardinal number n. Hence cardinality generalizes the notion of "number of elements." Let ω be the cardinality of the set of all natural numbers. That a set has this cardinality means that its elements can be numbered in such a way that each natural number is used precisely once. Such sets are said to *countable*. Examples are the even numbers, the odd numbers, the rational numbers, and all n-tuples of natural numbers. But the real numbers are not countable and it is easy to show that they have the same cardinality as the set of all infinite sequences of natural numbers. Call this cardinality ω^*. That $\omega^* > \omega$ can also be expressed by saying that all points of an interval or, to use an old word, a continuum, are many more than those with rational coordinates. Cantor's famous continuum hypothesis is simply the following: there is no cardinality strictly between ω and ω^*. The work by Gödel in 1938 and Paul Cohen in 1963 has shown that the continuum hypothesis has a position in mathematics similar to that of the axiom of parallels in classical geometry. It is an axiom which can be adjoined to the axioms of set theory to obtain perfectly consistent mathematics. But its contradiction could equally well be adjoined to these axioms, obtaining a perfectly consistent theory, too.

5.2 Limits of functions, continuity, open and closed sets

Limits

A sequence $a_1, a_2, \ldots$ of real numbers is nothing but a function from the natural to the real numbers, and it is easy to extend the notion of a limit to functions from real numbers to real numbers. That

$$\lim_{x \to x_0} f(x) = a \quad \text{or} \quad x \to x_0 \ \Rightarrow \ f(x) \to a$$

or, in words, that f has the limit a or tends to a as x tends to x_0 means that

$f(x)$ is arbitrarily close to a when x is sufficiently close to x_0. (8)

Here, as for sequences, it is useful to have a more technical version as follows: to every $\varepsilon > 0$ there is $\delta > 0$ such that

$$x \neq x_0 \quad \text{and} \quad |x - x_0| < \delta \ \Rightarrow \ |f(x) - a| < \varepsilon. \tag{9}$$

It is understood that (9) is required only for those x for which $f(x)$ is defined. To include limits of sequences we must also consider the case $x \to \infty$, interpreting the concluding part of (8) as "x is sufficiently large positive," and replacing (9) by

to every $\varepsilon > 0$ there exists an ω such that

$$x > \omega \ \Rightarrow \ |f(x) - a| < \varepsilon.$$

The case $x \to -\infty$ is dealt with in a similar way. In (9) we can also distinguish between limits from the right and from the left by replacing $x \neq x_0$ by $x > x_0$ and $x < x_0$ respectively. The four parts of Figure 5.1 illustrate the situations where f has a limit from the right, from the left, separate limits from both sides, and no limit from either side as $x \to x_0$.

The value of f at x_0 does not appear in (8) or (9) and may be anything. The function f need not even be defined at x_0, but if it is and

$$\lim f(x) = f(x_0) \quad \text{as} \quad x \to x_0$$

we say that the function is *continuous* at the point x_0. This does not occur in any of the four cases in the figure, but the functions drawn there are supposed to be continuous at all other points. When a function is discontinuous at too many points, it is hard to visualize. If we accept the general concept of a function, we must also accept that there can be a function f such that $f(x) = 0$ when x is not rational and $f(x) = 1/q$ when $x = p/q$ where p and q are integers which are relatively prime to each other. This function is continuous at x when x is irrational and discontinuous when x is rational. No drawing can do justice to this property.

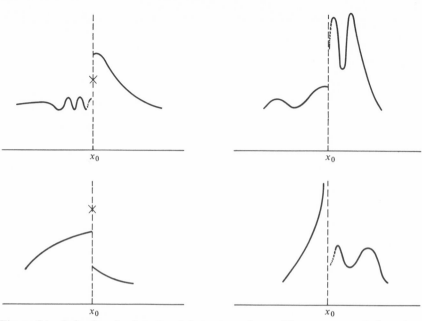

Figure 5.1   Behavior of a function f close to a point x_0. The crosses mark the value $f(x_0)$ of f at x_0 when it is supposed to be defined.

The rules for computing with continuous functions are the same as those for limits of sequences. If f and g are continuous at x_0 so are $f \pm g$ and fg and also $1/f$ provided $f(x_0) \neq 0$. If f is continuous at x_0 and g is continuous at $f(x_0)$, the composed function $g \circ f$ is continuous at x_0. The proofs are simple exercises in the use of (9).

Convergence of sequences of functions and uniform convergence

Let $f_1, f_2, \ldots$ be an infinite sequence of real functions with a common domain of definition D. When $\lim f_n(x)$ exists for every x in D we say that the sequence *converges pointwise* to the limit function $f(x) = \lim f_n(x)$. This seemingly innocent definition covers some unexpected phenomena. If, e.g., D is the interval $0 \leqslant x \leqslant 1$ and $f_n(x) = 0$ when $x = 0$ and when $n^{-1} \leqslant x \leqslant 1$ but f is otherwise arbitrary, it is clear that the sequence $f_1, f_2, \ldots$ converges pointwise to the limit function $f = 0$. The left part of Figure 5.2 illustrates this.

If, as in the figure, $f_n(x) > 1$ for every n and some x depending on n, most people would object to saying that the sequence tends to zero. But there is a more natural notion of convergence for sequences of functions, the uniform convergence. That a sequence $f_1, f_2, \ldots$ *converges uniformly* on a part D of the real axis to a function f means that to every $\varepsilon > 0$ there is an ω such that

$$n > \omega \quad \Rightarrow \quad |f_n(x) - f(x)| < \varepsilon \quad \text{for all } x \text{ in } D. \tag{10}$$

The right part of Figure 5.2 refers to this situation.

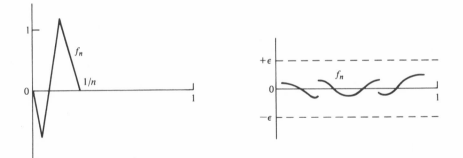

Figure 5.2   In the interval $0 \leqslant x \leqslant 1$ the sequence $f_1, f_2, \ldots$ to the left converges pointwise, and the sequence $f_1, f_2, \ldots$ to the right converges uniformly to the limit function $f = 0$.

Uniform convergence is useful because it preserves continuity. We have

Theorem 1. *A uniform limit of continuous functions is continuous, i.e.,* $f_1, f_2, \ldots$ *continuous in* $D, f_n \to f$ *uniformly in* $D \Rightarrow f$ *is continuous in* D.

We shall prove this theorem and at the same time demonstrate how the criterion (9) is used in a complicated situation. Let x_0 be in D. Adding and subtracting $f_n(x) - f_n(x_0)$ from $f(x) - f(x_0)$ and using the triangle inequality we get

$$|f(x) - f(x_0)| \leqslant |f_n(x) - f(x)| + |f_n(x_0) - f(x_0)| + |f_n(x) - f_n(x_0)|.$$

Let $\varepsilon > 0$ be given. According to (10) there is an ω_1 such that the first two terms on the right are $< \varepsilon/3$ when $n > \omega_1$ and this for all x in D. Fix such a number n. Now by (9) applied to the function f_n and with $a = f_n(x_0)$ there is a δ such that the third term is $< \varepsilon/3$ when $|x - x_0| < \delta$. Hence (9) holds for the function f and with $a = f(x_0)$. This means that f is continuous at x_0 and hence, since x_0 was arbitrary in D, f is continuous in all of D.

Several variables

Limits and continuity can be defined also for functions of several variables whose values have several components. In other words, we are now dealing with functions $f(x)$ of n real variables $x = (x_1, \ldots, x_n)$ having m components:

$$f(x) = (f_1(x), \ldots, f_m(x)).$$

We express this briefly by saying that f is a function from $\mathbf{R}^n$ to $\mathbf{R}^m$. The language used in (8) contains a clue. If we know what the word "near" means in $\mathbf{R}^p$ with p arbitrary, we know what limit and continuity means for our new functions. In $\mathbf{R}^p$ we shall use the terms limit and limit point interchangeably. One solution is to introduce a distance in $\mathbf{R}^p$ with all the usual properties including that the triangle inequality holds. We could, for instance, let

$$|y - z| = \max(|y_1 - z_1|, \ldots, |y_p - z_p|) \tag{11}$$

be the distance between y and z in $\mathbf{R}^p$. All we have said earlier about limits and continuity of functions including Theorem 1 is then meaningful and true. In the special case $n = m = 1$ we get back what we had before.

Open and closed subsets of $\mathbf{R}^p$

We could have used other distances than (11) without changing the definitions of limits and continuity. The citerion (9) just requires that to every $\varepsilon > 0$ there is a $\delta > 0$ such that (9) holds but says nothing about how this δ depends on ε. In order to avoid using a definite distance there have been efforts to free the definitions of limits and continuity of everything inessential. One way is to introduce the notion of *an open interval* in $\mathbf{R}^p$. Such an interval is defined by $2p$ numbers $a_1 < b_1, \ldots, a_p < b_p$ and consists of all $y = (y_1, \ldots, y_p)$ such that

$$a_1 < y_1 < b_1, \ldots, a_p < y_p < b_p.$$

If we interpret $y_1, \ldots, y_p$ as parallel coordinates, we get a line segment when $p = 1$, a parallelogram when $p = 2$, and a parallelepiped when $p = 3$. But we keep the word interval for all p. It is now easy to see that (9) is equivalent to

> to every open interval I around a there is an open
>
> interval J around x_0 such that $f(J) \subset I$.   (9′)

Here, naturally, $f(J) \subset \mathbf{R}^m$ is the image of the interval $J \subset \mathbf{R}^n$ under the map f. Using (9′) we now also have an alternative definition of continuity which takes a very simple form if we introduce the notion of an open set in $\mathbf{R}^p$. A subset A of $\mathbf{R}^p$ is said to be *open* if, whenever it contains a point it also contains an open interval around it; i.e.

$$A \in y \quad \Rightarrow \quad A \supset \text{an open interval around } y. \tag{12}$$

When f is a continuous function from all of $\mathbf{R}^n$ to $\mathbf{R}^m$, the set A defined by $f(x) \neq 0$ is open. For if $f(x_0) \neq 0$ there is an open interval I around $f(x_0)$ not containing the origin, and hence by (9′) an open interval J around x_0 in $\mathbf{R}^n$ where f does not vanish. That a function from an open part of $\mathbf{R}^n$ to $\mathbf{R}^m$ is continuous where it is defined can also be expressed so that

$$A \text{ open} \quad \Rightarrow \quad f^{-1}(A) \text{ open}$$

where $f^{-1}(A)$ is the set of all x in $\mathbf{R}^n$ such that $f(x)$ belongs to A. The proof is left to the reader as a logical exercise.

The complements $F = \mathbf{R}^p \setminus A$ of open sets A are said to be *closed*. In particular, if f is a continuous function from $\mathbf{R}^n$ to $\mathbf{R}^m$, the equation $f(x) = 0$ defines a closed subset of $\mathbf{R}^n$. A subset F of $\mathbf{R}^p$ is closed if and only if it contains its limit points, i.e., limits of convergent sequences contained in F. In fact, if F is closed its complement contains an open interval around each of its points and hence cannot contain any limit point of F. Conversely, if F contains its limit points and every open interval around a point y contains points in F, taking these intervals smaller and smaller we can construct a sequence of points in F tending to y which then must belong to F. This means that the complement of F must be open and hence that F is closed.

A subset B of $\mathbf{R}^p$ is said to be *bounded* if all components of $y = (y_1, \ldots, y_p)$ are bounded when y belongs to B. We have the following simple generalization of (6), called the Bolzano-Weierstrass principle and invented around 1870.

every bounded sequence in $\mathbf{R}^p$ contains a convergent subsequence.   (6')

The proof proceeds so that we first choose a subsequence of the given sequence where the first components form a convergent sequence, and then a subsequence of the subsequence where also the second components form a convergent sequence, and so on.

Uniform continuity and continuous functions from compact sets

Let f be a function to $\mathbf{R}^m$ from a part D of $\mathbf{R}^n$ and assume that f is continuous at all points of D. According to (9) this means that to every y in D and number $\varepsilon > 0$ there is a $\delta > 0$ such that $|f(x) - f(y)| < \varepsilon$ when x is in D and $|x - y| < \delta$. But this is not the end of all problems. The function may still be unbounded in D, as is $f(x) = x$ in $\mathbf{R}$ or $f(x) = x^{-1}$ in the interval $0 < x \leqslant 1$, and it need not be *uniformly* continuous in the sense that the number δ can be made to depend only on ε and not on y. If, e.g., $f(x) = x^2$, $D = \mathbf{R}$ and y is very large, $f(x) - f(y) = (x + y)(x - y)$ cannot be made small only by choosing $x - y$ small regardless of y. We can avoid these complications if we assume that D is *compact*, i.e., bounded and closed. In this case f has a lot of good properties listed in the following theorem.

Theorem 2. *Let f be a continuous function from a compact part K of $\mathbf{R}^n$ to $\mathbf{R}^m$. Then $f(K)$ is compact and f is uniformly continuous. If, in addition, f is a bijection, then the inverse function f^{-1} from $f(K)$ to K is also continuous.*

PROOF. If y is a limit point of $f(K)$, there is a sequence $(x^{(i)})$ in K such that $f(x^{(i)}) \to y$ as $i \to \infty$. According to (6') this sequence has a convergent subsequence, and since K is closed, its limit point x must be in K. But f is continuous and hence $f(x) = y$. This means that $f(K)$ is closed. If $f(K)$ were not bounded, we could choose a sequence $(x^{(i)})$ in K such that the sequence $(|f(x^{(i)})|)$ tends to infinity and this is not consistent with the continuity of f and the fact that $(x^{(i)})$ has a convergent subsequence. Hence $f(K)$ is compact. If f were not uniformly continuous, there would exist a number $\delta > 0$ and two sequences $(x^{(i)})$ and $(y^{(i)})$ in K such that $|x^{(i)} - y^{(i)}|$

$\to 0$ but $|f(x^{(k)}) - f(y^{(k)})| \geqslant \delta$ as $i \to \infty$. A passage to convergent subsequences gives a contradiction as before. If $(x^{(i)})$ is a sequence in K having two convergent subsequences with limits x' and x'' and $(f(x^{(i)}))$ converges to y in $f(K)$ we get $f(x') = f(x'')$. When f is a bijection, this means that $x' = x''$, i.e., all convergent subsequences of $(x^{(i)})$ have the same limit. It is then easy to see that the sequence itself must be convergent. Hence f^{-1} is continuous when f is a bijection.

Back to one variable

When $n = m = 1$, the preceding theorem can be made more precise. This depends on the following

Lemma. *Let $f(x)$ be a real continuous function defined when $a \leqslant x \leqslant b$. If c is between $f(a)$ and $f(b)$ there is a ξ between a and b such that $f(\xi) = c$.*

PROOF. We may assume that $f(a) < c < f(b)$. Let ξ be the least upper bound of the set of numbers $\eta < b$ such that $a < x \leqslant \eta \Rightarrow f(x) < c$. Since $f(x)$ is close to $f(a)$ when x is close to a, the set is not empty and hence ξ exists. Taking a sequence in the set tending to ξ and using the continuity of f, it follows that $f(x) \leqslant c$. The possibility that $f(\xi) < c$ is excluded by the definition of ξ. Hence $f(\xi) = c$ and, trivially, $a < \xi < b$.

A real function f is said to be (*strictly*) *monotone* if $f(b) - f(a)$ always has the same sign (plus or minus) when $b > a$. An interval on the real axis is said to be *open, half-open,* or *closed* according as it contains none, one, or two of its endpoints. The lemma and some reflection gives:

Theorem 3. *Let f be a real continuous function defined in an interval I. Then $f(I)$ is also an interval. Further, f is a bijection from I to $f(I)$ if and only if f is monotone. Finally, if f is monotone, it has a continuous inverse $f^{-1} : f(I) \to I$, and I and $f(I)$ are both either open, half-open, or closed.*

5.3 Topology

Topological spaces

Some of the terms that we have been using, like open, closed, and compact were not used in the same way by Dedekind and Cantor and their contemporaries. They come from a branch of mathematics called *general topology*, which is based on axiomatizations of the concepts of limit and continuity. General topology got its first form in a book by Hausdorff, *Set Theory*, from 1914. It has been the object of intensive study and it has changed its terminology many times. The modern version starts off by defining a *topological space* as a set E provided with a class of subsets which are said to be open. The minimal requirements on this class are

(i)   every union of open sets is open

(ii) finite intersections of open sets are open.

In addition, the set E itself and the empty set $\varnothing$ are open, and it is often required that

(iii)   the intersection of all open sets containing a point is the point itself.

Here "point" is the same as an element of E. A *topology* on a set E is, or is defined by a class of open sets containing E and $\varnothing$ and with the properties (i) and (ii). When (iii) holds we say that the topology *separates* points. *Closed* sets are of course defined as complements of open sets. The concept of compactness is more complicated. A topological space is *compact* if every open covering contains a finite covering. Here a *covering* of E is a family of open sets such that every point of E belongs to at least one of them; a covering *contains* another when its family of open sets contains the other family; and a covering is *finite* when its family has only finitely many open sets.

Letting E be a part of some $\mathbf{R}^n$ and taking as open sets of E the intersections $E \cap A$ with open sets A in $\mathbf{R}^n$, we get a lot of topological spaces with very different properties. That such an E is compact in the general sense above means precisely that E is a closed and bounded part of $\mathbf{R}^n$. This nontrivial fact was observed around 1900 by Borel and Lebesgue when they became interested in coverings of complicated sets by open intervals.

Neighborhoods, continuity and homeomorphisms

Let x be a point of a topological space E. Every subset of E containing an open set that contains x is said to be a *neighborhood* of x. A function f from a topological space E to another one, F, is said to be *continuous* at the point x of E if, to every neighborhood W of $f(x)$ in F there is a neighborhood V of x in E such that $f(V) \subset W$. A function f from E to F is then continuous everywhere if and only if it has the property (13).

This statement and the definitions preceding it must, of course, be supported by examples in order to be understood properly. They are here just as a breathtaking generalization of the concept of a continuous function. Using our definitions we can also say what *topology* is: it is the theory of topological spaces and continuous maps between them. The topological counterpart of the concept of an isomorphism is a *homeomorphism*, i.e., a bijection between topological spaces which is continuous in both directions. According to Theorem 3 above, two intervals of the real axis are homeomorphic if and only if they contain equally many endpoints, none, one or two.

Topology and intuition

General topology offers us mathematical models of, among other things, the everyday notions of path and connectivity. A path or a continuous curve in a topological space E from a point P to a point Q is defined as a continuous map f from the interval $0 \leqslant x \leqslant 1$ to E such that $f(0) = P$ and $f(1) = Q$. When f is injective, the path is simple in the sense that it does not intersect itself. When $P = Q$ but f is injective when $0 < x < 1$, we say that the path is a *simple closed curve* or a *Jordan curve* after a French mathematician. A topological space is said to be (*arcwise*) *connected* when there is at least one path between any two of its points.

Figure 5.3   A path from P to Q, and a Jordan curve in the plane.

These simple definitions are very intuitive but also too precise for us. Some topological results just escape intuition. In 1890 Peano constructed a bijection between an open interval and an open square which is continuous from the interval. Twenty years later Brouwer restored confidence in the concept of dimension by showing that an interval and a square are not homeomorphic. On the other hand, some results are intuitive trivialities. About 1890 Jordan proved that a Jordan curve in the plane divides the plane into two connected regions, the interior and the exterior. Anyone puzzled by this result is in good company. The mathematician Schwarz, a contemporary of Jordan, wrote some very good mathematics but simply did not want to hear about Jordan curves. Jordan's theorem has a counterpart in higher dimensions. For instance, a homeomorphic image in $E = \mathbf{R}^3$ of the sphere $S : x^2 + y^2 + z^2 = 1$ divides E into two connected parts. But most topological theorems are neither counterintuitive nor intuitive trivialities. There is, for example, Brouwer's fixed-point theorem. It says that every continuous map f from the ball $x_1^2 + \cdots + x_n^2 \leqslant 1$ in $\mathbf{R}^n$ to itself has at least one fixed point, i.e., a point P such that $f(P) = P$. When $n = 1$, in which case the "ball" is the interval $-1 \leqslant x \leqslant 1$, this follows from the fact that the continuous function $x \rightarrow f(x) - x$ is $\geqslant 0$ when $x = -1$ and $\leqslant 0$ when $x = 1$ and hence must vanish somewhere in the interval. The proof in the general case, and even when $n = 2$ and the ball is a disk, requires some deep thinking.

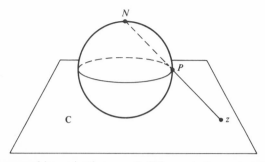

Figure 5.4 Stereographic projection $z \to P$. When z is far away, P is close to the north pole N. The complete plane $\hat{\mathbf{C}}$ is considered to have a point $z = \infty$ corresponding to N and is therefore homeomorphic to the sphere.

Since 1900 topologists have developed an algebraic machinery called algebraic topology which is a powerful tool for proving topological results. The ones we have sketched above are easy once this machinery has been constructed. It also yields plenty of results in higher dimensions where we cannot see, but can only reason. Some of them were actually obtained long before topology was given a firm foundation. We shall finish this chapter by giving two examples. Both have to do with polynomials in one or two complex variables. To begin with we remark that the complex plane $\mathbf{C}$ is homeomorphic to a pointed sphere, i.e., a sphere with one point removed. To see this one projects $\mathbf{C}$ on a sphere according to Figure 5.4. Its text explains what is meant by $\hat{\mathbf{C}}$, the plane $\mathbf{C}$ completed with a point at infinity. Let us also remark that the intuitive way of getting a homeomorphic image of a set in space is to imagine that it is made of rubber, and move it and squeeze it without tearing. To get the corresponding homeomorphism, just map the material points of the initial set to the corresponding material points of the deformed set. Just in case of the pointed sphere, these rules permit that a missing point is replaced by a hole. In this way, a sphere with two points removed is homeomorphic to a sphere with two closed caps removed and hence to a cylinder (Figure 5.5).

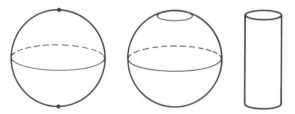

Figure 5.5 A sphere with two points removed is homeomorphic to a sphere with two closed caps removed and to a cylinder. None of the three sets is supposed to contain its boundary. The equators just mark that the figures represent spheres.

The fundamental theorem of algebra

In his doctoral thesis in 1851, Riemann says that he thinks of a polynomial $P(z)$ in one complex variable z as a map $z \to w = P(z)$ from a complex z-plane to a complex w-plane. When z moves around in the first plane, $P(z)$ moves around in the second plane. Had Riemann wanted to prove the fundamental theorem of algebra he could have done as follows. If $P(z) = z^n + a_{n-1}z^{n-1} + \cdots + a_0$ then $P(z) = z^n(1 + h(z))$ where $h(z) = a_{n-1}z^{-1} + \cdots + a_0 z^{-n}$ is very small when z is large. When z makes one turn around the circle $c_R : |z| = R$ in the z-plane then $w = P(z)$ makes n turns in the w-plane along a curve γ_R which is very close to the circle $|w| = R^n$. When we contract c_R to the origin, the curve γ_R contracts to the point $w = a_0$. Hence the image of the disk $|z| < R$ almost covers the disk $|w| < R^n$ and certainly the disk $|w| < R^n/2$ when R is big enough. But then the map $z \to P(z)$ is surjective from the z-plane to the w-plane and this is a way of expressing the fundamental theorem of algebra. When $n = 2$, the proof is illustrated by Figure 5.6.

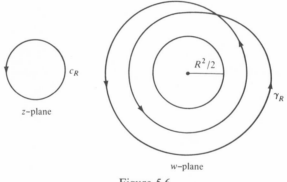

Figure 5.6

Algebraic curves

Let $P(z, w) = \Sigma a_{jk} z^j w^k$ be a polynomial in two variables z and w and not identically zero. When the coefficients and z and w are real, the equation $P(z, w) = 0$ defines a curve in $\mathbf{R}^2$, an algebraic curve. When z, w and the coefficients are allowed to be complex, the same equation defines a subset Γ of $\mathbf{C}^2$ which, by tradition, is also called an *algebraic curve*, although it is really something of dimension 2 lying in $\mathbf{C}^2$ which has dimension 4, the same as that of $\mathbf{R}^4$. Although we cannot see in four dimensions we would like to know what Γ looks like.

Let us first note that subsets of $\mathbf{C}^2$ consisting of points $(z, f(z))$ or $(g(w), w)$, where f and g are continuous and z and w belong to circular open disks in **C**, are homeomorphic to these disks. In fact, the corresponding bijections $(z, f(z)) \to z$ and $(g(w), w) \to w$ are continuous in both directions. For simplicity we shall refer to such subsets of $\mathbf{C}^2$ as disks. Returning to an algebraic curve Γ, we say that a point of Γ is regular when it has a neighborhood in Γ which is a disk. The reader is asked to accept that this happens when at least one of the partial derivatives $\partial P / \partial z$ and $\partial P / \partial w$ does not vanish at the point in question. In fact, close to a point where, for instance, the second one does not vanish, the equation $P(z, w) = 0$ can be shown to be equivalent to an equation $w = f(z)$ where f is continuous. Algebraic curves with only regular points are said to be regular. We now know what such a curve looks like locally, namely, a disk. To see what its global shape is we shall use a method developed by Riemann. His idea was to analyze Γ with the aid of the projection $(z, w) \to z$ from Γ to **C**. We say that the point (z, w) of Γ lies over or above z and think of it as lying somewhere above the z-plane.

As a first example let us choose $P(z, w) = w - Q(z)$ where Q is any polynomial. The projection $(z, Q(z)) \to z$ is then a bijection from Γ to **C** and it is also a homeomorphism. Hence the curve Γ is homeomorphic to a pointed sphere. Next, let $P(z, w) = z^2 + w^2 - 1$. Over a disk in the z-plane not containing the points $z = \pm 1$ there are then two disjoint disks of Γ where $w = \pm(1 - z^2)^{1/2}$. Hence, if γ is the path on Γ whose projection on **C** is the interval $I : -1 \leqslant x \leqslant 1$, γ is simple and closed and divides Γ into two disjoint parts. Both of them are homeomorphic to the plane **C** cut along I and hence to a pointed half-sphere. Hence Γ is homeomorphic to a sphere with two points removed (see Figure 5.7). Finally, consider the case $P = z^4 - w^2 - 1$. Over every disk in the z-plane not containing the points $1, -1, i, -i$ where $z^4 = 1$ there are two disjoint disks of Γ consisting of points where $w = \pm(1 - z^4)^{1/2}$. Letting γ_1 and γ_2 be the paths on Γ whose projections are, for instance, the straight lines I_1 from -1 to i and I_2 from $-i$ to 1, Γ minus these two paths consists of two disjoint parts both homeomorphic to **C** cut along I_1 and I_2 and hence to a cylinder with one point removed. Putting the two together, we see that Γ is homeomorphic to a torus with two points removed (see Figure 5.7).

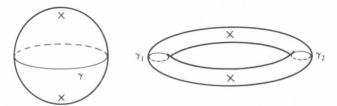

Figure 5.7 A sphere with two points removed separated by a simple closed path γ, and a torus with two points removed separated by two closed simple paths γ_1 and γ_2.

Reasoning as we have done here it is possible to prove that every algebraic curve with its singular points removed is homeomorphic to a sphere with a number of handles on it, also with a finite number of points removed. Note that one handle gives something homeomorphic to a torus. The number of handles is called the *genus* of the curve. This result, essentially obtained by Riemann in his thesis, is fundamental for the theory of algebraic curves. It is also a marvel of mathematical intuition.

5.4 Documents

Dedekind on Dedekind cuts

A passage from *Continuity and Irrational Numbers* (1872). The word continuous is used to mean unbroken, without gaps.

"... The comparison above of the rational numbers to a line has led to the latter being full of gaps, incomplete and discontinuous when we consider the line to be without gaps, complete and continuous. What does its continuity mean?

Richard Dedekind 1831–1916

The answer to this question must contain the scientific foundation on which we can base an investigation of all continuous regions. Vague talk about unbroken connection of the smallest parts will get us nowhere. We must get a precise definition of continuity which can be the basis of logical reasoning. For a long time I thought in vain about these things till I found what I was looking for. My discovery will be judged differently by different people but I believe that most of them will find it trivial. In the preceding paragraph it is pointed out that every point p on a straight line brings about a division of that line into two parts such that every point of one part lies to the left of every point of the other part. I maintain that the essence of continuity lies in the converse, i.e., in the following principle: If all points on a line belong to two classes such that every point of the first class lies to the left of every point in the other class, then there exists one and only one point which brings about this division of the line into two parts."

Poincaré on topology

In the preface to his article "Analysis situs" (from 1895) in which Poincaré laid the foundations of algebraic topology, he wrote

1854–1912

"... It has been said many times that Geometry is the art of correct reasoning supported by incorrect figures, but in order not to be misleading, these figures must satisfy certain conditions. The proportions may be altered but the relative positions of its various parts must not be changed. The use of figures is there mainly to clarify the relations between the objects that we study and these relations belong to a branch of Geometry called Analysis situs which describes the relative positions of lines and surfaces without regard to their size. There are relations of the same kind in hyperspace; as shown by Riemann and Betti, there is also an Analysis situs in spaces of more than three dimensions. Through this branch of science we know these kinds of relations although our knowledge cannot be intuitive; our senses fail us. In certain cases it gives us the service that we demand from geometric figures...."

Literature

First Concepts of Topology, by Chinn and Steenrod (Random House, 1966), is an elementary text on the level of this chapter. Modern standard works like *Lectures on Algebraic Topology*, by Dold (Springer-Verlag, 1972) are not accessible without the sort of background to be found, for instance, in *Introduction to Topology* by Lefschetz (Princeton, 1949) or *Algebraic Topology, An Introduction*, by Massey (Harbrace College Mathematics Series, 1967), or (in French) *Topologie des surfaces* by Gramain (Presses Univ. de France, 1971).

6
THE HEROIC CENTURY

The theory of differentiation and integration is called *infinitesimal calculus*, which means computation with infinitely small entities. It dates from the seventeenth century, which saw the birth of modern mathematics. For some time the strict proofs of the Greeks had been abandoned in favor of heuristic reasoning. Audaciously exploring new approaches, mathematicians surpassed everything that had been done before. The century had long religious wars, severe crop failures, and serious outbreaks of the plague, but for science and mathematics it was a time of unprecedented discoveries. The development was rapid. The works of Galilei (from 1604) on accelerated motion are almost childish compared to what Leibniz and the brothers Bernoulli did at the end of the century, solving many different problems of infinitesimal calculus and variational calculus with modern methods and notation.

Infinitesimal calculus was born out of efforts to compute areas of plane figures bounded by curved lines and the volumes of bodies bounded by curved surfaces. Certain of these computations, which are easy calculus exercises, had given the Greeks a great deal of trouble. In fact, almost everything that Archimedes wrote had to do with such problems, and his results mark the high tide of Greek mathematics. Before him Eudoxus from Chnidos had computed the volume of the cone and the pyramid, but Archimedes computed the volume and the area of the sphere, the area of the parabolic segment, the center of gravity of the triangle, and the area enclosed by a certain spiral now called the spiral of Archimedes. Of these, the volume of the cone and the pyramid, the area of the parabolic segment, the center of gravity of the triangle, and the area of the segment of the spiral all depend on the same integral, namely $\int x^2 \, dx$, but there are no signs that Archimedes saw the connection between these problems. Every one of them he solves with different and often very ingenious methods. In his work on the spiral he even says that the problem treated there has no connection with certain other problems, e.g., the volume of the paraboloid segment which he has just mentioned and which leads to the integral $\int x \, dx$. Archimedes' proofs are very strict. He uses the "method of exhaustion" of Eudoxus, in which figures with known areas are circumscribed and inscribed into the figure whose area one wants to compute. There is a similar method for arc lengths, and Archimedes used it to prove that the number π lies between $3\frac{1}{7}$ and $3\frac{10}{71}$. The proof is perfectly rigorous and seventeenth century mathematicians often used Archimedes as a model when they wanted to give absolutely convincing proofs. But the rigor also made the proofs cumbersome and difficult. It is reasonable to assume that in most cases Archimedes had known the solutions of his problems before he worked out the proofs. Through a letter from Archimedes to Eratosthenes of Alexandria, discovered in 1906, we even know his method for this. He called it the mechanical method. The gist of it was to consider, e.g., a plane figure as something with a weight composed of straight lines each one with a weight proportional to its length. The actual work is then done by balancing various geometric figures against each other and looking for centers of gravity. About this Archimedes wrote:

> "This procedure is, I am persuaded, no less useful even for the proofs of the theorems themselves; for certain things became clear to me by the mechanical method, although they had to be demonstrated by geometry afterwards because their investigation by the said method did not furnish an actual demonstration . . . and I deem it necessary to expound the method partly because I have spoken of it and I do not want to be thought to have uttered vain words, but equally because I am persuaded that it will be of no little service to mathematics; for I apprehend that some, either of my contemporaries or of my successors will, by means of the method when once established, be able to discover other theorems in addition, which have not yet occurred to me."

But it was a long time before these prophetic remarks came true. There are several reasons why Archimedes' work was not continued by others, e.g., Archimedes' superior gifts and the sterilizing effect that the Roman conquest had on Greek science in general. But the main cause is probably the Greek geometric method itself. It is impeccable but it does not reveal the right connections and therefore makes new discoveries difficult. The progress predicted by Archimedes did not occur until the seventeenth century.

The main works of Greek literature and philosophy were printed in Italy before 1520, and the first edition of Archimedes' work in Basel in 1544. Contemporary mathematics, represented by Cardano, was then rather algebraic, and Archimedes' influence came later with Galilei and Kepler, both of whom were astronomers and physicists more than they were mathematicians. But from their time, the early seventeenth century, until about 1670, mathematicians are constantly quoting Archimedes. He is translated and commented upon, and everybody declares him a paragon and a source of inspiration.

In the beginning of the seventeenth century, scientists had to work under very primitive conditions, but later these improved enormously through the consolidation of the universities and the foundation of scientific societies and periodicals. Mathematicians had no periodical before 1665, when the newly founded Royal Society started publishing *Philosophical Transactions*. Before this time they had to write letters to each other or print books, often at their own expense when no patron of science was available. Publishers and printers who could do such a job were scarce, and sometimes not very honest. After the printing a new ordeal awaited the author. The general uncertainty about the foundations of infinitesimal calculus made it very easy for the rivals to find the weak points and criticize them. There were many bitter controversies of this kind, fought in bad faith by both parties. It is not surprising that many preferred to work in peace and quiet, just telling good friends about new results. Certain mathematical dilettantes, e.g., Mersenne in Paris and Collins in London, conducted a large correspondence supplying the mathematical news service with excerpts from letters. Students traveled a lot and in this way new ideas spread efficiently but perhaps in not a very orderly fashion. The random contacts and the fact that all mathematicians were working on essentially the same problems made priority fights very common. The quarrel between Newton and Leibniz about who invented infinitesimal calculus was well-known at the time, even to nonmathematicians.

Roughly speaking, there are two major periods of the seventeenth century, the time before and that after 1670. The most important names of the first period are the Italians Galilei (1564–1642) and Cavalieri (1598–1647), the German astronomer Kepler (1571–1630), the Frenchmen Fermat (1601–1663), Descartes (1596–1650), and Blaise Pascal (1623–1662), the Dutchman Huygens (1629–1695), and the Englishmen Wallis (1616–1703) and Barrow (1630–1677). All these did work preparatory to infinitesimal calculus proper which then, in a brief period after 1670, was created by Newton (1642–1727), the German Leibniz (1646–1716), and the Scotsman James Gregory (1638–1675). Before going into details we should perhaps say a few words about the actors in the drama. Galilei, professor in Pisa and Padua, broke with Aristotelian physics and discovered the laws of falling bodies. He constructed telescopes and made some fundamental astronomical discoveries. He believed in the theory of Copernicus that the earth and the planets move around the sun, but the Church forced him to deny this heresy in 1633. Cavalieri was professor of mathematics in Bologna and a friend of Galilei's. Kepler succeeded the Danish astronomer Tycho Brahe as imperial mathematician in Prague; he deduced, from Brahe's observations, that the planets move in elliptic orbits around the sun. Fermat, a lawyer in Toulouse, worked with number theory and analysis, and he corresponded with Pascal and Descartes. The second of these last two was a nobleman, a soldier, a philosopher and a teacher of royalty. He died in Stockholm at the court of Queen Christina. Descartes' great mathematical discovery was analytic geometry. At the age of 16, Pascal discovered a fundamental theorem about conic sections. Later, he wrote about probability theory and computed areas and centers of gravity. He had several religious crises, and his contemporaries knew him more as a philosopher and religious writer than as a mathematician. Huygens studied law and thought of becoming a diplomat but soon made a name for himself as a scientist. His analysis of progressive waves and the refraction of light is valid even today. Huygens was elected member of the French Academy of Sciences in 1666 and lived in Paris for a long time. There he met Leibniz and got him interested in the new mathematics. Wallis started as a theologian and became professor of mathematics in Oxford in 1649. Barrow, Newton's teacher, was professor of mathematics in Cambridge and later retired to a parsonage. He left his chair to Newton, and it is likely that Newton had some influence on Barrow's work. Gregory was professor of mathematics at St. Andrews in Scotland. With this background, we are ready now for the two main characters, Newton and Leibniz.

Newton came to Cambridge in 1660, at the age of 17. Nine years later he succeeded Barrow and planned to publish a treatise on derivatives and series containing the fundamental theorems of infinitesimal calculus. It remained in manuscript, but was printed after his death and became known as the theory of fluxions. Newton considered the derivative as a velocity, and called it a *fluxion*. In his main work, *Philosophie Naturalis Principia Mathematica*, known as *Principia* and printed in 1687, Newton proved that the movements of celestial bodies can be deduced from the law of motion (the force equals the time derivative of momentum) and the law of gravitation. *Principia* was the first big success for the combination of physics and mathematics, and it has been followed by many others for almost 300 years. The firmly rooted prestige now enjoyed by this couple started with Newton's work. Its first unparalleled success has led to sometimes exaggerated hopes that mathematics in combination with, e.g., biology or economics will yield the same brilliant results.

Most of Newton's contemporaries thought that comets were the work of God or the Devil, and were portentous signs of coming events. After *Principia* educated people could no longer have this faith, but philosophy and religion soon adjusted to the fact that the movements of celestial bodies are as predictable as those of the wheels of a clockwork. According as new planets were discovered it was difficult to maintain that God intended them to be five in order to join the sun and the moon in a sacred seven, but *Principia* did not shake God's position as the Creator. On the contrary, Creation appeared as an even greater miracle than before. Politically, and in religious matters, Newton was a conservative and he had a firm belief in God. Among his unpublished manuscripts there are long investigations of religious chronology and the topography of hell. The spirit of the times was such that they are consistent with *Principia*. After 1690 Newton served for some time as director of the Royal Mint and an M.P. for Cambridge University.

Leibniz started his career as a precocious student in Leipzig, and after 1676 earned his living as diplomat, genealogist, and librarian to the house of Hanover. One of its members became king of England in 1714, under the name of George I. He was supported by the Whig party. Since Newton was a fanatical Tory, he and Leibniz were political adversaries, which is assumed to explain some of the animosity between them. The contact with Huygens in Paris in 1673 was the beginning of Leibniz's career as a mathematician. He visited London many times and exchanged letters with Newton, Collins, Huygens, and many others. Leibniz founded the academies of science of Leipzig and Berlin, and published most of his mathematical papers in *Acta Eruditorum*, the journal of the Leipzig Academy. He was a pioneer in symbolic logic, but was also a philosopher in the classical tradition, who occupied himself with explaining the universe and proving God's existence. His most important philosophical work remained in manuscript. What he published was more or less tailored to the taste of ruling princes. In any case they had no difficulty in accepting his famous dictum that we live in the best of all possible worlds.

Here we leave the personalities and return to mathematics. We shall follow the development through an analysis of three themes: mathematical rigor versus heuristic reasoning, connections between problems, and the balance between geometry and algebra.

Among the Greeks philosophical and logical aspects and mathematical rigor were dominant, and there was almost nothing left for heuristic arguments. There was a large gap between the initial idea and the artfully executed, polished proof, and this must have had a frustrating effect on ingenious mathematicians. To a large extent, progress in the seventeenth century is due to the fact that mathematical rigor was neglected in favor of heuristic reasoning. Archimedes' mechanical method was not known, but mathematicians started arguing as he had done. They talked about "infinitely small quantities" and "sums of infinitely small quantities." A plane figure, for instance, was considered to be composed of parallel line intervals and—now comes the meaningless but useful point of view—its area as the sum of the areas of corresponding infinitely narrow rectangles. Aided by such arguments it is easy to convince oneself that if the figure is doubled in every direction, its area gets four times as large. In the same way, if a body is doubled in every direction, its volume is multiplied by eight. This observation in general form is due to Cavalieri and was called *Cavalieri's principle*. Before coming to believe it Cavalieri checked it against all the areas and volumes computed by Archimedes.

Similar arguments were used to treat the second main problem of infinitesimal calculus, the determination of tangents and the calculation of arclength. The existence of tangents was postulated and the curve itself was thought of as composed of infinitely small line segments sometimes considered as parts of tangents, sometimes as cords. The arclength was supposed to be the sum of the lengths of these infinitely small parts.

Not everybody was content with these arguments and Fermat, for instance, generally took great care to provide strict proofs in every special case. Others did not, but all had the feeling of being on safe ground. "It would be easy," Fermat says somewhere, "to give a proof in the manner of Archimedes, but I content myself by saying this once in order to avoid endless repetitions." Pascal assures his readers that the two methods —Archimedes and the principle of infinitesimals—only differ in the manner of speaking, and Barrow remarks nonchalantly that "it would be easy to add a long proof in the style of Archimedes, but to what purpose?" As time passes the references to Archimedes tend to become mere formalities, often used to give an air of respectability to methods which Archimedes himself would certainly not have endorsed.

Some 100 hundred years later, all these consideration were taken care of by the concept of a limit. It is also not difficult to find quotations from, e.g., Pascal or Newton, where this concept occurs more or less explicitly, but we have only to read them in context to realize that the time was not yet ripe for a systematic theory. Instead, the really significant steps forward were made when Newton and Leibniz turned away from the past, justifying infinitesimal calculus by its fertility and coherence rather than by rigorous proofs.

With this we come to the second theme, the connection between problems. Nowadays, we compute volumes, areas, and arclengths using a single operation, *integration*; and we treat problems about tangents, maxima, and minima using another one, *differentiation*. These two, on the other hand, are connected via the main theorem of integral calculus. The Greeks had different methods for all these geometric problems, and not even Archimedes saw that they are connected when seen from a higher and more abstract point of view. But the seventeenth century continued where Archimedes had left off. Within a period of 50 years, Cavalieri, Fermat, Huygens, Barrow, and Wallis succeeded in reducing many computations of volumes, areas, and arclengths to the problem of integrating (or, using their own term, *finding the quadrature* of) certain simple functions, e.g., entire or even fractional powers of x. Some of these quadratures were found and others were guessed at. Using special methods they had also found the tangents of certain curves. The decisive steps were then taken by Newton and Leibniz, both of whom introduced a special notation for the derivative of a function, and by Leibniz, who did the same for the integral and gave the algebraic formulas governing the use of these notations. This made all previous work obsolete. The new calculus had lucid formulas and simple procedures for computing volumes, areas, arclengths, and tangents using only the two basic operations, integration and differentiation. Leibniz's notations became universally accepted. The first calculus textbook, published shortly before 1700 by the Marquess de l'Hospital, was an adaptation of a manuscript by Johann Bernoulli, a pupil of Leibniz. Together with infinitesimal calculus, a new tool of analysis was invented, the power series. In 1668 Mercator had made the sensational discovery that the logarithm could be developed into a power series, and Newton, Gregory, and Leibniz competed to find power series for the basic functions.

Our third theme is the balance between geometry and algebra. The mathematics of the Greeks was geometric. In the works of Archimedes there is not a single formula; everything is expressed in words and figures.

The seventeenth century found this geometric method to be a straitjacket and finally got rid of it. The beginnings had already been made. The Arabic numerals, based on the positional system, had proved themselves superior in practice to the Roman numerals, computations with letters had been gradually accepted, and algebra and the theory of equations as we know them now had been studied intensively in Italy in the sixteenth century. The usefulness of algebra had become obvious. Using simple school algebra it is child's play to prove, for instance, the basic lemma of Archimedes' treatise on spirals, whereas without algebra it is a feat. Galilei sticks to geometry but Fermat already uses algebra rather freely, and analysis gets less and less geometric until Leibniz creates the new calculus giving analysis an algebraic form. But the geometric tradition put up a stubborn resistance. The most striking example of this is *Principia*, where the terminology and the proofs are geometric, although Newton himself had achieved his results using calculus. Only afterwards did he give them a geometric form.

After its breakthrough around 1700, infinitesimal calculus was to consolidate, grow, and find a host of new applications. But the heroic time was over—the rapid development, the great discoveries, and the hard fights.

Documents

Galilei on uniformly accelerated motion

In this passage from *Two New Sciences* (1638), Galilei uses Euclidean geometry to make a diagram illustrating uniform and uniformly accelerated motion.

Galileo Galilei 1564–1642

Theorem I, Propos. I. The time it takes a body with uniform acceleration to move a certain distance equals the time it takes for the same body to move the same distance with a uniform velocity equal to half the sum of the least and the largest of its velocities in the uniformly accelerated motion. In fact, let AB represent the time it takes for the body to move from C to D with uniform acceleration. On AB we represent the increasing velocities, the last one by the line EB. We draw AE and several lines parallel to EB supposed to represent the increasing velocities. We half EB in F and draw the parallels FG to BA and GA to FB. The parallellogram $AGFB$ is equal to the triangle AEB. In fact,...

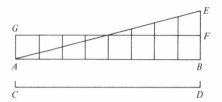

Figure 6.1   Uniform and uniformly accelerated motion according to Galilei.

Newton on fluxions

In this passage from *Treatise on the Quadrature of Curves* (1704), Newton disassociates himself from Leibniz's infinitesimals and asserts his priority. His fluxions are derivatives and his fluents primitive functions. At the end of the quoted text he is dealing with the main theorem of integral calculus—in modern language, the fact that the derivative of the function $x \to \int_a^x f(t)\, dt$ equals $f(x)$ when f is continuous.

"1. I consider that mathematical quantities are given by continuous movement, not that they consist of very small parts. Lines are not composed by putting utterly small parts together, they are generated by moving points, surfaces by moving lines, bodies by moving surfaces, angles by rotation of half-rays, times by steady flow and the same in other cases. These things occur in Nature and we can observe them daily when watching moving bodies.

Isaac Newton 1643–1727

"2. As I considered that, in the same time interval, quantities with large velocities become larger than quantities with small velocities, I tried to determine a quantity from the velocity with which it moves. These velocities I have called fluxions and the quantities generated by them fluents and in 1665 and 1666 I discovered and worked out the method of fluxions which I have here used for the quadrature of curves.

"3. The fluxions behave like the increments of the fluents in utterly small time intervals. More precisely, they are directly proportional to them.

"4. Let the arcs *ABC* and *ABDG* be generated by the ordinates *BC* and *BD* moving on the basis *AB* with the same uniform velocity. Then the fluxions of these areas are to each other as the ordinates *BC* and *BD* and we can think of them as represented by these ordinates since they are to each other as the just beginning increments of the areas...."

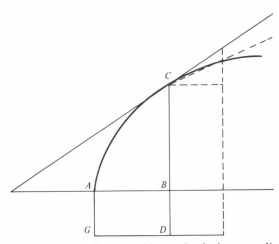

Figure 6.2   The fundamental theorem of integral calculus according to Newton.

Leibniz on differentials

Between 1672 and 1677 Leibniz created his basic formalism of infinitesimal calculus. Already from the beginning he tried to turn the complicated geometric diagrams of his predecessors into algebraic formulas. One of his early efforts was the formula

$$\text{omn. } xl = x \text{ omn. } l - \text{omn. omn. } l \qquad (*)$$

which refers to areas. Later he replaced the letters omn. meaning "sum of" by the integral sign $\int$. He also tried to find an inverse of integration using a symbol d with the loose meaning of "difference." In his algebraic experiments with this symbol he made false starts, like $d(u/v) = du/dv$, but in 1677 he got the correct formulas. In terms of these, the mysterious (*) above takes the familiar form $\int x\,dy = xy - \int y\,dx$, the formula for integration by parts. Here is how Leibniz himself wrote some of the rules of differential calculus (MS. July 1677, translated by J. M. Child in *The Early Mathematical Manuscripts of Leibniz* (1920).) Note that a means a constant and the bars parentheses.

> "*Addition and Subtraction.* Let $y = v \pm w \pm a$, then $\overline{dy}$ will be equal to $\overline{dv} \pm \overline{dw} \pm 0$. *Multiplication.* Let y be equal to avw, then $\overline{dy}$ or $\overline{davw}$ or $\overline{advw}$ will be equal to $avdw + awdv$."

In *History and Origin of Differential Calculus*, written about 1714 to answer the attacks of the followers of Newton, Leibniz—talking of himself in the third person—says, among other things:

1646–1716

"Now it certainly never entered the mind of anyone else before Leibniz to institute the notion peculiar to the new calculus by which the imagination is freed from the perpetual reference to diagrams.... But now by the calculus of Leibniz the whole of geometry is subjected to analytical computation, and those transcendent lines that Descartes called mechanical [curves whose construction does not belong to Euclidean geometry] are also reduced to equations chosen to suit them by considering the differences $dx,\ldots$."

Literature

This survey of the seventeenth century follows a section entitled "Calcul infinitesimal" of *Eléments d'histoire des mathématiques*, by N. Bourbaki (Hermann, 1969). This article also has a list of references. Bertrand Russell's *A History of Western Philosophy* (Simon and Schuster 1945) has a section on Leibniz, and James Newman's anthology *The World of Mathematics* (Simon and Schuster 1956) contains a biography of Newton written by C. Andrade.

7
DIFFERENTIATION

7.1 *Derivatives and planetary motion*. Derivatives and rules of differentiation. Ballistic motion and planetary motion. 7.2 *Strict analysis*. The mean value theorem and Taylor's formula. The differentiability of inverse functions. 7.3 *Differential equations*. Systems of ordinary differential equations. Existence and uniqueness. 7.4 *Differential calculus of functions of several variables*. Partial derivatives. Taylor's formula. The chain rule and the Jacobi matrix. Smooth local bijections. Implicitly defined functions. General coordinates. 7.5 *Partial differential equations*. Elasticity. Heat conduction. Distributions. History. 7.6 *Differential forms*. Products of vectors and the Grassmann algebra. Differentials in $\mathbf{R}^n$. Differential forms. 7.7 *Differential calculus on a manifold*. Hypersurfaces and curves in $\mathbf{R}^n$, tangents and normals. Manifolds in $\mathbf{R}^n$. Charts and atlases. Exact and closed forms, the de Rham complex, and the cohomology of a manifold. 7.8 *Document*. Riemann on physics and partial differential equations.

After a short presentation of what a derivative is and the rules of differentiation, differential calculus is illustrated by the modern version of Newton's deduction of planetary motion from the physical postulates. In principle, high school mathematics suffices for this, but it turns out that a complete understanding of the things that happen in the mathematical model requires the mean value theorem and the theorem of differentiability of inverse functions. Short but complete proofs are given and now the text requires some background in college mathematics. Then follows a section about systems of ordinary differential equations. There are not many examples but this section is not difficult, and the material is very important for everyone who wants to understand and use mathematics. Those who went through the contraction theorem in Banach spaces of Chapter 4 get an existence and uniqueness proof which is certain to impress everyone who sees it for the first time. After this, the scene changes to differential calculus in several variables. An introductory section states the simplest properties of partial derivatives, Taylor's formula, smooth bijections (here the contraction theorem is used again), and implicitly defined functions. Partial differential equations are treated in passing. Finally, there is a section on differential forms, followed by a section on differential calculus on manifolds. It begins with the simplest facts about tangents and normals of curves, surfaces, and hypersurfaces and ends with a short presentation of manifolds and the de Rham complex intended to whet the reader's appetite for differential topology.

7.1 Derivatives and planetary motion

Derivatives and rules of differentiation

Let us consider a curve $y = f(x)$ in a coordinate system and draw the tangent T to the curve at the point $x = a$ (Figure 7.1). Here we have taken the word tangent in its original geometric meaning. The derivative of a function is the analytical counterpart of this concept. Writing the equation of the tangent as $y = f(a) + (x - a)c$, we can transfer our geometrical intuition to the formula

$$x \quad \text{close to } a \quad \Rightarrow \quad f(x) = f(a) + (x - a)c + \text{small} \tag{1}$$

where the word small indicates something that is small compared to $x - a$. It is convenient to write this as $(x - a)o(1)$ where the symbol "$o(1)$," pronounced "little o of 1," means a quantity that tends to zero as x tends to a. We can then write (1) as

$$x \to a \quad \Rightarrow \quad f(x) = f(a) + (x - a)c + (x - a)o(1) \tag{2}$$

or as

$$c = \frac{f(x) - f(a)}{x - a} + o(1),$$

which is the same as

$$c = \lim_{x \to a} \frac{f(x) - f(a)}{x - a}. \tag{3}$$

If the curve has a tangent in the analytical sense (2) then the limit of the right side of (3) exists, and conversely. The limit need not exist, but if it does, the function f is said to be *differentiable* at the point a with the derivative $c = f'(a)$. Correspondingly, the derivative of f at the point x is

$$f'(x) = \lim_{h \to 0} \frac{f(x + h) - f(x)}{h} \tag{4}$$

when the limit exists. In that case the function is of course continuous at

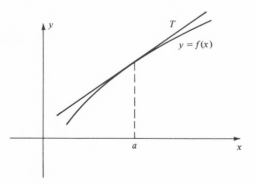

Figure 7.1   A curve and its tangent.

the point x. However, it is easy to construct functions which are not differentiable at certain selected points, for instance, because the function has a corner there (as $|x|$ at $x = 0$), or because the quotient of the right side of (4) oscillates between two fixed values as $h \to 0$ (as for $f(x) = x \sin x^{-1}$ at $x = 0$). In 1872 Weierstrass constructed a continuous function which is not differentiable at any point. This provoked a considerable outcry but the agitation has now subsided and we shall find it quite natural that some functions are differentiable and others are not.

When $f(x) \geqslant f(a)$ $(f(x) \leqslant f(a))$ for all x close to a, f is said to have a *local minimum* (*maximum*) at a. It follows from (3) that the derivative $f'(a)$ vanishes in such a point provided it exists and f is defined in an interval around a. In fact, $f'(a)$ is then at the same time the limit of numbers $\geqslant 0$ and numbers $\leqslant 0$.

One way of looking at the derivative used by Newton is to interpret it as a velocity. We then change our notations so that t is time and x the coordinate of a point P that moves along a straight line L. If P has the coordinate $x(t)$ at the time t, the function $t \to x(t)$ describes the movement of P. Its velocity at time t is then $x'(t)$ or, with Newton's notation, $\dot{x}(t)$. The acceleration is defined as the derivative of $\dot{x}(t)$ and will be denoted by $x''(t)$ or $\ddot{x}(t)$. When the movement is uniform, then $x(t) = x_0 + vt$ where x_0 and v are constant so that $\dot{x}(t) = v$ and $\ddot{x}(t) = 0$ for all t.

The classical rules for computing the derivatives of sums, products, quotients, and composed functions are easy consequences of (4) and run as follows:

$$H = f + g \quad \Rightarrow \quad H'(x) = f'(x) + g'(x)$$
$$H = fg \quad \Rightarrow \quad H'(x) = f(x)g'(x) + f'(x)g(x)$$
$$H = f/g \quad \text{and} \quad g(x) \neq 0 \quad \Rightarrow \quad H'(x) = (g(x)f'(x) - g'(x)f(x))/g(x)^2$$
$$H = f \circ g \quad \Rightarrow \quad H'(x) = f'(g(x))g'(x) \quad \text{(the chain rule).} \tag{5}$$

Here it is assumed that f and g are differentiable at x and the implied equalities mean that $H'(x)$ exists with the value stated. The last rule is proved by the following computation:

$$f(g(x + h)) = f(g(x)) + f'(g(x))(g(x + h) - g(x))$$
$$+ (g(x + h) - g(x))\varepsilon_1 = f(g(x)) + f'(g(x))g'(x)h + h\varepsilon_2$$

where $\varepsilon_1 \to 0$ when $g(x + h) - g(x) \to 0$ and $\varepsilon_2 \to 0$ when $h \to 0$. By various simple arguments it is also possible to compute the derivatives of the traditional elementary functions, e.g.,

$$f(x) = x^a, \quad \text{where} \quad x > 0 \quad \Rightarrow \quad f'(x) = ax^{a-1},$$
$$f(x) = e^x \quad \Rightarrow \quad f'(x) = e^x,$$
$$f(x) = \log|x|, \quad \text{where} \quad x \neq 0 \quad \Rightarrow \quad f'(x) = 1/x,$$
$$f(x) = \cos x \quad \Rightarrow \quad f'(x) = -\sin x,$$
$$f(x) = \sin x \quad \Rightarrow \quad f'(x) = \cos x.$$

Combining these formulas with the general rules (5), we can compute the derivatives of a large class of functions explicitly. In particular, for polynomials one has

$$H(x) = a_0 + a_1 x + \cdots + a_n x^n$$

$$\Rightarrow \quad H'(x) = a_1 + 2a_2 x + \cdots + na_n x^{n-1}. \tag{6}$$

The art of computing with derivatives is called differential calculus. In the section that follows we shall demonstrate the magic of this calculus by using it to prove the main result of Newton's *Principia*: the orbits of planets and comets spinning around the sun are conic sections. We shall then see that the algebraic side of the differential calculus represented by the rules (5) is not enough. We have to add theorems representing the analytical side, for instance, the fact that only constants have the derivative zero. Neither is it possible to pretend that all functions are differentiable. We shall work with functions f defined in open intervals such that the k successive derivatives $f^{(1)} = f', f^{(2)} = (f')' = f'', \ldots, f^{(k)} = (f^{(k-1)})'$ exist and are continuous. For simplicity we shall call them C^k functions. We shall mainly deal with C^1 and C^2 functions. Repeated use of the rules (5) shows that sums, products, quotients with nonvanishing denominators, and compositions of C^k functions are C^k functions.

Ballistic motion and planetary motion

Let us start with the simple ballistic motion analyzed in the beginning of the seventeenth century by Galilei. In his mathematical model a point-sized body K moves in a vertical plane where we have a rectangular system of coordinates x, y with vertical y-axis. If we let $x = x(t), y = y(t)$ denote the position of the body at time t, the movement is such that

$$\ddot{x}(t) = 0, \qquad \ddot{y}(t) = -g \tag{7}$$

where g is a positive number, the same for all bodies. These equations also follow from Newton's law of motion,

mass times acceleration = force,

and the law of gravitation: the earth attracts bodies on its surface by a vertical force of size mg where m is the mass of the body. The rules for taking derivatives show that the functions

$$x = x_0 + ut, \qquad y = y_0 + vt - gt^2/2 \tag{8}$$

are solutions of (7) for all values of the constants x_0, y_0, u, v. Here the pair (x_0, y_0) is the position and the pair (u, v) the velocity of the body at $t = 0$. A small computation shows that the orbit is a parabola with vertical axis. It follows that every solution of (7) has the form (8) if we can show that

$$\dot{w}(t) = 0 \text{ in an interval} \quad \Rightarrow \quad w(t) \text{ is a constant there.} \tag{i}$$

In fact, then (7) implies, for instance, that $\dot{w}(t) = 0$ where $w(t) = \dot{y}(t) + gt$ so that $\dot{y}(t) + gt = \text{constant} = v$, etc.

To study planetary motion we shall make a mathematical model of a body attracted by the sun. The body and the sun are assumed to be points with masses m and M where M is much larger than m. By the law of gravitation, the body is subject to a force of size $amMr^{-2}$ directed towards the sun. Here a is a constant and r the distance to the sun. The sun is also attracted by the body with a force of the same size and the opposite direction. But the sun has a very large mass and is influenced very little by this force so we decide to neglect it. If we introduce a rectangular system of coordinates x, y, z with its center in the sun, then $r = (x^2 + y^2 + z^2)^{1/2}$, and if the position of the body at time t is denoted by $x = x(t), y = y(t)$, $z = z(t)$, the law of motion says that

$$m\ddot{x} = -amMxr^{-3}, \qquad m\ddot{y} = -amMyr^{-3}, \qquad \ddot{z} = -amMzr^{-3}.$$

The left sides are the three components of mass times acceleration and the right sides the three components of a force of size $amMr^{-2}$ directed towards the origin. Dividing by m and assuming a suitable time scale we can assume that $aM = 1$, and then we get the three equations

$$\ddot{x} = -xr^{-3}, \qquad \ddot{y} = -yr^{-3}, \qquad \ddot{z} = -zr^{-3} \qquad (9)$$

where $r > 0$ and x, y, z are twice differentiable functions of t. Since the right sides then are continuous functions of t, it follows that x, y, z are also C^2 functions. Our mathematical machinery is not yet good enough for a complete analysis of (9) and we shall write down the missing points as we go along. They will be dealt with later. We have to start with such an item, namely

$$w = \dot{w} = 0 \quad \text{for one} \quad t \quad \Rightarrow \quad w = 0 \quad \text{for all } t \qquad \text{(ii)}$$

when $w = ax + by + cz$ is a linear combination, with constant coefficients, of our functions x, y, z. Note that, given t_0, $w(t_0) = 0$, $\dot{w}(t_0) = 0$ is just a system of two linear equations for three unknowns a, b, c and hence it has a solution which we can normalize so that $a^2 + b^2 + c^2 = 1$. Choosing $ax + by + cz$ as a new z coordinate, the meaning of (ii) is then simply that $z(t) = 0$ for all t. Hence we can disregard the third equation (9) and assume that the movement takes place in the plane $z = 0$. The first two equations then imply the following ones

$$\dot{x}\ddot{x} + \dot{y}\ddot{y} = -(x\dot{x} + y\dot{y})r^{-3}, \qquad x\ddot{y} - \ddot{x}y = 0. \qquad (10)$$

We can also see that

$$\dot{r} = 0 \quad \Rightarrow \quad \dot{x}^2 + \dot{y}^2 - r^{-1} = 0. \qquad (10')$$

In fact, $\dot{r} = 0$ implies $x\dot{x} + y\dot{y} = r\dot{r} = 0$ so that, taking the time derivative, $\dot{x}^2 + \dot{y}^2 + x\ddot{x} + y\ddot{y} = 0$. Insertion of $\ddot{x}$ and $\ddot{y}$ according to (9) gives the desired result. Now in (10), all members are derivatives since

$$h = 2^{-1}(\dot{x}^2 + \dot{y}^2) \quad \Rightarrow \quad \dot{h} = \dot{x}\ddot{x} + \dot{y}\ddot{y}$$
$$h = x\dot{y} - \dot{x}y \quad \Rightarrow \quad \dot{h} = x\ddot{y} - \ddot{x}y$$
$$h = r^{-1} = (x^2 + y^2)^{-1/2} \quad \Rightarrow \quad \dot{h} = -(x\dot{x} + y\dot{y})r^{-3}.$$

Supposing as before that (i) holds, (10) shows then that there are constants E and c such that

$$2^{-1}(\dot{x}^2 + \dot{y}^2) - r^{-1} = E, \qquad x\dot{y} - \dot{x}y = c. \tag{11}$$

Both these numbers have physical significance. The product mE is called the *energy* of the body. It is the sum of the kinetic energy $2^{-1}m(\dot{x}^2 + \dot{y}^2)$, where the last factor is the square of the velocity, and the potential energy $- mr^{-1}$. To see what c means we introduce polar coordinates $x = r\cos\theta$, $y = r\sin\theta$. In order to proceed we then have to assume that

$$r \text{ and } \theta \text{ are } C^2 \text{ functions of } t. \tag{iii}$$

Then the rules (5) show that

$$\dot{x} = \dot{r}\cos\theta - \dot{\theta}r\sin\theta, \qquad \dot{y} = \dot{r}\sin\theta + \dot{\theta}r\cos\theta$$

and a small computation shows that (11) can be written as

$$2^{-1}(\dot{r}^2 + r^2\dot{\theta}^2) - r^{-1} = E, \qquad r^2\dot{\theta} = c \tag{12}$$

and (10') as

$$r \text{ constant} \quad \Rightarrow \quad r^2\dot{\theta}^2 - r^{-1} = 0. \tag{12'}$$

It follows from the second equation of (12) that $c/2$ is the sector velocity, i.e., the area swept per unit time by a straight line from the origin to the body. That the sector velocity is constant is one of the three laws that Kepler deduced from astronomical observations. The two others say that the orbits of the planets are conic sections with the sun at one of the foci, and that the squares of their periods are proportional to the cubes of the averge distances to the sun. We shall come back to these laws but for the moment we return to (12).

If, to start with, r is constant, (12) and (12') show that

$$r = c^2, \qquad \dot{\theta} = c^{-3}, \qquad E = -1/2c^2.$$

This means that the body moves with angular velocity c^{-3} in a circle with radius c^2. This also gives a solution of (9). Another special case is $c = 0$ implying $\dot{\theta} = 0$ so that the body moves on a straight line through the sun. Disregarding these two simple cases it remains to study the equations (12) in an interval of time where r is not a constant and when $c \neq 0$. To determine the orbit we shall try to find out how r varies with θ and then we need to know that

$$\dot{\theta} \neq 0 \quad \Rightarrow \quad t \text{ is a } C^2 \text{ function of } \theta. \tag{iv}$$

Putting $t = h(\theta)$ it follows that also $r = r(h(\theta))$ is a C^2 function of θ and we get $r' = \dot{r}(t)h'(\theta) = \dot{r}/\dot{\theta}$. In fact, $t = h(\theta)$ gives $1 = h'(\theta)\dot{\theta}$ when we take the derivative with respect to t. Now $r^2\dot{\theta} = c$ and inserting this, i.e. $\dot{r} = cr'r^{-2}$ and $\dot{\theta} = cr^{-2}$, into the first equation of (12) shows that

$$2^{-1}(r'^2 r^{-4}c^2 + r^{-2}c^2) - r^{-1} = E.$$

To simplify this formula, replace r by the function $u = r^{-1}$. It is a C^2 function of θ and, since $u' = -r'r^{-2}$, we get

$$2^{-1}c^2(u'^2 + u^2) - u = E.$$

With $v = cu - c^{-1}$, $v' = cu'$, we get something still simpler, namely

$$v'^2 + v^2 = 2E + c^{-2} > 0. \tag{13}$$

The right side must be positive since v is not a constant. Suppose now that we knew that

$$w'^2 + w^2 = 1 \text{ in an interval} \quad \Rightarrow \quad w = \cos(\theta + \theta_0) \text{ there} \tag{v}$$

when $w = w(\theta)$ is a nonconstant C^2 function. In that case (13) implies that

$$v = (2E + c^{-2})^{1/2} \cos\theta$$

provided we choose a suitable origin for θ. A computation of $r = r(\theta)$ gives us the final result

$$r = p(1 + e\cos\theta)^{-1} \quad \text{where} \quad p = c^2, \quad e = (2Ec^2 + 1)^{1/2}.$$

According to the last section of chapter 4, this is the equation of a conic section with one focus at the origin, the eccentricity e, and the parameter p. When $0 < e < 1$ the orbit is an ellipse, when $e = 1$ a parabola, and when $e > 1$ a branch of a hyperbola. When $e = 0$ we get back the circular orbit. In all cases it is possible to get solutions of (9) from the orbits by simple calculations.

We have now verified two of Kepler's laws and we shall finish this section by verifying the third one as well. Since the sector velocity $c/2 = r^2\dot\theta/2$ is constant, the period T of an elliptical orbit is $2Y/c$ where Y is the area of the ellipse. This area is πab where a and b are the semi-axes of our ellipse. A little computation shows that $a = p(1 - e^2)^{-1}$ and $b = p(1 - e^2)^{1/2}$ so that $Y = \pi p^2(1 - e^2)^{-3/2}$ and hence

$$T = 2\pi p^{3/2}(1 - e^2)^{-3/2}.$$

On the other hand, the average distance m to the sun equals

$$m = 2^{-1}\left(p(1 + e)^{-1} + p(1 - e)^{-1}\right) = p(1 - e^2)^{-1}$$

so that $T = 2\pi m^{3/2}$ and this proves the third law.

7.2 Strict analysis

Differential calculus is both algebra and analysis. We shall now present the analytical side by stating and proving some of its most important theorems. Aided by them we shall then be able to fill the gaps (i) to (v) of the previous section.

The mean value theorem and Taylor's formula

The assertion (i) is an immediate consequence of the mean value theorem of differential calculus expressed by the formula

$$f(b) - f(a) = (b - a)f'(\xi).$$

Here f is assumed to be continuous when $a \leqslant x \leqslant b$, and possess a derivative when $a < x < b$, and the formula says that there is a ξ between a and b where the equality holds. In the special case when $f(a) = f(b) = 0$, the theorem follows from the fact that if $f \neq 0$ somewhere between a and b, then there is a point ξ between a and b where f is positive and assumes its largest value or f is negative and assumes its least value. In both cases $f'(\xi) = 0$. In the general case, this result is applied to the function

$$h(x) = (b - a)f(x) - (x - a)f(b) - (b - x)f(a)$$

for which $h(a) = h(b) = 0$. That $h'(\xi) = 0$ then gives the desired result. The geometrical meaning of the mean value theorem is illustrated by Figure 7.2 and its legend.

The mean value theorem has many applications. We can for instance show that if f is differentiable close to a point a and $f''(a)$ exists, then f has a local strict minimum (maximum) at a when $f''(a) > 0$ ($f''(a) < 0$) and $f'(a) = 0$. Figure 7.3 illustrates this. In fact, $f'(\xi) = f'(\xi) - f'(a) = (\xi - a)f''(a) + (\xi - a)o(1)$ then has the same sign as the first term on the

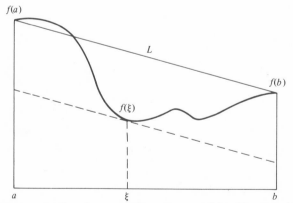

Figure 7.2   The mean value theorem. The tangent at ξ has the same inclination as the line L.

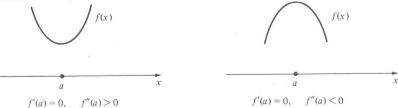

Figure 7.3   The second derivative f'' and strict maxima and minima.

141

right if $\xi - a \neq 0$ is small. Hence $f(x) - f(a) = (x - a)f'(\xi)$ has the same sign as $f''(a)$ when $x - a \neq 0$ is small and ξ lies between x and a.

Another application of the mean value theorem is the following statement: if there is a number c such that

$$|f''(x)| \leqslant c|f(x)| + c|f'(x)| \tag{14}$$

when x is close to a and $f(a)$ and $f'(a)$ both vanish, then $f = 0$ close to a. Of course, we have to assume here that f'' exists close to a and this implies that f and f' are continuous there. According to (14), $|f''(x)|$ then has a finite upper bound $h(\delta)$ when $|x - a| \leqslant \delta$ and $\delta > 0$ is small. By the mean value theorem, $|f'(x)| = |f'(x) - f'(a)| = |(x - a)f''(\xi_1)| \leqslant \delta h(\delta)$ so that also $|f(x)| = |f(x) - f(a)| = |(x - a)f'(\xi_2)| \leqslant \delta^2 h(\delta)$ when $|x - a| \leqslant \delta$. Inserting this into (14) gives $h(\delta) \leqslant c\delta(1 + \delta)h(\delta)$ so that $h(\delta) = 0$ when $c\delta(1 + \delta) < 1$. Hence $f'' = 0$ close to a. But then $f' = \text{constant} = f'(a) = 0$ close to a and hence also $f = \text{constant} = f(a) = 0$ close to a.

We can now prove (ii) and (v). According to (9), the function $w = ax + by + cz$ satisfies the equation $\ddot{w} = Fw$ where $F(t) = -r^{-3}$ is a continuous function of t. Hence $|\ddot{w}| \leqslant c|w|$ when t lies in an interval and c is the maximum of $|F|$ in this interval. Hence (ii) follows from (14). In (v) we have to do with a C^2 function w of θ such that $w'^2 + w^2 = 1$ and hence, by differentiation, $w'(w'' + w) = 0$. Since w is not a constant, $w' \neq 0$ somewhere and then also in an interval. There $w'' + w = 0$, and the same holds at the endpoints of the interval because w is assumed to be a C^2 function. If $w' = 0$ at an endpoint, then $w = \pm 1$ and hence $w'' = \pm 1$ there. But then $w' \neq 0$ on both sides of the endpoint. This means that $w'' + w$ vanishes in the entire interval where w is defined. Assume for simplicitly that this interval contains 0, and find a θ_0 such that $w(0) = \cos \theta_0$ and $w'(0) = \sin \theta_0$. This is possible since $w'(0)^2 + w(0)^2 = 1$. The function $u(\theta) = w(\theta) - \cos(\theta + \theta_0)$ then has the property that $u'' + u = 0$ everywhere where w is defined and $u(0) = u'(0) = 0$. But then, according to (14), $u = 0$ everywhere. This proves (v). The proof was not so simple but this is perhaps not surprising since (v) happens to be wrong for C^1 functions. Putting, for instance, $w(\theta) = 1$ when $\theta \leqslant 0$ and $w(\theta) = \cos\theta \geqslant 0$ we have $w'^2 + w^2 = 1$ everywhere. But w is not a C^2 function, only a C^1 function. It does not have a second derivative at the origin.

The mean value theorem can also be used to prove Taylor's formula with a remainder term, more precisely, the formula

$$f(x) = f(a) + (x - a)f'(a) + \cdots + \frac{(x - a)^{n-1}}{(n - 1)!} f^{(n-1)}(a)$$

$$+ \frac{(x - a)^n}{n!} f^{(n)}(\xi). \tag{15}$$

Here f is assumed to be n times differentiable in an interval around a and, precisely as for the mean value theorem, the formula holds for some suitably chosen ξ between a and x. The last term on the right is called *Lagrange's remainder term*. It vanishes when p is a polynomial of degree $< n$. The point of Taylor's formula is that the difference between $f(x)$ and the polynomial

$$F(x, a) = f(a) + (x - a)f'(a) + \cdots + \frac{(x - a)^{n-1}}{(n - 1)!} f^{(n-1)}(a)$$

is much less than the power $(x - a)^{n-1}$ of the last term of the polynomial when $x - a$ is small. In other words, close to a point, a function which is many times differentiable behaves very much like the polynomial given by the first terms of its Taylor series. To prove (15) one considers the function

$$t \rightarrow h(t) = f(x) - F(x, t) - \frac{(x - t)^n}{n!} K$$

where t ranges between x and a and $K = K(x, a)$ is chosen such that $h(a) = 0$. We also have $h(x) = 0$, and some computation gives

$$h'(t) = \frac{(x - t)^{n-1}}{(n - 1)!} \left(K - f^{(n)}(t) \right).$$

Since, by the mean value theorem, $h'(\xi) = 0$ for some ξ between a and x, this proves (15).

The differentiability of inverse functions

We shall prove a theorem about inverse functions. It says that if f is a real function defined in an open interval I and $f' \neq 0$ everywhere, then $f(I)$ is also an open interval and f a bijection whose inverse is differentiable and continuously differentiable as many times as f is. To see this we note that the mean value theorem, $f(b) - f(a) = (b - a)f'(\xi)$, and the assumption that $f' \neq 0$ imply that $f(b) \neq f(a)$ when $b \neq a$, so that f is a bijection. Let $g : f(I) \rightarrow I$ be the inverse of f. According to Theorem 3 of Chapter 5, $f(I)$ is an open interval and g a continuous function. We shall see that g is differentiable. Let $y \in f(I)$ and choose $\eta \in f(I)$ close to y. Let $x = g(y)$ and $\xi = g(\eta)$ be the corresponding points of the interval I so that also $y = f(x)$, $\eta = f(\xi)$. Then

$$\frac{f(\xi) - f(x)}{\xi - x} = \frac{\eta - y}{g(\eta) - g(y)}.$$

Since f and g are continuous, ξ tends to x at the same time as η tends to y. Hence, the common limit of both sides is $f'(x)$ and this means that g is differentiable, with the derivative $g'(y) = 1/f'(x) = 1/f'(g(y))$. The rules for taking derivatives then show that g has as many derivatives as f and that g is a C^k function when f is.

We can now prove (iii). In fact, $r = (x^2 + y^2)^{1/2}$ is a C^2 function of t when x and y are and $r > 0$. We also have $x/r = \cos \theta$ and $y/r = \sin \theta$.

Since the derivatives $-\sin\theta$ and $\cos\theta$ of the right sides do not vanish at the same time, θ, considered in small intervals, is a C^k function of either x/r or y/r and this for every k. Hence θ is a C^2 function of t.

7.3 Differential equations

Systems of ordinary differential equations

A *differential equation* is usually described as a relation between a function and its derivatives. It is said to be *ordinary* when the function depends on just one variable. We shall consider systems of such equations, as many as there are unknown functions. They are commonly written as

$$u'_1 = f_1(t, u_1, \ldots, u_n), \ldots, u'_n = f_n(t, u_1, \ldots, u_n). \tag{16}$$

Here t is one real variable and $f_1(t, v), \ldots, f_n(t, v)$ functions of t and n real variables $v = (v_1, \ldots, v_n)$ defined when t lies in an open interval I and v in an open part V of $\mathbf{R}^n$. By a solution of the system is meant n differentiable functions $u = u(t) = (u_1 = u_1(t), \ldots, u_n = u_n(t))$ such that the equations (16) hold when t lies in some interval. Putting $u_1 = x$, $u_2 = y$, $u_3 = z$, $u_4 = \dot{x}$, $u_5 = \dot{y}$, $u_6 = \dot{z}$, Newton's equations (9) can be written as

$$\dot{u}_1 = u_4, \qquad \dot{u}_2 = u_5, \qquad \dot{u}_3 = u_6,$$
$$\dot{u}_4 = -u_1 r^{-3}, \qquad \dot{u}_5 = -u_2 r^{-3}, \qquad \dot{u}_6 = -u_3 r^{-3}$$

where $r = (u_1{}^2 + u_2{}^2 + u_3{}^2)^{1/2}$, and then are a special case of (16) with $V = \mathbf{R}^6$ minus the part where $v_1 = 0$, $v_2 = 0$, $v_3 = 0$.

If we interpret the vector $u(t)$ as the state of a system at time t, and the derivative as the rate of change of the state, we can express (16) in words as

> the rate of change of the system at a given time depends only on
> the time and the state at that time. (17)

This is of course a very general situation and its matheamatical formulation (16) fits a lot of time-dependent processes. Planetary motion is just one example. The entire solar system satisfies such a system of equations and, generally speaking, every mechanical system. Flows of air, water, and electricity are other examples. A very great number of physical, chemical and economic processes fit into the frame given by (16). In every case, the mathematical analysis gives information, in principle complete, about the processes when the right sides of (16) are known. As we have seen, planetary motion can be expressed in terms of simple functions but such favorable circumstances are an exception. In general, the mathematical analysis must be implemented by numerical calculations. In order for all this to make sense it is of course necessary that

> the state at one time determines the states at subsequent times. (17′)

The corresponding requirement in the mathematical model (16) is uniqueness: if two solutions are equal when $t = t_0$, they are also equal in a subsequent interval. In the model we are also forced to show that solutions really exist. We formulate both of these requirements as

through every point in $I \times V$ there passes

a uniquely determined solution of (16).   (18)

Next we shall state hypotheses about the right sides of (16) from which (18) follows. It is for instance not sufficient that they be continuous. In fact, the function $u = 1$ and $u = \cos t$ are both equal to 1 when $t = 0$ but they are different solutions of the differential equation $u' = -(1 - u^2)^{1/2}$ when $0 \leqslant t \leqslant \pi$.

Existence and uniqueness

In order to proceed we need a little bit of integration theory, more precisely the fact that if $g(t)$ is a continuous function, then its integral

$$G(t) = \int_{t_0}^{t} g(s) \, ds$$

is a C^1 function with the derivative g such that $G(t_0) = 0$. In particular, by the mean value theorem,

$$|G(t)| \leqslant |t - t_0| \max |g(s)|$$

where the maximum is taken over all s between t and t_0. Using this, we can rewrite (16) and the condition that the function $t \to (t, u(t))$ passes through a given point (t_0, u_0), i.e., that $u(t_0) = u_0$, as

$$u_k(t) = u_{0k} + \int_{t_0}^{t} f_k(s, u_1(s), \ldots, u_n(s)) \, ds, \qquad (k = 1, \ldots, n), \quad (16^*)$$

provided $f_1, \ldots, f_n$ are continuous functions, and this we assume. The next step is to use that contraction theorem in Banach spaces which was proved at the end of Chapter 4. The Banach spaces B_δ that we shall use consist of functions $v(t) = (v_1(t), \ldots, v_n(t))$ with continuous components defined when $|t - t_0| \leqslant \delta$ and with the norm

$$|v| = \max |v_k(t)| \quad \text{when} \quad |t - t_0| \leqslant \delta \quad \text{and} \quad 1 \leqslant k \leqslant n.$$

That this space is complete follows from Theorem 1 of Chapter 5. We now define a function T from B_δ to B_δ by letting the right sides of (16*) be the components of $T(u)$. Then this formula simply means that $u = T(u)$. By choosing δ small we shall arrange things so that T becomes a contraction fulfilling the requirements of the contraction theorem. This means that (16*) has a unique solution in B_δ and hence that (18) holds. A simple hypothesis making this program go through is the following one

(A)   the functions $f_1, \ldots, f_n$ are continuous in $I \times V$ and have the property that there is a number a such that

$$|f(t, v) - f(t, w)| \leqslant a|v - w|$$

when t, v and t, w are sufficiently close to t_0, u_0.

145

In fact, this and the definition of T gives

$$|T(u) - T(v)| \leqslant a\delta|u - v|$$

when $|u - u_0|$ and $|v - u_0|$ are small enough. Since $|u_0 - T(u_0)| \leqslant \text{const } \delta$ when δ is small enough, it follows that there are positive numbers r, δ_0, c_0 such that

$$\delta < \delta_0 \quad \Rightarrow \quad |T(u) - u_0| \leqslant c_0\delta, \qquad |T(u) - T(v)| \leqslant c_0\delta|u - v|$$

when $|u - u_0| = r$, $|v - u_0| \leqslant r$. A look at the assumptions of the contraction theorem then shows that we have the desired result $(A) \Rightarrow (18)$ when δ is small enough. The condition (19) is called a *Lipschitz condition*, after its originator, and will be commented on in the next section.

7.4 Differential calculus of functions of several variables

Partial derivatives

Keeping all variables fixed except one, a function $f(x)$ of n real variables $x = (x_1, \ldots, x_n)$ gives rise to n functions $x_k \to f_k(x_k) = f(x)$ of one real variable. Their derivatives, denoted by

$$\partial_k f(x) = \lim_{h \to 0} (f_k(x_k + h) - f_k(x_k))/h$$

are called the partial derivatives of f. They can be computed from the values of f on n straight lines through x parallel to the coordinate axes. For this reason, the existence of the partial derivatives tell us relatively little about the function itself (when $n > 1$) unless we restrict the class of functions that we consider. A simple way out is to keep to C^1 functions, i.e., continuous functions with continuous partial derivatives defined in open subsets of $\mathbf{R}^n$. Such a function is differentiable in the sense that

$$f(x + h) - f(x) = \sum_1^n \partial_k f(x)h_k + o(1)|h|$$

where $|h| = \max(|h_1|, \ldots, |h_n|)$ and $o(1) \to 0$ as $|h| \to 0$. This assertion is proved by the mean value theorem and a small computation. For $n = 2$ it is the following one

$$f(x_1 + h_1, x_2 + h_2) - f(x_1, x_2)$$
$$= f(x_1 + h_1, x_2 + h_2) - f(x_1, x_2 + h_2) + f(x_1, x_2 + h_2) - f(x_1, x_2)$$
$$= \partial_1 f(x_1 + \theta_1 h_1, x_2 + h_2)h_1 + \partial_2 (f(x_1, x_2 + \theta_2 h_2)h_2$$

where θ_1 and θ_2 are numbers between 0 and 1. The proof in the general case is left to the reader.

Replacing x and h in (20), by $x + sy$ and ty, and letting t tend to zero shows that the function $s \to f(x + sy)$ has the derivative $\sum \partial_k f(x + sy)y_k$. In

146

particular, the Lipschitz condition (19) holds when $f_1, \ldots, f_n$ are C^1 functions, since the mean value theorem applied to the functions $s \to f_j(t, sv + (1-s)w)$ gives

$$f_j(t, v) - f_j(t, w) = \sum \partial_k f_j(t, \theta v + (1-\theta)w)(v_k - w_k)$$

where θ lies between 0 and 1.

When f and all its partial derivatives are C^1 functions, we say that f is a C^2 function. For a C^2 function of two variables the mean value theorem gives

$$f(x_1 + h_1, x_2 + h_2) - f(x_1 + h_1, x_2) - f(x_1, x_2 + h_2) + f(x_1, x_2)$$
$$= g(x_2 + h_2) - g(x_2) = \partial_2 g(x_2 + \theta_2 h_2) h_2$$
$$= (\partial_2 f(x_1 + h_1, x_2 + \theta_2 h_2) - \partial_2 f(x_1, x_2 + \theta_2 h_2)) h_2$$
$$= \partial_1 \partial_2 f(x_1 + \theta_1 h_1, x_2 + \theta_2 h_2) h_1 h_2 + o(1) h_1 h_2$$

where $g(x_2) = f(x_1 + h_1, x_2) - f(x_1, x_2)$ and $o(1) \to 0$ as h_1 and $h_2 \to 0$. The same computation with $g(x_1) = f(x_1, x_2 + h_2) - f(x_1, x_2)$ gives the same result with $\partial_2 \partial_1 f$ in the last member. Hence, a division by $h_1 h_2$ and a passage to the limit shows that $\partial_1 \partial_2 f = \partial_2 \partial_1 f$. From this we see that $\partial_j \partial_k f = \partial_k \partial_j f$ for all j and k and every C^2 function f. In other words, a second order derivative does not depend on the order between the two differentiations.

The classes of C^k functions are defined recursively when $k > 2$. That f is a C^k function means exactly that all its partial derivatives are C^{k-1} functions. Hence a C^k function can be differentiated k times. We denote its repeated partial derivatives by

$$\partial^\alpha f(x) = \partial_1^{\alpha_1} \ldots \partial_n^{\alpha_n} f(x) \tag{21}$$

where $\alpha = (\alpha_1, \ldots, \alpha_n)$ with integral components ≥ 0. The number $|\alpha| = \alpha_1 + \cdots + \alpha_n$ is called the *order* of the derivative. When $|\alpha| \leq k$ and f is a C^k function, the right side of (21) exists and is independent of the order between the derivatives. When f is a C^k function with a suitable k we say that f is a *smooth* function. This terminology is of course a little vague, but it is very convenient. When f is a C^k function for all k, it is said to be a C^∞ function. A function from $\mathbf{R}^n$ to $\mathbf{R}^p$ is said to be a C^k function or smooth when all its components have the corresponding property.

Taylor's formula

According to (20), the function $g(t) = f(a + t(x - a))$, where $a \in \mathbf{R}^n$ is differentiable with the derivative $\sum \partial_k f(a + t(x - a))(x_k - a_k)$ when f is a C^1 function. Repeated use of this formula, Taylor's formula for g, and some thought gives Taylor's formula for a smooth function f of n variables

x, namely

$$f \in C^{k+1} \implies f(x) = \sum_{|\alpha| \leq k} \partial^\alpha f(a) \frac{(x-a)^\alpha}{\alpha!}$$

$$+ \sum_{|\alpha| = k+1} \partial^\alpha f(a + \theta(x-a)) \frac{(x-a)^\alpha}{\alpha!}. \qquad (22)$$

The left side means that f is a C^{k+1} function, the θ on the right is an appropriately chosen number between 0 and 1 and

$$(x-a)^\alpha = (x_1 - a_1)^{\alpha_1} \dots (x_n - a_n)^{\alpha_n}, \qquad \alpha! = \alpha_1! \dots \alpha_n!.$$

The last term of (22) is at most a constant times $|x-a|^{k+1}$ when x is closed to a and hence much smaller than the previous terms (when they do not vanish). Using Taylor's formula we can see how a smooth function behaves near a given point. If, e.g., f is a C^2 function and $\partial_1 f(a) = 0, \dots, \partial_n f(a) = 0$ but the quadratic form

$$Q(h) = \sum_{|\alpha|=2} \partial^\alpha f(x) h^\alpha / \alpha! = 2^{-1} \sum_{j,k=1}^n \partial_j \partial_k f(a) h_j h_k$$

is positive definite, i.e., > 0 when $h \neq 0$, then f has a local strict minimum at the point a.

The chain rule and the Jacobi matrix

It follows from the rules of differentiation that sums, products, and quotients with nonvanishing denominators of C^k functions are C^k functions. Also, the composition of two C^k functions is a C^k function. This follows from the chain rule: if $g_1, \dots, g_n$ are differentiable at the point $y \in \mathbf{R}^p$ and f is differentiable at the point $x = g(y) = (g_1(y), \dots, g_n(y))$, then $H = f \circ g$ is differentiable at the point y and

$$\partial_k H(y) = \sum_{k=1}^n \partial_j f(x) \partial_k g_j(y). \qquad (23)$$

The proof follows from the formula

$$f(g(y+h)) =$$

$$= f(g(y)) + \sum_{j=1}^n \partial_j f(x)(g_j(y+h) - g_j(y)) + \varepsilon_1 |g(y+h) - g(y)|$$

$$= f(g(y)) + \sum_{j=1}^n \sum_{k=1}^p \partial_j f(x) \partial_k g_j(y) h_k + \varepsilon_2 |h|$$

where $\varepsilon_1 \to 0$ when $g(y+h) - g(y) \to 0$ and $\varepsilon_2 \to 0$ when $h \to 0$.

In the chain rule there appears the *Jacobi matrix*

$$g'(y) = (\partial_k g_j(y))$$

of n functions with respect to p variables $y_1, \dots, y_p$. Taking j as a column index and k as a row index, it is of type $n \times p$. If $f = (f_1, \ldots, f_q)$ are q functions of x, the chain rule (23) applied to every component of f can also be written in matrix form as

$$(f \circ g)'(y) = f'(g(y)) g'(y). \tag{24}$$

When $n = p = q = 1$, this is the rule for computing the derivative of a composed function, and hence we are free to use the same notation for the derivative and the Jacobi matrix.

EXAMPLES. The Jacobi matrix of polar coordinates in the plane,

$$x_1 = r \cos \theta, \qquad x_2 = r \sin \theta \tag{25}$$

is

$$\begin{pmatrix} \partial_r x_1 & \partial_\theta x_1 \\ \partial_r x_2 & \partial_\theta x_2 \end{pmatrix} = \begin{pmatrix} \cos \theta & -r \sin \theta \\ \sin \theta & r \cos \theta \end{pmatrix} \tag{26}$$

where ∂_r and ∂_θ denote partial derivatives with respect to the corresponding variable. For spherical polar coordinates in space,

$$x_1 = r \cos \theta \cos \varphi, \qquad x_2 = r \cos \theta \sin \varphi, \qquad x_3 = r \sin \theta \tag{27}$$

the Jacobi matrix is

$$\begin{bmatrix} \partial_r x_1 & \partial_\theta x_1 & \partial_\varphi x_1 \\ \partial_r x_2 & \partial_\theta x_2 & \partial_\varphi x_2 \\ \partial_r x_3 & \partial_\theta x_3 & \partial_\varphi x_x \end{bmatrix} = \begin{bmatrix} \cos \theta \cos \varphi & -r \sin \theta \cos \varphi & -r \cos \theta \sin \varphi \\ \cos \theta \sin \varphi & r \sin \theta \sin \varphi & r \cos \theta \sin \varphi \\ \sin \theta & r \cos \theta & 0 \end{bmatrix}. \tag{28}$$

Smooth local bijections

Conceptually, theoretically, numerically, and practically the following question is very important. Let $h(y)$ be a function from a part of $\mathbf{R}^n$ to another. When is it possible to compute y as a function of $x = h(y)$ at least locally? When $n = 1$ we know one answer: it suffices that h is a C^1 function and that $h' \neq 0$, and in that case $y = h^{-1}(x)$ is a C^1 function. For many variables, the answer is the same provided we interpret $h' \neq 0$ as saying that the Jacobi matrix is invertible. We formulate this as a theorem.

The bijection theorem for smooth functions. *Let h be a C^k function from $\mathbf{R}^n$ to $\mathbf{R}^n$ defined close to some y_0 and assume that $k > 0$ and that the Jacobi matrix $h'(y_0)$ is invertible. Then there are open neighborhoods V of y_0 and W of $x_0 = h(y_0)$ such that h^{-1} is a C^k function from W to V.*

EXAMPLES. The Jacobi matrix (26) is singular only when $r = 0$. When $r > 0$ we can get $r = (x_1^2 + x_2^2)^{1/2}$ and $\theta = \arctan x_2/x_1$ or $\theta = \text{arc cot } x_1/x_2$ as functions of x at least locally. When $r = 0$, the angle θ ceases to be a function of x. The Jacobi matrix (28) is seen to be singular only when

$r \cos \theta = 0$, i.e., when $x_3 = \pm r$ (north pole and south pole of a ball of radius r; θ is the latitude), and then φ (the longitude) ceases to be function of x. When $r \cos \theta \neq 0$ we can compute r, θ, and φ as functions of x at least locally.

PROOF. We shall use the contraction theorem of Chapter 4. Let $A = (a_{jk})$ be a square nonsingular matrix and put $g_k(z) = \Sigma a_{kj} z_j + y_{k0}$. Then $z \to y = g(z)$ and $y \to z = g^{-1}(y)$ are C^k bijections between $\mathbf{R}^n$ and $\mathbf{R}^n$. Hence it suffices to prove the theorem for the composite function $h \circ g$. The chain rule (24) gives $(h \circ g)'(0) = h'(y_0)A$. If we choose A so that the right side is the unit matrix E and change y for z, it is clear that it suffices to show the theorem when $y_0 = 0$ and $h'(0) = E$. We may of course also assume that $h(0) = 0$. This means that $h(y) = y + H(y)$, where H is a C^k function such that $H(0) = H'(0) = 0$. In particular,

$$|y| \leqslant \delta, \quad |z| \leqslant \delta \quad \Rightarrow \quad |H(y) - H(z)| \leqslant c(\delta)|y - z|$$

where $0 \leqslant c(\delta) \to 0$ as $\delta \to 0$. In fact, by the mean value theorem,

$$H_k(y) - H_k(z) = \Sigma \, \partial_j H_k (\theta y + (1 - \theta)z)(y_j - z_j)$$

where θ lies between 0 and 1 and the derivatives $\partial_j H_k$ are small near the origin. We can then apply the contraction theorem with $U = \mathbf{R}^n$, $T = H$, $u_0 = 0$ and $T(u_0) = u_0$, and $c(\delta) < 1$. The result is that h is a bijection from every sufficiently small ball V around the origin, that h^{-1} is continuous, and that $h(V)$ contains a ball around the origin. The differentiability of h^{-1} and that fact that h and h^{-1} are C^k functions at the same time are then easy to prove, but this will not be done here.

Implicitly defined functions

In analysis it happens very frequently that one wants to solve an equation of the type $f(t, x) = 0$ where $x = (x_1, \ldots, x_n)$ are given real numbers and t is unknown. In other words, we want to get $t = t(x)$ as a function of x. If f is a C^1 function and t_0, x_0 are such that $f(t_0, x_0) = 0$ and $\partial_t f(t_0, x_0) \neq 0$, then the equation has precisely one solution $t = t(x)$ close to t_0 when x is close to x_0 and equal to t_0 when $x = x_0$. In addition, this solution is a C^1 function of x. These assertions are called the *theorem about implicitly defined functions*, and follow from the bijection theorem. In fact, if, e.g., $h_0 = f(t, x)$, $h_1 = x_1, \ldots, h_n = x_n$, and $h = (h_0, \ldots, h_n)$ it is easy to see that the matrix $h'(t_0, x_0)$ is invertible. Hence, there are C^1 functions $g_0, \ldots, g_n$ of $n + 1$ variables $y = (y_0, \ldots, y_n)$ such that the systems $y_0 = h_0(t, x), \ldots, y_n = h_n(t, x)$ and $t = g_0(y)$, $x_1 = g_1(y), \ldots, x_n = g_n(y)$ have the same solutions when t, x is close to t_0, x_0. Since $h_1 = x_1$, etc., it follows that $g_1 = y_1$, etc., and hence $y_0 = f(t, x)$ and $t = g_0(y_0, x)$ are equivalent equations. Hence, finally, the function $t(x) = g_0(0, x)$ has the desired properties. It is a C^k function when f is.

EXAMPLE. The equation $f(t, x) = t^2 - x = 0$ has the solution $t = x^{1/2}$ when $x > 0$ but the solution is not differentiable when $x = 0$. When $x = t = 0$ then $\partial_t f = 0$.

General coordinates

Polar coordinates (25) and (27) are special cases of general coordinates in $\mathbf{R}^n$, i.e., functions $y = h(x)$ of $x \in \mathbf{R}^n$ such that h is a smooth bijection between open parts of $\mathbf{R}^n$. When the components of h are first degree polynomials we have the parallel coordinates of linear algebra. When $n = 2$, the equations $h_1 = $ constant and $h_2 = $ constant represent two families of curves intersecting transversally (Figure 7.4).

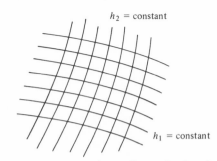

Figure 7.4   General coordinates in the plane.

7.5 Partial differential equations

A *partial differential equation* is a relation between a function of many variables and its partial derivatives, and *a system* of partial differential equations is a set of relations between a number of functions of many variables and their partial derivatives. A definition as general as that obviously requires examples and we shall give some.

The most important examples of systems of partial differential equations occur in the mathematical models of classical physics describing the movement of elastic bodies, flows of fluids and gases, flows of heat, and how electric and magnetic fields change with time. We shall touch upon two of these categories. In both of them, the state of the physical medium at time t and at the point $x = (x_1, x_2, x_3)$ in space is described by some smooth function $u(t, x)$ with one or several components. The system expresses its rate of change $\partial_t u$ at t, x in terms of the spatial derivatives of u at t, x. In other words, how the state function u changes with t at a point x depends only on the behavior of u close to x at time t. In fact, the spatial derivatives are given by this information. Sometimes the acceleration $\partial_t^2 u$ is expressed in terms of $\partial_t u$ and the spatial derivatives. This case reduces to the previous one if we incorporate $\partial_t u$ into the state function, i.e., introduce a new state function $t, x \rightarrow (u, v)$ and add the equations $\partial_t u = v$ to the system.

151

Elasticity

Let us assume that a homogeneous elastic string is stretched initially between two points a and b on an x-axis, and that the string starts moving a little in a x, y-plane under the influence of outside forces. Let $u(t, x)$ be the deviation in the y direction at the point x (Figure 7.5). In the simplest model for the movement of a string, its deviation is a smooth function u satisfying the differential equation

$$a < x < b \quad \Rightarrow \quad c^{-2}\partial_t^2 u - \partial_x^2 u = 0 \tag{29}$$

where $c > 0$ is a number characteristic of the string. When f and g are C^2 functions, then

$$u(t, x) = f(x + ct) + g(x - ct) \tag{30}$$

is a solution of the equation and it is not difficult to prove that every C^2 solution has that form. The functions $v(t, x) = f(x + ct)$ and $w(t, x) = g(x - ct)$ are solutions of the simpler differential equations

$$\partial_t v - c\partial_x v = 0 \quad \text{and} \quad \partial_t w + c\partial_x w = 0.$$

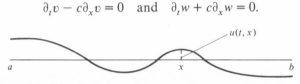

Figure 7.5   The deviation of a string from its position of rest.

They represent movements of the string that can be described very simply as waves moving with the velocity c to the right and to the left respectively (Figure 7.6). The equation (29) does not determine the movement uniquely. We must add to it initial conditions, i.e., the position and the velocity of the string at some given time, and boundary conditions, i.e., something about the movement at the end points. We get the simplest case when the

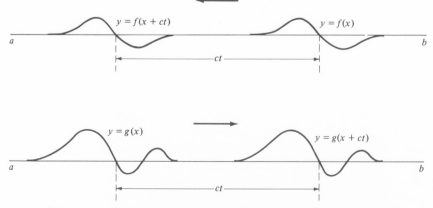

Figure 7.6   Waves along a string before they attain the endpoints.

string does not move at all, i.e., when $u = h(x)$ depends only on x. The equation (29) then reduces to $h'' = 0$ so that $u = Ax + B$ is a first degree polynomial. The string is then straight and uniquely determined by its positions $h(a)$ and $h(b)$ at the end points. In case of a string of infinite length, the function

$$u(t, x) = 2^{-1}(v(x + ct) + v(x - ct))$$

solves the initial problem $u(0, x) = v(x)$, $\partial_t u(0, x) = 0$. The interpretation is that we keep the string in the position v at time 0 and then let it loose. The result is that v gives rise to two waves going in opposite directions with the velocity c. With the aid of the formula (30) it is possible to answer in an elementary way a large number of questions of how the string moves under various boundary conditions but this would carry us too far afield and will not be done here.

Going from one space variable x to two, x_1, x_2, or three, x_1, x_2, x_3, (29) becomes, respectively,

$$c^{-2}\partial_t^2 u - \partial_1^2 u - \partial_2^2 u = 0 \quad \text{and} \quad c^{-2}\partial_t^2 u - \partial_1^2 u - \partial_2^2 u - \partial_3^2 u = 0.$$

A physical situation described by the first equation is, for instance, the movement of a thin homogeneous elastic membrane stretched over a region V of the x_1, x_2 plane. The second equation is the famous wave equation describing the propagation of both light and sound waves. In both cases c is the velocity of propagation. Time-independent solutions of them satisfy Laplace's equation in two and three variables, $\partial_1^2 u + \partial_2^2 u = 0$ and $\partial_1^2 u + \partial_2^2 u + \partial_3^2 u = 0$, and are called *harmonic functions*. The problem of finding a harmonic function in an open part V of $\mathbf{R}^2$ or $\mathbf{R}^3$ taking given values on the boundary of V is called *Dirichlet's problem*. When the boundary of V and the boundary values are smooth, Dirichlet's problem has a unique solution. There are several proofs of this statement and all of them are tied to some decisive forward step in the development of mathematical analysis. The modern functional analysis version of Riemann's proof of 1851 is described at the end of Chapter 4.

Heat conduction

Let us imagine an interval $a < x < b$ of the x axis as a homogeneous heat-conducting rod and let $u(t, x)$ be the temperature of the rod at the point x and at time t. In the simplest model of heat conduction, u then satisfies the partial differential equation

$$a < x < b \quad \Rightarrow \quad \partial_t u - c\partial_x^2 u = 0,$$

called *the heat equation*. Here $c > 0$ is a constant measuring the thermal conductivity of the rod. Taking for simplicity $c = 1$,

$$u(t, x) = (2\pi t)^{-1/2} e^{-x^2}/2t$$

is a solution of the equation for all x when $t > 0$. For small $t > 0$, the temperature is high near $x = 0$ and small far away. As t increases heat propagates out from the origin and the difference is evened out. The function u also appears in probability theory as the frequency function of a normal distribution with the variance t.

The wave equation and the heat equation are linear in the sense that they have the form $F = 0$ where F is a first degree polynomial in the unknown function and its derivatives. The differential equations of hydrodynamics, which we have to pass over here, are not linear. If they are written in the form $F = 0$, then F contains products of the unknown functions and their derivatives. Linear equations are in general easier to treat than nonlinear ones.

Distributions

The simplest mathematical models of physics are very strongly idealized. Point-sized masses are accepted as a matter of course, and when it comes to wave propagation one would not hesitate to replace the smooth waves of Figure 7.6 by triangle-shaped ones. Then the functions f and g are no longer differentiable everywhere. The equation (30) stands but (29) loses its meaning. This is an example of the constant temptation in differential calculus to differentiate functions that do not have derivatives. We have taken the security measure of only working with smooth functions, but there are other ways. It is possible to extend differential calculus from smooth functions to a kind of generalized functions called distributions in such a way that essentially all the algebraic rules still hold. This theory was worked out by Laurent Schwartz in the forties. Its starting point is the formula for integration by parts, and we will touch upon it in the next chapter.

History

The fundamental differential equations of elasticity and hydrodynamics are from the period 1750–1850. Heat conduction was treated by Fourier in 1822 in a famous book entitled *Théorie Analytique de la Chaleur*. The fundamental equations of electricity got their final form through Maxwell in 1860. The equations for the movement of electrically loaded fluids and gases are used in astronomy and plasma physics, and stem from the thirties. The many initial value problems and boundary problems that can be stated in these models have been very important in the development of analysis. Hydrodynamical problems of this kind have proved to be particularly hard. The reason is not only that the analysis is difficult, but that it is not always easy to relate the model to reality.

In analysis, partial differential equations are also studied for their own sake. Before the age of specialization set in, in the beginning of this century, most great mathematicians contributed to the subject, e.g., Gauss, Cauchy, Riemann, Poincaré, and Hilbert. The general theory has had a steady growth and a rapid development since 1950 when distributions appeared as a new tool in analysis.

7.6 Differential forms

So far we have kept away from Leibniz's version of differential calculus. He wrote the derivative as a quotient,

$$f'(x) = df(x)/dx.$$

Behind this lay the idea that the quotient is the limit of a quotient of the differences $\Delta f(x) = f(x + h) - f(x)$ and $\Delta x = (x + h) - x$ as Δx goes to zero, and that the differences $\Delta f(x)$ and Δx tend to the differentials $df(x)$ and dx. The formula $df(x) = f'(x)\, dx$ is a consequence of this point of view. The free manipulation with differentials gives many nice formulas, but the concept of a differential as an infinitely small entity which does not vanish is not tenable and has been abandoned. In the present-day version of differential calculus, the differentials are vectors which can be multiplied and then give rise to differential forms, i.e., linear combinations of products of differentials. After a preparatory section on products of vectors we shall show how this is done.

Products of vectors and the Grassmann algebra

Let $e_1, \ldots, e_n$ be a basis of a real vector space L. That two vectors $u = a_1 e_1 + \cdots + a_n e_n$ and $v = b_1 e_1 + \cdots + b_n e_n$ are linearly dependent can then be expressed as

$$a_j b_k - a_k b_j = 0 \tag{31}$$

for all j and k. In fact, if (for instance) $u = cv$ is a multiple of v, then $a_j = cb_j$ for all j and (31) follows. And if (31) holds, then $b_k u = a_k v$ for all k and hence u and v are linearly dependent. The condition (31) can also be expressed in terms of a vector product $u \wedge v$, i.e., a map from $L \times L$ to a so-far unspecified linear space $L \wedge L$ with the following properties: the functions $u \to u \wedge v$ and $v \to u \wedge v$ are linear, $v \wedge u = - u \wedge v$ and all $e_j \wedge e_k$ with $j < k$ constitute a basis of $L \wedge L$ when $e_1, \ldots, e_n$ is a basis of L. In fact, then $u \wedge v = (a_1 e_1 + \cdots + a_n e_n) \wedge (b_1 e_1 + \cdots + b_n e_n)$ is a sum of n^2 terms $a_j b_k e_j \wedge e_k$ and hence equal to

$$\sum_{j<k} (a_j b_k - a_k b_j) e_j \wedge e_k.$$

Consequently, that $u \wedge v = 0$ is equivalent to (31). If we continue as above to form products of more than two vectors and their linear combinations we get a mathematical object $G(L)$ invented around 1840 by Grassmann and now called the *Grassmann algebra over L*. It consists of linear combinations of products $u_1 \wedge \cdots \wedge u_q$ of $q \leqslant n$ factors called *q-vectors*, and lying in an unspecified linear space $\wedge^q L = L \wedge \cdots \wedge L$ (q factors) having the property that all functions $u_k \to u_1 \wedge \cdots \wedge u_k \wedge \cdots \wedge u_q$ are linear, that a product changes its sign when two factors are permuted, and that all $e_{i_1} \wedge \cdots \wedge e_{i_q}$ with $i_1 < \cdots < i_q$ constitute a basis of $\wedge^q L$ when

$e_1, \ldots, e_n$ is a basis of L. The proof that every linear space L has a Grassmann algebra and that the multiplication is associative and distributive is not difficult but a bit long and we must refrain from giving it. We show instead that if the vectors

$$u_1 = \sum a_{1k} v_k, \ldots, u_q = \sum a_{qk} v_k$$

are q linear combinations of q other vectors $v_1, \ldots, v_q$, then

$$u_1 \wedge \cdots \wedge u_q = \det(a_{jk}) v_1 \wedge \cdots \wedge v_q \tag{32}$$

where the determinant $\det(a_{jk})$ is defined by the formula (2) of Chapter 4 with n replaced by q. The proof is an excercise in the use of the summation sign. Writing $u_j = \sum a_{jk_j} v_{k_j}$ with summation over $k_j = 1, \ldots, q$ then

$$u_1 \wedge \cdots \wedge u_q = \sum a_{1k_1} \cdots a_{qk_q} v_{k_1} \wedge \cdots \wedge v_{k_q}$$

with summation over $k_1, \ldots, k_q$. The vector product on the right vanishes when two summation indices are equal. When they are all different, $k_1, \ldots, k_q$ is a permutation of $1, \ldots, q$ and hence the product equals $\varepsilon v_1 \wedge \cdots \wedge v_q$ where $\varepsilon = \pm 1$ according as the permutation is even or odd. This proves the assertion. Let us remark finally that q vectors $u_1, \ldots, u_q$ are linearly independent if and only if $u_1 \wedge \cdots \wedge u_q \neq 0$. In fact, if, e.g., $u_1 = b_2 u_2 + \cdots + b_q u_q$ is a linear combination of the others, then the product is a linear combination of products $v \wedge u_2 \ldots \wedge u_q$ where v is one of the vectors $u_2, \ldots, u_q$ and hence it vanishes. Conversely, if $u_1, \ldots, u_q$ are linearly independent they can be completed to a basis of L, and then $u_1 \wedge \cdots \wedge u_q$ is a member of a basis of $\wedge^q L$ and hence not zero. From all this it follows, for instance, that all n-vectors $\neq 0$ are proportional to each other.

Differentials in $\mathbf{R}^n$

The modern concept of the differential of a function f from an open part V of $\mathbf{R}^n$ to $\mathbf{R}$ is the following one. To every x in V we attach an unspecified linear space V_x with the basis $dx_1, \ldots, dx_n$, and the differential $df(x)$ of a differentiable function f is a vector in V_x defined by

$$df(x) = \partial_1 f(x) \, dx_1 + \cdots + \partial_n f(x) \, dx_n. \tag{33}$$

Hence, as x varies, the differentials lie in different linear spaces but that need not disturb us. The rules for differentiation can now be written very briefly as

$$d(f+g) = df + dg, \qquad d(fg) = f \, dg + g \, df,$$

$$f \neq 0 \quad \Rightarrow \quad d(1/f) = -(1/f^2) \, df.$$

These formulas appear in the very first textbooks of infinitesimal calculus. In the sequel we only consider smooth functions. The chain rule (23) can be written as a relation between differentials at the point y, namely

$$x = g(y) \quad \Rightarrow \quad d(f \circ g)(y) = \sum \partial_j f(x) \, dg_j(y)$$

or, if we want, as

$$d(f \circ g) = (df) \circ g \tag{34}$$

where the right side means that we shall write $g(y) = (g_1(y), \dots, g_n(y))$ instead of x in $df(x) = \sum \partial_j f(x) \, dx_j$. Letting S be this operation, namely the substitution $x \to g(y)$, we can also write this as $dSf = Sdf$ or, discarding f, as $dS = Sd$. We shall state our observation as

Leibniz's lemma. *The differential operator d commutes with substitutions.*

We give some examples. If $f(x) = x_1^2 + x_2^2$ so that $df(x) = 2x_1 dx_1 + 2x_2 dx_2$ and we put $x_1 = \cos t$, $x_2 = \sin t$, then $f(x) = 1$ and $df(x) = 2 \cos t \, d \cos t + 2 \sin t \, d \sin t = 2(-\sin t \cos t + \sin t \cos t) \, dt = 0$ and it is indeed true that $d1 = 0$. From $dt^2 = 2t \, dt$ and the substitution $t = f(x)$ we get $df(x)^2 = 2f(x) \, df(x)$ and this is also true.

According to Leibniz's lemma, when computing with differentials we may at any time replace our variables by functions of other variables and still keep all signs of equality. It is this property that makes the algebra of differentials so very useful. The lemma is a monument to Leibniz's intuition and ingenuity. Here is one application. That the Jacobi matrix $g'(x)$ of n functions $g(x) = (g_1(x), \dots, g_n(x))$ of n variables $x = (x_1, \dots, x_n)$ is nonsingular at x means that the differentials $dg_1(x), \dots, dg_n(x)$ are linearly independent. In fact, this linear independence means precisely that the rows of the matrix are linearly independent. Every differential $df(x)$ is then a linear combination $a_1(x) \, dg_1(x) + \dots + a_n(x) \, dg_n(x)$ and we can assert that the coefficients $a_1, \dots, a_n$ are nothing but the partial derivatives of the function f with respect to the variables $y_1 = g_1(x), \dots, y_n = g_n(x)$. In fact, considering $x = g^{-1}(y)$ as a function of y we get $df(x) = a_1(x) \, dy_1 + \dots + a_n(x) \, dy_n$.

We have defined differentials and their linear combinations as vectors in linear spaces V_x. Hence we can also form vector products of differentials. A product of n of them is then a multiple of $dx_1 \wedge \dots \wedge dx_n$, i.e., we have

$$df_1(x) \wedge \dots \wedge df_n(x) = J(x) \, dx_1 \wedge \dots \wedge dx_n.$$

According to (32), the factor $J(x)$ is just the *Jacobian*, defined as the determinant of the Jacobi matrix. Jacobians appear also in integration theory. For future use and in order to show how differential calculus works we shall now compute the Jacobians of polar coordinates according to (25) and (27). First,

$$x_1 = r \cos \theta, \qquad x_2 = r \sin \theta \quad \Rightarrow \quad dx_1 \wedge dx_2 = r \, dr \wedge d\theta. \tag{35}$$

In fact, (note that $dr \wedge dr = 0$ and $d\theta \wedge d\theta = 0$)

$$dx_1 \wedge dx_2 = (\cos \theta \, dr - r \sin \theta \, d\theta) \wedge (\sin \theta \, dr + r \cos \theta \, d\theta)$$

$$= \cos^2\theta \, r \, dr \wedge d\theta - \sin^2\theta \, r \, d\theta \wedge dr = r \, dr \wedge d\theta.$$

The corresponding formula in three variables is

$$x_1 = r \cos \theta \cos \varphi, \quad x_2 = r \cos \theta \sin \varphi, \quad x_3 = r \sin \theta$$
$$\Rightarrow \quad dx_1 \wedge dx_2 \wedge dx_3 = -r^2 \cos\theta \; dr \wedge d\theta \wedge d\varphi \tag{36}$$

and it is proved by the calculations

$$x_1 = u \cos \varphi, \quad x_2 = u \sin \varphi \quad \Rightarrow \quad dx_1 \wedge dx_2 = u \; du \wedge d\varphi,$$
$$u = r \cos \theta, \quad x_3 = r \sin \theta \quad \Rightarrow \quad du \wedge dx_3 = r \; dr \wedge d\theta,$$

noting that

$$dx_1 \wedge dx_2 \wedge dx_3 = u \; du \wedge d\varphi \wedge dx_3 = -u \; du \wedge dx_3 \wedge d\varphi.$$

Differential forms

Expressions

$$\omega(x) = a_1(x) \; dx_1 + \cdots + a_n(x) \; dx_n \tag{37}$$

where $a_1, \ldots, a_n$ are smooth functions from some open part V of $\mathbf{R}^n$ are called *Pfaffian forms*, after Pfaff who wrote about them in the beginning of the nineteenth century. Leibniz's magic symbol d can be applied also to Pfaffian forms. We simply put

$$d\omega(x) = da_1(x) \wedge dx_1 + \cdots + da_n(x) \wedge dx_n \tag{38}$$

where the right side is a 2-vector in $V_x \wedge V_x$. When $\omega(x) = df(x)$ is a differential this gives

$$d^2 f(x) = \sum_k d\partial_k f(x) \wedge dx_k = \sum_{j,\,k} \partial_j \partial_k f(x) \; dx_j \wedge dx_k = 0. \tag{39}$$

The last sum vanishes since $\partial_j \partial_k f = \partial_k \partial_j f$ but $dx_j \wedge dx_k + dx_k \wedge dx_j = 0$. In the chapter on integration we shall prove the converse of (39), which asserts that every Pfaffian form ω such that $d\omega = 0$ is the differential of a function. The following instances of (38) appear in the theory of line integrals:

$$\omega = P \; dx + Q \; dy \quad \Rightarrow \quad d\omega = (Q_x - P_y) \; dx \wedge dy \tag{40}$$
$$\omega = P \; dx + Q \; dy + R \; dz \quad \Rightarrow$$
$$d\omega = (R_y - Q_z) \; dy \wedge dz + (P_z - R_x) \; dz \wedge dx + (Q_x - P_y) \; dx \wedge dy. \tag{41}$$

In the first formula, x, y are real variables and P, Q smooth functions of them, and an index x or y indicates the corresponding derivative. In the other formula x, y, z are real variables and P, Q, R smooth functions. To prove (40) we make the computation

$$d\omega = (P_x \; dx + P_y \; dy) \wedge dx + (Q_x \; dx + Q_y \; dy) \wedge dy$$
$$= P_y \; dy \wedge dx + Q_x \; dx \wedge dy = (Q_x - P_y) \; dx \wedge dy$$

and the reader can then prove (41) by himself. The sums of (35) to (41) are examples of differential forms, i.e., finite sums

$$\omega(x) = \sum a(x) \; df_1(x) \wedge \cdots \wedge df_q(x),$$

the sum running over a finite number of functions $a, f_1, \ldots, f_q$. When q is constant, ω is said to be q-*form*. Functions are 0-forms and Pfaffians are 1-forms. Every q-form can be written uniquely as

$$\omega(x) = \sum a_{i_1 \ldots i_q}(x)\, dx_{i_1} \wedge \cdots \wedge dx_{i_q}$$

with summation over $i_1 < \cdots < i_q$. Its differential d is then defined by

$$d\omega(x) = \sum da_{i_1 \ldots i_q}(x) \wedge dx_{i_1} \wedge \cdots \wedge dx_{i_q}. \tag{42}$$

One example that occurs in integration theory and is recommended as an exercise is the following one

$$d(P\, dy \wedge dz + Q\, dz \wedge dx + R\, dx \wedge dy) = (P_x + Q_y + R_z)dx \wedge dy \wedge dz \tag{43}$$

with notations as in (41).

It is a very important fact that Leibniz's lemma holds also when d operates on differential forms. We prove this for 1-forms given by (37). According to (38), we have $d(f\omega) = df \wedge \omega + f\, d\omega$ for functions f. Hence it follows from (39) that $d(f\, dg) = df \wedge dg$ when f and g are functions. Now let $x = g(y)$. Then

$$d(\omega \circ g) = d\big(a_1 \circ g(y)\, dg_1(y) + \cdots\big)$$
$$= d(a_1 \circ g)(y) \wedge dg_1(y) + \cdots = (d\omega) \circ g$$

which is the assertion. The computation is the same for q-forms.

7.7 Differential calculus on a manifold

With all the basic tools of differential calculus at our disposal we shall make some applications, as old as the calculus itself, to the computation of tangents, tangent planes, and normals of smooth curves and surfaces. Working with hypersurfaces, i.e., manifolds in $\mathbf{R}^n$ defined by a single equation, this can be done in one stroke, at the same time providing an insight into the geometry of $\mathbf{R}^n$. Hypersurfaces will also serve to introduce the concept of a smooth manifold, easy to handle when the bijection theorem for smooth maps is available. The full force of differential calculus becomes apparent only when it is done on manifolds. Here we can also make a connection with topology. A sketch of the link between the two, the de Rham complex, concludes the chapter.

Hypersurfaces and curves in $\mathbf{R}^n$, tangents and normals

Let f be a real smooth function from $\mathbf{R}^n$, let $y \in \mathbf{R}^n$ be fixed, and assume that $f(y) = 0$. What is the geometric meaning of the equation $f(x) = 0$ when x is close to y? If $df(y) \neq 0$, the answer is given by the theorem on implicitly defined functions. In fact, when one of the partial derivatives $\partial_1 f(y), \ldots, \partial_n f(y)$ does not vanish, and assuming that $\partial_1 f(y) \neq 0$, the equation $f(x) = 0$ with x close to y is equivalent to an equation $x_1 = h(x_2, \ldots, x_n)$, where h is a smooth function defined close to $(y_2, \ldots, y_n)$ and equal to y_1 at that point. This is the equation of a curve when $n = 2$, a surface when $n = 3$, and something called a hypersurface in the general case. We note in passing that hypersurfaces are also given by equations

$$x = h(t) = (h_1(t), \ldots, h_n(t)) \tag{44}$$

where $t = (t_2, \ldots, t_n)$ runs over an open subset of $\mathbf{R}^{n-1}$, $t \to h(t)$ is a bijection, and precisely $n - 1$ of the differentials $dh_1, \ldots, dh_n$ are linearly independent there. This gives nothing new, for if, e.g., $dh_2, \ldots, dh_n$ are linearly independent at a point s corresponding to $y \in \mathbf{R}^n$, then, by the bijection theorem, $t_2, \ldots, t_n$ are smooth functions of $x_2, \ldots, x_n$ close to $y_2, \ldots, y_n$ and then (44) amounts to x_1 being a smooth function of $x_2, \ldots, x_n$ close to y.

Now let $f(x) = 0$ be the equation of a hypersurface close to y where $f(y) = 0$, $df(y) \neq 0$ and consider the formula

$$(x_1 - y_1)\partial_1 f(y) + \cdots + (x_n - y_n)\partial_n f(y) = 0. \tag{45}$$

When $n = 2$, this is the equation of a straight line, when $n = 3$ the equation of a plane, and in the general case the equation of a hyperplane, all passing through y and said to be tangent to the hypersurface at y. To motivate this, note that if a straight line, $x = y + at$, where $a \neq 0$ is in $\mathbf{R}^n$ and t is a real variable, lies in this hyperplane, then $\sum_1^n a_k \partial_k f(y) = 0$, so that, by Taylor's formula,

$$f(y + at) = f(y) + t \sum_1^n a_k \partial_k f(y) + t^2 H(t) = t^2 H(t)$$

where H is bounded. Vectors a with the property above are said to be *tangent* to the hypersurface at y (see Figure 7.7). As an example of all this, let $f(x) = c_1 x_1^2 + \cdots + c_n x_n^2 - 1$ where at least one of the numbers c_k is positive. Since it reduces to a conic when $n = 2$, the hypersurface $f(x) = 0$ will also be called a *conic*. In this case, (45) can be written as

$$c_1 x_1 y_1 + \cdots + c_n x_n y_n = 1$$

which is the classical formula for the tangent hyperplane of a conic.

Since (45) is the equation of the tangent hyperplane, all multiples of the *gradient* vector $\partial f(y) = (\partial_1 f(y), \ldots, \partial_n f(y))$ are said to be *normal* to the hypersurface $f(x) = 0$ at the point y (see Figure 7.7). A unit normal is one of length 1. There are two of them, opposite to each other. In case the hypersurface is given in the form (44), normals can be computed in another way. Consider the differential forms

$$\sigma_k = \sigma_k(x) = (-1)^k \, dx_1 \wedge \cdots \wedge dx_{k-1} \wedge dx_{k+1} \wedge \cdots \wedge dx_n.$$

In particular, $dx_k \wedge \sigma_k(x) = dx_1 \wedge \cdots \wedge dx_n$ for all k. On the hypersurface, i.e., when $x = h(t)$, all σ_k are multiples of $dt_2 \wedge \cdots \wedge dt_n$ so that

$$\sigma_k = J_k(x) \, dt_2 \wedge \cdots \wedge dt_n, \qquad k = 1, \ldots, n.$$

The vector $J = (J_1, \ldots, J_n) \neq 0$ is normal to the hypersurface at the point x. In fact, since precisely $n - 1$ of the differentials $dx_1, \ldots, dx_n$ are linearly independent on the hypersurface, J does not vanish, and since $df(x) = \partial_1 f(x)\, dx_1 + \cdots + \partial_n f(x)\, dx_n = 0$ on the hypersurface, taking products by $dx_1 \wedge \cdots$ with dx_p and dx_q left out, we get $\partial_p f(x) J_q(x) - \partial_q f(x) J_p(x) = 0$ for all p and q so that J is a multiple of the gradient vector $\partial f(x)$.

Curves in $\mathbf{R}^n$ are defined by equations (44) where t now is just one real variable and $h'(t) \neq 0$. The straight line $x = h(u) + sh'(u)$, where s varies over $\mathbf{R}$, is then tangent to the curve at the point $y = h(u)$. All multiples of the vector $h'(u)$ are said to be *tangent* to the curve at y (see Figure 7.7). The unit tangents are those of length 1. There are two of them, opposite to each other.

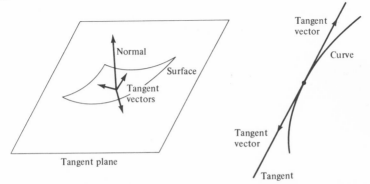

Figure 7.7   Tangent planes, tangents, and normals to a surface. Tangents to a curve.

Application of differential calculus to the study of curves and surfaces is called differential geometry. What we have given here is just a beginning, little more than a conceptual frame. In its extension lies, for instance, the theory of curvature containing deep results by Gauss and Riemann and also contemporary mathematicians. Differential geometry is interesting not only mathematematically. It is one of the main ingredients of the theory of general relativity.

Manifolds in $\mathbf{R}^n$

The 1-dimensional manifolds in $\mathbf{R}^n$ are the curves, the $(n - 1)$-dimensional ones the hypersurfaces. The formal definition of a manifold of dimension p in $\mathbf{R}^n$ runs as follows.

Let $h(t) = (h_1(t), \ldots, h_n(t))$ be n smooth real functions of $p \leqslant n$ real variables $t = (t_1, \ldots, t_p)$ defined in some open part V of $\mathbf{R}^p$ and consider the function

$$t \to x = h(t) \tag{46}$$

from V to $\mathbf{R}^n$. We say that the image $h(V)$ is a (*smooth*) *manifold* of dimension p or, shorter, a *p-manifold*, if h is a bijection and, moreover,

161

precisely p of the differentials $dh_1(t), \ldots, dh_n(t)$ are linearly independent and this for every t in V (since they are linear combinations of $dt_1, \ldots, dt_p$, p is the largest possible number). More generally, a p-manifold M in $\mathbf{R}^n$ is defined as a part of $\mathbf{R}^n$ which is the union of at most countably many such pieces $h(V)$ put together in such a way that any two of them are disjoint or overlap, each one meeting at most finitely many others.

Informally speaking, a p-manifold is something lying in some $\mathbf{R}^n$ with $n \geqslant p$ (in general n is much larger than p) which might be very complicated globally but locally looks just like an open piece of $\mathbf{R}^p$. Objects of this kind are natural generalizations of curves and surfaces, but they also appear in classical mechanics, more precisely in numerical descriptions of states of physical systems. Then the components of $x = (x_1, \ldots, x_n)$ are positions, directions, and perhaps also velocities describing a state of the system, the manifold M is the collection of all possible states, and its dimension is the number of degrees of freedom of the system. A simple example is the space of positions of a rod with one end fixed at the origin. If l is the length of the rod, a corresponding manifold is, for instance, the sphere $x_1{}^2 + x_2{}^2 + x_3{}^2 = l^2$, a 2-manifold in $\mathbf{R}^3$. When the rod is not fixed, the space of its positions is a 5-manifold in $\mathbf{R}^6$ defined by $(x_1 - y_1)^2 + (x_2 - y_2)^2 + (x_3 - y_3)^2 = l^2$.

Locally, manifolds are intersections of hypersurfaces. In fact, if $f_{p+1}, \ldots, f_n$ are smooth real functions from $\mathbf{R}^n$ vanishing at y, and their differentials are linearly independent there, a slight extension of our arguments in connection with hypersurfaces shows that if x is close to y, the equations $f_{p+1}(x) = 0, \ldots, f_n(x) = 0$ are equivalent to a set of equations (46) where $t_1, \ldots, t_p$ are p of the variables $x_1, \ldots, x_n$ chosen so that their differentials together with those of $f_{p+1}, \ldots, f_n$ are linearly independent. Conversely, if M is a p-manifold given by (46) close to $h(s) = y \in M$ and, for instance, $dh_1, \ldots, dh_p$ are linearly independent, then by the bijection theorem, close to $s = (s_1, \ldots, s_p)$, $t_1, \ldots, t_p$ are smooth functions of $x_1, \ldots, x_p$ and hence, close to y, (46) is equivalent to $n - p$ equations $x_{p+1} - g_{p+1}(x_1, \ldots, x_p) = 0, \ldots, x_n - g_n(x_1, \ldots, x_p) = 0$ where the g's are smooth functions.

Charts and atlases

Let us enliven our definition of a manifold M with some suggestive terminology. A pair h, V where h is a map from V to M as described above is called a *chart*, the variables t are called *coordinates* (and sometimes *parameters*) of the point $x = h(t)$, and $h(V)$ will be called a *charted part* of M. Sometimes V will be called a chart for short. An *atlas* of M is an at most countable collection of charts such that all its charted regions cover M and each one of them meets at most finitely many others. Figure 7.8 shows two charts whose charted regions overlap. Note that every x in the intersection $N = h(V) \cap h'(V')$ has two sets of coordinates t and t', and

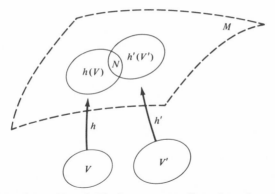

Figure 7.8   Two charts V and V', the corresponding charted regions $h(V)$ and $h'(V')$, and their intersection N.

that the equality $x = h(t) = h'(t')$ gives a bijection between the parts $h^{-1}(N)$ and $h'^{-1}(N)$ of the two charts V and V'. This bijection is smooth. In fact, according to the bijection theorem, t is a smooth function of p of the variables $x_1 = h_1(t), \ldots, x_n = h_n(t)$ if suitably chosen and hence is a smooth function of t'.

Here follow some examples of charts and atlases. First consider the upper half of the ellipse $(x/a)^2 + (y/b)^2 = 1$ where $y > 0$. It has a chart h, V given by

$$x = h_1(t) = t, \qquad y = h_2(t) = b\left(1 - (t/a)^2\right)^{1/2}$$

where V is the interval $-a < t < a$. Note that h_2 ceases to be a smooth function when $t = \pm a$. We can give a short description of this chart by saying that x has been taken as a coordinate on the upper half of the ellipse. The variable x is also a coordinate on the lower half (change the sign of h_2). On the left and right parts of the ellipse y can be taken as a coordinate. All this gives us an atlas of 4 charts. Trigonometric functions give other charts, e.g., h, V where

$$x = h_1(t) = a \cos t, \qquad y = h_2(t) = b \cos t$$

and t lies in an open interval V of length $< 2\pi$. Two such charts give an atlas of the ellipse if, together, they cover a closed interval of length at least 2π.

The ellipsoid $(x/a)^2 + (y/b)^2 + (z/c)^2 = 1$ can be treated in the same way. Using x, y, z as coordinates we get an atlas of six charts corresponding to the parts of the ellipsoid where one coordinate is < 0 or > 0. Trigonometric functions give other charts, e.g. (we now make the presentation less formal by not writing out h and V).

$$x = a \cos \theta \cos \varphi, \qquad y = b \cos \theta \sin \varphi, \qquad z = c \sin \theta$$

where, e.g., $-\pi/2 < \theta < \pi/2$ and $0 < \varphi < 2\pi$. This chart covers the entire ellipsoid except the curve $x = a \cos \theta, y = 0, z = c \sin \theta, -\pi/2 < \theta < \pi/2$, and the points $(0, 0, -c)$ and $(0, 0, c)$. Putting here $\cos \theta = (1 + t^2)^{-1/2}$,

$\sin \theta = t(1 + t^2)^{-1/2}$ and $a = b = c = 1$, we get a chart h, V of the unit sphere except the poles and a half-circle between them where V is the band $0 < \varphi < 2\pi$, t arbitrary real, in the φ, t-plane. In geography this is the chart of the world associated with Mercator's projection (~ 1675). The words chart and atlas have been chosen precisely because their everyday meanings correspond very well to the mathematical ones.

Exact and closed forms, the de Rham complex, and the cohomology of a manifold

A 1-form $\omega = f\, dx + g\, dy$ in an open region of the plane is said to be *closed* if $d\omega = 0$, and *exact* if there is a function h from V such that $\omega = dh$. (According to (40), ω is closed if and only if $\partial f/\partial y - \partial g/\partial x = 0$.) Since $d^2h = 0$ for all functions, exact $\Rightarrow$ closed but whether the converse, closed $\Rightarrow$ exact, is true or not depends on V. It is true when V is a rectangle and this will be shown in the next chapter, section 2. But if, e.g., V is $\mathbf{R}^2$ minus the origin (or a circular ring $0 \leqslant b < x^2 + y^2 < a$), it is not true. In fact, the differential form

$$d\theta = (y\, dx - x\, dy)/(x^2 + y^2) = d \text{ arc tan } y/x$$

where θ is the angle of polar coordinates, is closed but not exact in this region. Locally, θ is a smooth function from V but it becomes many-valued when continued around the origin. The Figure 7.9 illustrates this.

Transferring the notions of exact and closed forms from the plane to differential forms on a p-manifold M gives an important piece of mathematical machinery called the de Rham complex. Let us first consider a smooth function f from M to $\mathbf{R}$. For every chart h, V of the manifold we get a smooth function from V to $\mathbf{R}$ defined by $f(x) = f_V(t)$ when $x = h(t) \in h(V)$. To every $x = h(t)$ in $h(V)$ we then construct a linear space M_x with basis $dt_1, \ldots, dt_p$ and agree to put

$$dt'_j = \sum_1^p (\partial t'_j/\partial t_k)\, dt_k$$

when $x = h(t) = h'(t')$ and h, V and h', V' are overlapping charts. (Incidentally, the space M_x is called the *cotangent space* of M at x.) We now define the differential $df(x)$ at x of our smooth function f by putting

$$df(x) = df_V(t) = \sum_1^p (\partial f_V(t)/\partial t_k)\, dt_k$$

where $x = h(t)$. The right side is in M_x and the chain rule shows that it does not depend on the choice of chart.

Starting from the differentials we now construct differential forms on M by forming vector products, multiplying by smooth functions, and adding. We then get finite sums

$$\omega(x) = \sum a(x)\, df_1(x) \wedge \cdots \wedge df_q(x)$$

which, for fixed q, are called *q-forms* and form a linear space which will be

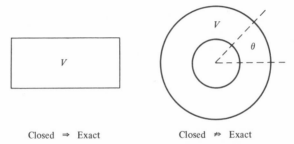

Closed ⇒ Exact   Closed ↛ Exact

Figure 7.9   Whether closed ⇒ exact depends on the region V.

denoted by $\bigwedge^q M$. In particular, $\bigwedge^0 M$ is the space of smooth functions. When $q > p$, $\bigwedge^q M$ contains only the zero element. It turns out that

$$\omega(x) \to d\omega(x) = da(x) \wedge df_1(x) \wedge \cdots \wedge df_q(x)$$

defines a linear map d from $\bigwedge^q M$ to $\bigwedge^{q+1} M$ such that $d^2\omega = 0$ for all ω. The *de Rham complex* is, by definition, the chain of maps

$$0 \to \bigwedge^0 M \xrightarrow{d} \bigwedge^1 M \xrightarrow{d} \cdots \to \bigwedge^p M \to 0.$$

Its importance for topology was pointed out by de Rham in 1930.

A q-form ω on a manifold is said to be *closed* if $d\omega = 0$ and *exact* if $\omega = d\sigma$ for some $(q-1)$-form σ. Among the functions only 0 is considered to be exact. The fact that closed ⇒ exact when $q = 1$ and M is an interval is a special case of a lemma by Poincaré: closed ⇒ exact for all forms when M is smoothly bijective to an interval. The philosophical implication of this lemma is that the difference between closed and exact is a global property. The size of this difference is measured by the *cohomology groups* of M, i.e., the quotient spaces

$$H^q(M) = Z_q / C_q$$

where Z_q and C_q are, respectively, the space of closed and exact q-forms on M, or, in other words, the kernel of the map from $\bigwedge^q M$ and the image of the map to $\bigwedge^q M$ in the de Rham complex. The elements of H^0 are the locally constant functions from M, and hence dim H^0 equals the number of connected pieces of M. By Poincaré's lemma, $H^q(M) = 0$ when $q > 0$ and M is smoothly bijective to an interval. In particular, $H^0(\mathbf{R}^n) \approx \mathbf{R}$ and $H^q(\mathbf{R}^n) = 0$ when $q > 0$ ($\approx$ means isomorphic to). Let S^n denote the so-called *n-sphere*, given by the equation $x_0^2 + x_1^2 + \cdots + x_n^2 = 1$ in $\mathbf{R}^{n+1}$. Then S^1 is just the unit circle, and it is easy to see that $H^1(S^1) \approx \mathbf{R}$. More generally, it can be proved that $H^q(S^n) = 0$ unless $q = 0$ or n and that $H^n(S^n) \approx \mathbf{R}$. When M is compact, all its cohomology groups have finite dimension.

These simple examples given an idea what the cohomology groups look like. Their most important property is that they behave very nicely under maps between manifolds. In fact, a smooth map F from a manifold M to another one, M', gives a linear substitution map $\omega(x') \to \omega(F(x))$ from differential forms on M' to differential forms on M. According to Leibniz's lemma, this map carries closed forms into closed forms and exact forms into exact forms. Hence F induces a linear map

$$F_* : H^q(M') \to H^q(M)$$

of the cohomology groups. When F is a bijection, so is F_*. In particular the dimensions of the corresponding cohomology groups must then be the same. Certain important properties of the map F can also be read off from the induced maps above, but to go into this would carry us too far afield.

7.8 Document

Riemann on physics and partial differential equations

Riemann took a lively interest in physics. His lectures on partial differential equations given around 1860 and published in 1882 are still a model of clarity and simplicity. Here are some passages from his introduction.

"As is well known, physics became a science only after the invention of differential calculus. It was only after realizing that natural phenomena are continuous that attempts to construct abstract models were successful. The task is two-fold: to devise simple basic concepts referring to time and space and to find a method to deduce from the processes which can be checked against experiments.

"The first such basic concept, that of an accelerating force, is due to Galilei. He found it to be the simple time independent cause of motion in a free fall. Newton took the second step: he found the concept of an attracting center, a simple cause of force. Contemporary physics still works with these two concepts, an accelerating force and an attracting or repellent center...all attempts to bypass these concepts have failed.

"But the methods—differential calculus—through which one passes from the concepts to the processes have been improved in an essential way. In the first period after the invention of differential calculus, only certain abstract cases were treated: in a free fall, the mass of the body was considered to be concentrated at its center of gravity, the planets were mathematical points,...so that the passage from the infinitely near to the finite was made only in one variable, the time. In general, however, this passage has to be done in several variables. For the basic concepts deal only with points in time and space and the processes describe actions over finite times and distances. Such passages lead to partial differential equations. [Riemann then reviews the work of d'Alembert and Fourier, both having dealt with such equations.]

"Since then, partial differential equations are the basis of all physical theorems. In the theory of sound in gases,

Bernhard Riemann 1826–1866

liquids and solids, in the investigations of elasticity, in optics, everywhere partial differential equations formulate basic laws of nature which can be checked against experiments. It is true that in most cases, these theories start from the assumption of molecules subject to certain forces. The constants of the partial differential equations then depend on the distribution of molecules and how they influence each other at a distance. But one is far from being able to draw definite conclusions from these distributions.... In all physical theories and by all phenomena explained by molecular forces, partial differential equations constitute the only verifiable basis.

These facts, established by induction, must also hold *a priori*. True basic laws can only hold in the small and must be formulated as partial differential equations. Their integration provides the laws for extended parts of time and space."

Note that this was written before the work of Maxwell and before relativity theory. Riemann's vision is only partly true in quantum mechanics but on the whole it has held up remarkably well. What he says about classical physics and partial differential equations can be said today.

Literature

The basic material of this chapter is in every calculus book. *Lectures on Ordinary Differential Equations*, by Hurewicz (MIT Press, 1964) is a charming and not too difficult account of the subject. The classic on partial differential equations is *Methods of Mathematical Physics*, vols. I, II, by Courant and Hilbert (Interscience, 1953 and 1964). *Topology from the Differentiable Viewpoint*, by Milnor (Virginia University Press, 1967), and *Differentiable Topology*, by Guillemin and Pollack (Prentice-Hall, 1974) are very readable introductions to manifolds.

8
INTEGRATION

The theory of integration has developed from simple computations of length, area, and volume to some very abstract constructions. We start with some intuitive calculations of area and volume giving, for instance, Archimedes' formulas for the volume and area of a ball, but after a reprimand from A. himself, we become more serious and turn to the Riemann integral. With both differentiation and integration at our disposal, we can then go through some of the fundamental results of analysis at a brisk pace. By means of a series of well-chosen examples we shall at the same time prove the basic properties of the Fourier transform, one of the most versatile tools of mathematics. After this there are sections on the Stieltjes integral and on integration on manifolds. Only the simplest properties of manifolds and differential forms are used. I chose this way in order to be able to give the right proof of Green's formula, and at the same time demonstrate one of the magic tricks of analysis, Stokes's formula in all its generality.

8.1 Areas, volumes, and the Riemann integral

Areas and volumes

Let $f(x)$ be a real and continuous function of a real variable x and let $S(x)$ be the area bounded by the curve $y = f(x)$, the x-axis, and verticals through the points a and x on the x-axis. Figure 8.1 shows that the area $S(x + h) - S(x)$, bounded by the x-axis, the curve $y = f(x)$, and verticals through x and $x + h$, lies between mh and Mh where m is the least and M the largest value of f in the interval between x and $x + h$. Since f is

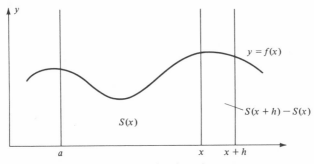

Figure 8.1 The derivative of an area.

continuous, both m and M tend to $f(x)$ as h tends to zero. Hence

$$S'(x) = \lim_{h \to 0} \frac{S(x + h) - S(x)}{h} = f(x) \qquad (1)$$

for all x. Behind this "hence" there is the tacit understanding that the area $S(x)$ is positive over the x-axis and negative under it, and positive to the right and negative to the left of a. Even so, the entire argument is extremely clear and convincing. Let us say that a function F is *a primitive function* or an *integral* of another function f in an interval if $F' = f$ in the interval. Using this terminology we can now give our result in abstract form as follows.

Theorem. *Every continuous function f has an integral F and every integral F has the property that $F(x_2) - F(x_1)$ is the area bounded by the x-axis, the curve $y = f(x)$, and verticals through x_1 and x_2.*

In fact, we have seen that $S(x)$ is an integral, and if F is another one then $F' = S'$ and hence, by the mean value theorem, $F - S$ is a constant. In the sequel we shall also use Leibniz's notation (from 1675)

$$\int_a^b f(x)\, dx, \qquad (2)$$

in words, "the integral of f from a to b (with respect to x)," for the area $F(b) - F(a)$. The function f is called the *integrand*.

The insight contained in our theorem is as old as infinitesimal calculus itself. To show its power we shall combine it with the following list of primitive functions

f	F
$x^{\alpha - 1}, \quad \alpha \neq 0$	x^α / α
x^{-1}	$\log |x|$
$(1 + x^2)^{-1}$	$\arctan x.$

Here arc tan x is the angle between $-\pi/2$ and $\pi/2$ whose tangent is x. According to the first row of the table

$$\int_a^1 x^{\alpha-1}\,dx = (1-a^\alpha)/\alpha, \qquad \int_1^b x^{\alpha-1}\,dx = (b^\alpha-1)/\alpha,$$

and according to the second one,

$$\int_a^1 x^{-1}\,dx = \log a^{-1}, \qquad \int_1^b x^{-1}\,dx = \log b,$$

and according to the third one

$$\int_{-\tan\theta}^{\tan\theta}(1+x^2)^{-1}\,dx = 2\theta,$$

when $a, b > 0$. Letting $b\to\infty$, $a\to0$ and $\theta\to\pi/2$ we get, respectively,

$$\alpha > 0 \quad \Rightarrow \quad \int_0^1 x^{\alpha-1}\,dx = \alpha^{-1}, \qquad \alpha > 0 \quad \Rightarrow \quad \int_1^\infty x^{-\alpha-1}\,dx = \alpha^{-1} \quad (3)$$

$$\int_0^1 x^{-1}\,dx = \infty, \qquad \int_1^\infty x^{-1}\,dx = \infty \tag{4}$$

$$\int_{-\infty}^{+\infty}(1+x^2)^{-1}\,dx = \pi. \tag{5}$$

The integrals (3) and (5) are said to be *convergent*, and the integrals (4) *divergent*. The areas under $y = 1/x$ from 0 to 1 and from 1 to ∞ are not finite.

It is also easy to show Archimedes' result (250 B.C.) that the volume of a ball of radius R is $4\pi R^3/3$, and that the area of its surface is $4\pi R^2$. In fact, by Figure 8.2, if $F(x)$ is the volume of the part of the ball denoted by B_x, then $F'(x) = \pi(R^2 - x^2)$ and $F(0) = 0$ so that $F(x) = \pi R^2 x - 3^{-1}\pi x^3$ and $F(R) = 2\pi R^3/3$. Hence $V(R) = 2F(R) = 4\pi R^3/3$ is the volume of the ball and, by Figure 8.2 again, its derivative $V'(R) = 4\pi R^2$ is the area of the sphere.

Let us now imagine that Archimedes has risen from the dead so that we can show him our proofs, translated, of course, into Greek by some

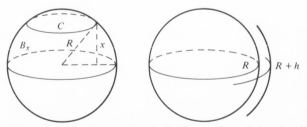

Figure 8.2   The volume and area of a ball. The radius of the circular disc C is $(R^2 - x^2)^{1/2}$, its area is $\pi(R^2 - x^2)$. The figure on the right indicates that if a ball of radius R has the volume $V(R)$, its area is $V'(R)$. In fact, $V(R+h) - V(R)$ is close to this area times h when h is small.

competent interpreter. In a conversation afterwards A. then tells us that he himself could have done something similar although he is surprised at the ease of our calculations. But he is far from satisfied. He doubts that it is possible simply to ignore the method of exhaustion used by him and based on a comparison between inscribed and circumscribed figures whose volumes and areas are known for certain. With this somewhat caustic remark Archimedes disappears, leaving us to contemplate the situation.

The Riemann integral

As long as mathematicians were dealing only with smooth functions, they had no reason to question the notion of area implicit in the definition (2) of the integral. But in 1854, when Riemann wrote about trigonometric series, he was forced to be precise because the functions he had to deal with were not smooth. Riemann's terse definition can be rendered as follows: divide the interval from a to b in smaller intervals separated by points $x_0 = a, x_1, \ldots, x_{n+1} = b$ such that $x_0 < x_1 < \ldots$ when $a < b$ and $x_0 > x_1 > \ldots$ when $a > b$ and consider sums

$$\sum_0^n f(\xi_j)(x_{j+1} - x_j) \tag{6}$$

where ξ_j is a number between x_j and x_{j+1}. The integral (2) is then defined as the limit of these sums when the subdivision gets infinitely fine in the sense that the largest of the numbers $|x_{j+1} - x_j|$ tends to zero.

We suppose here that the limit exists. Note that this is also a precise definition of the area $S(b)$ and that we get (2) from (6) by the suggestive substitutions

$$f(\xi_j) \text{ becomes } f(x), \quad x_{j+1} - x_j \text{ becomes } dx, \quad \sum_0^n \text{ becomes } \int_a^b.$$

The integral sign is actually a handwritten variant of the sign of summation invented by Leibniz. We can say that (2) at the same time denotes a number and the history of its computation. Note that the variable x of (2) represents a summation and that it can be changed to other symbols without changing the integral.

Let m_j and M_j be the infimum and supremum respectively of f in the interval between x_j and x_{j+1}. The Riemann sum (6) then lies between the sums

$$\sum_0^n m_j(x_{j+1} - x_j) \quad \text{and} \quad \sum_0^n M_j(x_{j+1} - x_j) \tag{7}$$

and it is easy to see that it tends to the integral (2) if and only if these sums do. Further, every term in them is the area of a rectangle with base $x_{j+1} - x_j$. Riemann's definition would have satisfied Archimedes.

The Riemann integral applies only to bounded functions and intervals. A bounded function for which the integral (2) exists is said to be *integrable* or, more precisely, *Riemann integrable* in the interval between a and b. Continuous functions are integrable, for by the theorem on uniform continuity, the largest of the numbers $M_j - m_j$ of (7) tends to zero with the largest of the numbers $|x_{j+1} - x_j|$. We can also allow f to have points of discontinuity provided they can be enclosed in finite collections of intervals whose total length is arbitrarily small.

Here are the basic properties of the integral. It is monotone, i.e.,

$$f \leqslant g, \quad a < b \quad \Rightarrow \quad \int_a^b f(x)\, dx \leqslant \int_a^b g(x)\, dx \tag{8}$$

(when $a > b$ the inequality is reversed), and it is linear, i.e.,

$$h = Af + Bg \quad \Rightarrow \quad \int_a^b h(x)\, dx = A \int_a^b f(x)\, dx + B \int_a^b g(x)\, dx. \tag{9}$$

In particular, since $-|f(x)| \leqslant f(x) \leqslant |f(x)|$,

$$b > a \quad \Rightarrow \quad \left| \int_a^b f(x)\, dx \right| \leqslant \int_a^b |f(x)|\, dx. \tag{10}$$

Here A and B are numbers and f and g are supposed to be integrable, in which case $Af + Bg$ and $|f|$ are also integrable. All this follows easily from the corresponding properties of the Riemann sums (7). We also have

$$\int_a^b f(x)\, dx + \int_b^c f(x)\, dx = \int_a^c f(x)\, dx. \tag{11}$$

Here, whenever two of the integrals exist so does the third one and the equality holds. When $a = c$ the right side vanishes. From (8) it follows that

$$m \leqslant f(x) \leqslant M \quad \text{when } a \leqslant x \leqslant b$$
$$\Rightarrow \quad m(b - a) \leqslant \int_a^b f(x)\, dx \leqslant M(b - a), \tag{12}$$

together with (10) one of the most frequently used inequalities of analysis.

If (f_p) is a family of integrable functions depending on one or several parameters p, and $f_p \to f$ uniformly in the interval from a to b as $p \to p_0$, then f is integrable and

$$p \to p_0 \quad \Rightarrow \quad \int_a^b f_p(x)\, dx \to \int_a^b f(x)\, dx.$$

This follows from the definition of the integral and (10) applied to the functions $f_p - f$. The more precise inequality (12) combined with an easy argument shows that

$$F(x) = \int_a^x f(t)\, dt, \ f \text{ continuous at } x \quad \Rightarrow \quad F'(x) = f(x). \tag{13}$$

This result, sometimes called *the fundamental theorem of calculus*, is what we started with. From (13) it follows that

$$\int_a^b F'(x)\, dx = F(b) - F(a) \tag{14}$$

when the derivative F' is continuous in the closed interval between a and b.

With this we have created an algebraic machinery for the integral and also liberated it from the notion of area. We can now allow ourselves to integrate complex-valued functions simply by putting

$$\int_a^b f(x)\,dx = \int_a^b \operatorname{Re} f(x)\,dx + i\int_a^b \operatorname{Im} f(x)\,dx$$

when $\operatorname{Re} f$ and $\operatorname{Im} f$ are integrable. It is then a simple exercise to show that the properties (9), (10), (11), (13), and (14) are true for complex functions and complex linear combinations of them. We shall also consider integrals

$$\int f(x)\,dx = \int_{\mathbf{R}} f(x)\,dx$$

over the real line $\mathbf{R}$ of C_0 functions f, i.e., such continuous functions which vanish outside bounded intervals. The integral (2) is then independent of a and $b > a$ when f vanishes outside the corresponding interval, and there is no need to write out the limits of integration.

Figure 8.3 A C_0 function. It is continuous and vanishes outside some bounded interval. In other words, its *support*, defined as the complement of the largest open set where the function vanishes, is compact.

Orientation

When integrating a function over an interval I between a and b and using the definition (2) we have to decide whether we shall integrate from a to b or from b to a. It follows from (11) with $c = a$ that the two choices differ in sign only. A way of looking at this is to say that (2) is an integral over an oriented interval. It is also possible to integrate over a nonoriented interval I. We shall then write the integral as

$$\int_I f(x)\,dx$$

fixing the sign so that the integral is $\geqslant 0$ when $f \geqslant 0$.

8.2 Some theorems of analysis

Traditionally, mathematical analysis consists of the branches of mathematics where infinitesimal calculus is used in some form or other. It has been and still is a very rich mine of mathematical knowledge. We shall now review some basic theorems of analysis using differentiation and integration. Various aspects of the theory of the Fourier transform appear as illustrations. At the end, having proved the Fourier inversion formula, we can go into the philosophy of harmonic analysis.

Differentiation under the integral sign

Let $f(t, x)$ be a continuous function defined in a product $W = I \times J$ where I is an interval on the t-axis and J a compact interval $a \leqslant x \leqslant b$. The integral

$$F(t) = \int_a^b f(t, x) \, dx \tag{15}$$

is then a continuous function of t. In fact,

$$F(t') - F(t) = \int_a^b (f(t', x) - f(t, x)) \, dx \tag{16}$$

and since f is uniformly continuous on compact parts of W, the integrand tends to zero as t' tends to t, uniformly when $a \leqslant x \leqslant b$. We shall also see that if the partial derivative $\partial_t f(t, x)$ exists and is continuous, then the derivative

$$F'(t) = \partial_t F(t) = \int_a^b \partial_t f(t, x) \, dx \tag{17}$$

also exists and is a continuous function of t. Combining (15) and (17) we get

$$\partial_t \int_a^b f(t, x) \, dx = \int_a^b \partial_t f(t, x) \, dx \tag{18}$$

so that the derivative has gone right through the integral sign. To show this, divide both sides of (16) by $t' - t$. By the mean value theorem, the new integrand equals $\partial_t f(\tau, x)$ where τ lies between t and t'. In particular, the points (τ, x) and (t, x) are uniformly close when t' is close to t. The function $s, x \to \partial_t f(s, x)$ being continuous and hence uniformly continuous on compact parts of $I \times J$, it follows that $\partial_t f(\tau, x)$ tends to $\partial_t f(t, x)$ as $t \to t'$, uniformly when $a \leqslant x \leqslant b$. Hence (17) follows and we have just seen that the right side is a continuous function of t. Besides, it is easy to see that

$$F(t, u, v) = \int_u^v f(t, x) \, dx$$

is a continuous function of its three arguments t, u, v in the product $I \times J \times J$ when f is continuous in $I \times J$. Moreover, (13) shows that $\partial_u F(t, u, v) = -f(t, u)$, $\partial_v F(t, u, v) = f(t, v)$ and, if $\partial_t f$ is continuous, $\partial_t F(t, u, v)$ is given by (17) with u and v on the right side. Hence, in this case, F is a C^1 function.

The *Fourier transform*

$$F(t) = \int e^{-ixt} f(x) \, dx \tag{19}$$

of a C_0 function f provides an example of (17). In fact, the function $t, x \to e^{-ixt} f(x)$ and all its derivatives with respect to t are continuous everywhere. Hence F is a C^k function for all k and, since the function $t \to e^{ct}$ has the derivative ce^{ct} for every complex number c, we get

$$F^{(k)}(t) = \int e^{-ixt} (-ix)^k f(x) \, dx. \tag{20}$$

Repeated integrations

Let us consider an integral

$$F(t) = \int_a^b f(t, x) \, dx$$

where f is continuous when t lies between α and β and x between a and b. We then know F to be continuous and we shall see that

$$\int_\alpha^\beta F(t) \, dt = \int_a^b \left(\int_\alpha^\beta f(t, x) \, dt \right) dx, \tag{21}$$

or, inserting F,

$$\int_\alpha^\beta \left(\int_a^b f(t, x) \, dx \right) dt = \int_a^b \left(\int_\alpha^\beta f(t, x) \, dt \right) dx. \tag{22}$$

In other words, the result of the two integrations is independent of the order in which they are performed. The proof is very simple. A Riemann sum

$$\sum_0^n F(\tau_j)(t_{j+1} - t_j), \qquad \tau_j \text{ between } t_j \text{ and } t_{j+1},$$

for the left side of (21) can also be written as

$$\int_a^b R(x) \, dx$$

where the Riemann sum

$$R(x) = \sum_0^n f(\tau_j, x)(t_{j+1} - t_j)$$

differs from the integral

$$S(x) = \int_\alpha^\beta f(t, x) = \sum_0^n \int_{t_j}^{t_{j+1}} f(t, x) \, dt$$

by at most the length $|\beta - \alpha|$ of the interval times the maximum δ of $|f(\tau_j, x) - f(t, x)|$ when t lies between t_j and t_{j+1} and x between a and b. Since f is continuous, δ tends to zero and hence $R(x)$ uniformly to $S(x)$ when the largest of the numbers $|t_{j+1} - t_j|$ tends to zero. This proves (21).

Again, consider the Fourier transform (19) of a C_0 function f. If g is another C_0 function, (22) shows that

$$\int F(t) g(t) \, dt = \int G(x) f(x) \, dx \tag{23}$$

175

where

$$G(x) = \int e^{-ixt} g(t)\, dt \qquad (24)$$

is the Fourier transform of g.

Exact differentials

We are going to apply the formula (22) to the following problem. Given C^1 functions $f_1, \ldots, f_n$ of n real variables $x = (x_1, \ldots, x_n)$, when is there a C^2 function h such that $f_k = \partial_k h$ for all k? Here $\partial_k h$ is the partial derivative $\partial h(x)/\partial x_k$. In terms of differentials the same question runs as follows: when is a differential form $\omega = f_1\, dx_1 + \cdots + f_n\, dx_n$ exact in the sense that $\omega = dh$ for some function h? Since $\partial_j \partial_k h = \partial_k \partial_j h$ for all j and k, the problem is solvable only if $\partial_j f_k = \partial_k f_j$ for all j and k. According to the formula (38) of Chapter 7,

$$d\omega = \sum_{j<k} (\partial_j f_k - \partial_k f_j)\, dx_j \wedge dx_k,$$

this condition is equivalent to $d\omega = 0$. We shall see that this necessary condition is also sufficient provided we move in an open interval $I \subset \mathbf{R}^n$.

To begin with, consider the case $n = 2$ and suppose that I contains the origin, which is no restriction. We introduce the two functions

$$h_1(x) = \int_0^{x_2} f_2(0, y_2)\, dy_2 + \int_0^{x_1} f_1(y_1, x_2)\, dy_1$$

$$h_2(x) = \int_0^{x_1} f_1(y_1, 0)\, dy_1 + \int_0^{x_2} f_2(x_1, y_2)\, dy_2$$

and remark in passing that, obviously, the right sides can be interpreted as integrals $\int_{\gamma_1} \omega(y)$ and $\int_{\gamma_2} \omega(y)$ where $\omega(y) = f_1(y)\, dy_1 + f_2(y)\, dy_2$ and γ_1 and γ_2 are the two curves of Figure 8.4. For, if (y_1, y_2) lies on the first part of γ_1, then $y_1 = 0$ so that $\omega(y) = f_2(0, y_2)\, dy_2$ where y_2 runs between 0 and x_2. And so on for the other parts.

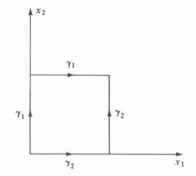

Figure 8.4   Integration along different paths.

The difference between h_1 and h_2 we can write as

$$h_1(x) - h_2(x) = -\int_0^{x_2}(f_2(x_1 y_2) - f_2(0, y_2))\, dy_2$$
$$+ \int_0^{x_1}(f_1(y_1, x_2) - f_1(y_1, 0))\, dy_1$$

and hence according to (14) as

$$-\int_0^{x_1}\left(\int_0^{x_2}\partial_1 f_2(y)\, dy_2\right) dy_1 + \int_0^{x_2}\left(\int_0^{x_1}\partial_2 f_1(y)\, dy_1\right) dy_2.$$

By virtue of (22) this vanishes when $\partial_1 f_2 = \partial_2 f_1$ and in that case $h_1 = h_2$ in the entire two-dimensional interval I. It follows from the text preceding (19) that h_1 and h_2 are C^1 functions, and (13) shows that $\partial_1 h_1 = f_1$ and $\partial_2 h_2 = f_2$. Hence $h = h_1 = h_2$ is a C^2 function such that $\partial_1 h = f_1$ and $\partial_2 h = f_2$. An extension of the proof to more variables does not introduce new difficulties, only more bookkeeping, and we have

Theorem (Exact differentials). *A 1-form ω of class C^1 in an interval is exact, i.e., the differential of a function, if and only if it is closed, i.e., $d\omega = 0$.*

Replacing the interval by an arbitrary open set makes the theorem false. See the end of Chapter 7.

Generalized integrals

As remarked above, the Riemann integral applies only to bounded functions and bounded intervals, but it is easy to extend it to, e.g., continuous functions in open intervals $a < x < b$, allowed to extend to infinity. We simply put

$$\int_a^b f(x)\, dx = \lim \int_\alpha^\beta f(x)\, dx \quad \text{when } \beta \uparrow b \text{ and } \alpha \downarrow a \tag{25}$$

provided the limit exists. If $f(x) \geqslant 0$ everywhere, the integral on the right increases as β increases and α decreases and hence the limit exists but may be $+\infty$. Applying the inequality (10) to intervals close to a and b and using Cauchy's principle of convergence shows that the limit (25) exists and is finite, provided

$$\int_a^b |f(x)|\, dx < \infty$$

in which case the integral of the left side of (25) is said to be *absolutely convergent* and the function *absolutely integrable*. The basic properties (8), (9), (10) extend to generalized integrals.

It follows from (3) that a continuous function f from the real axis is absolutely integrable if the function $x^{1+\varepsilon}f(x)$ is bounded for some $\varepsilon > 0$. In particular, since $x^k e^{-x}$ is bounded in the interval $x \geqslant 0$ when $k > 0$, all

177

products $P(x)e^{-a|x|}$ or $P(x)e^{-ax^2}$ where $a > 0$ and P is a polynomial have this property. When f is continuous and $\int |f(x)|\, dx$ is finite, its Fourier transform

$$F(t) = \int e^{-ixt} f(x)\, dx$$

is defined, for $|e^{is}| = 1$ for all real s. Another important example is Euler's Gamma function $\Gamma(\alpha)$, for $\alpha > 0$ defined by the integral

$$\Gamma(\alpha) = \int_0^\infty x^{\alpha-1} e^{-x}\, dx, \tag{26}$$

seen by (3) to be absolutely convergent.

Change of integration variable

Let $h(y)$ be a function such that $x = h(y)$ varies between $a = h(\alpha)$ and $b = h(\beta)$ as y varies between α and β. Then

$$\int_a^b f(x)\, dx = \int_\alpha^\beta f(h(y))\, dh(y) = \int_\alpha^\beta f(h(y)) h'(y)\, dy. \tag{27}$$

This is a simple way of stating the rule for changing variables of integration. For integrals over nonoriented intervals it says that

$$\int_{h(J)} f(x)\, dx = \int_J f(h(y))|h'(y)|\, dy. \tag{28}$$

Both formulas are correct when f is integrable and h is a C^1 bijection, for if $y_0, \ldots, y_{n+1}$ is a subdivision of the interval J from α to β and if $x_0 = h(y_0), \ldots, x_{n+1} = h(y_{n+1})$ is the corresponding subdivision of the interval $I = h(J)$ from a to b, if, by the mean value theorem, η_j between y_j and y_{j+1} is chosen so that $x_{j+1} - x_j = h'(\eta_j)(y_{j+1} - y_j)$ and, finally, if $\xi_j = h(\eta_j)$, then the two sides of the identity

$$\sum_0^n f(\xi_j)(x_{j+1} - x_j) = \sum_0^n f(h(\eta_j)) h'(\eta_j)(y_{j+1} - y_j)$$

are Riemann sums for the corresponding integrals of (27), and (28) follows from (27).

A simple case of (27) is

$$\int_a^b f(x)\, dx = \int_{a-c}^{b-c} f(y + c)\, dy.$$

Hence, letting $a \to -\infty$ and $b \to +\infty$ we get

$$\int f(x)\, dx = \int f(x + c)\, dx$$

with integration over the real axis $\mathbf{R}$ provided f is continuous and absolutely integrable. In particular,

$$\int e^{-it(x+y)} f(x + y)\, dx = \int e^{-itx} f(x)\, dx$$

so that, multiplying by e^{ity},

$$\int e^{-itx} f(x+y)\, dx = e^{ity} \int e^{-itx} f(x)\, dx \tag{29}$$

for all real y. In other words, the Fourier transform of the function $x \to f(x+y)$ is $e^{ity} F(t)$ where F is the Fourier transform of f.

Using (27) it is easy to see that if f is a C^{k+1} function close to the origin, then $(f(x) - f(0))/x$ is a C^k function there. In fact, the substitution $t = xu$ gives

$$\frac{f(x) - f(0)}{x} = x^{-1} \int_1^x f'(t)\, dt = \int_0^1 f'(xu)\, du$$

where k differentiations under the last integral sign are permitted, resulting in continuous functions of x. Finally, by the substitution $x = \sin^2\theta$,

$$\int_0^1 (x(1-x))^{-1/2}\, dx = \int_0^{\pi/2} 2 d\theta = \pi \tag{30}$$

for $x(1-x) = \sin^2\theta \cos^2\theta$ and $dx = 2\sin\theta\cos\theta\, d\theta$. But this conclusion might be too hasty since the integral on the left is generalized. However, integrating between $\sin^2\varepsilon$ and $\sin^2(2^{-1}\pi - \varepsilon)$ where $\varepsilon > 0$ is small and passing to the limit saves the situation.

Passage to the limit under the integral sign and dominated convergence

It is very important in analysis to have conditions permitting us to pass to the limit under the integral sign,

$$\lim_{p \to p_0} \int_a^b f_p(x)\, dx = \int_a^b \lim_{p \to p_0} f_p(x)\, dx$$

where (f_p) is a collection of functions depending on one or several parameters p. When the interval is finite, the functions f_p continuous when $a \leqslant x \leqslant b$, and the convergence uniform we know already that the formula is true. We are going to give an easy but important extension of this result. The family (f_p) is said to *converge locally uniformly* in an interval I when $p \to p_0$ if there is a function f such that $f_p \to f$ as $p \to p_0$ uniformly on every compact subinterval of I. In particular, if the f_p are continuous so is the limit function f. This follows from Theorem 1 of Chapter 5.

Theorem (Dominated convergence). *If a family (f_p) of continuous functions in an open interval $a < x < b$ converges locally uniformly to a limit function $f(x)$ as $p \to p_0$ and if there is a continuous function $g(x) \geqslant 0$ such that $\int_a^b g(x)\, dx$ is finite and $|f_p(x)| \leqslant g(x)$ for all x, then (27) holds so that*

$$\lim_{p \to p_0} \int_a^b f_p(x)\, dx = \int_a^b f(x)\, dx.$$

PROOF. Since f is continuous and $|f(x)| \leqslant g(x)$, all integrals involved exist. The inequality

$$\left| \int_a^b (f_p(x) - f(x))\, dx \right|$$

$$\leqslant \int_\alpha^\beta |f_p(x) - f(x)|\, dx + 2\int_\beta^b g(x)\, dx + 2\int_a^\alpha g(x)\, dx,$$

true for all $\alpha < \beta$ between a and b, shows that its left side tends to zero as $p \to p_0$. In fact, given $\varepsilon > 0$, we can fix α and then β so that the sum of the two last terms on the right is $< \varepsilon/2$ and then choose p so close to p_0 that the first term is also $< \varepsilon/2$.

We shall apply this theorem several times, first to show that the Fourier transform F of f is a C^k function with derivatives given by (20) when f is continuous and $x^k f(x)$ is absolutely integrable. In fact,

$$F(t + h) - F(t) = \int e^{-ixt} g(xh) f(x)\, dx$$

where $g(y) = e^{-iy} - 1$ is bounded and tends to zero as $y \to 0$. This proves that F is continuous. Dividing by h and rewriting $g(xh)/h$ as $(g(xh)/xh)x$ where $g(y)/y = (e^{-iy} - 1)/y$ is bounded and tends to $-i$ as $y \to 0$, proves that F is a C^1 function and that (20) holds when $\int |xf(x)|\, dx < \infty$. Iteration gives the general result for all derivatives of F.

Let us also note that (23) holds when f and g are continuous and absolutely integrable and F and G are their Fourier transforms. In fact, let $h(x)$ be a C_0 function equal to 1 when $|x| < 1$ and such that $|h(x)| \leqslant 1$ everywhere and put $h_\varepsilon(x) = h(\varepsilon x)$. Then, as $\varepsilon \to 0$, the function $f_\varepsilon(x) = h_\varepsilon(x) f(x)$ tends to $f(x)$ locally uniformly and $|f_\varepsilon(x)| \leqslant |f(x)|$ everywhere. At the same time its Fourier transform $F_\varepsilon(t) = \int e^{-ixt} f_\varepsilon(x)\, dx$ is uniformly bounded independently of ε and tends uniformly to $F(t)$ as $\varepsilon \to 0$. This follows from the estimates

$$|F_\varepsilon(t)| \leqslant \int |f(x)|\, dx, \ |F_\varepsilon(t) - F(t)| \leqslant \int |(1 - h(\varepsilon x))| |f(x)|\, dx$$

and dominated convergence. Now since (23) holds for C_0 functions, it holds when f and g are replaced by f_ε and $g_\varepsilon = h_\varepsilon g$ with Fourier transforms F_ε and G_ε. Since $F_\varepsilon(t) g_\varepsilon(t)$ tends to $F(t) g(t)$ locally uniformly and its absolute value is bounded by a constant times $|g(t)|$ and analogously for $f_\varepsilon(x) G_\varepsilon(x)$, the desired result follows by dominated convergence. Finally, let us prove that

$$\varepsilon \downarrow 0 \Rightarrow \varepsilon^{-1} \int e^{-x^2/2\varepsilon^2} f(x)\, dx \to f(0) \int e^{-x^2/2}\, dx \tag{31}$$

when f is continuous and absolutely integrable. In fact, when f vanishes for

$|x| \leqslant 1$, this follows from the inequality

$$\left| \varepsilon^{-1} \int e^{-x^2/2\varepsilon^2} f(x)\, dx \right| \leqslant \varepsilon^{-1} e^{-2\varepsilon^{-2}} \int |f(x)|\, dx$$

and when f is a C_0 function it follows by dominated convergence from the formula

$$\varepsilon^{-1} \int e^{-x^2/2\varepsilon^2} f(x)\, dx = \int e^{-x^2/2} f(\varepsilon x)\, dx,$$

obtained by changing x to εx in the first integral. In the general case, let h be the C_0 function used above and write $f = (1-h)f + hf$. Here $(1-h)f$ vanishes when $|x| \leqslant 1$ and hf is a C_0 function so that (31) is established.

Integration by parts

The formula for integration by parts,

$$\int_a^b f(x) g'(x)\, dx + \int_a^b f'(x) g(x)\, dx = f(b) g(b) - f(a) g(a), \quad (32)$$

where f and g are supposed to be C^1 functions, results from (14) and (9) if we put $F = fg$ in (14) and note that $(fg)' = fg' + f'g$. A good way to remember (32) is to write $h'(x)\, dx$ as $dh(x)$ when h is a C^1 function. The formula then becomes

$$\int_a^b f(x)\, dg(x) + \int_a^b g(x)\, df(x) = f(b) g(b) - f(a) g(a) \quad (33)$$

and we can write (14) as

$$\int_a^b dF(x) = F(b) - F(a). \quad (34)$$

An important application is to Taylor's formula for C^{n+1} functions,

$$f(x) = f(a) + (x-a) f'(a) + \cdots + \frac{(x-a)^n}{n!} f^{(n)}(a) + R_n(x)$$

with the remainder term

$$R_n(x) = \int_a^x \frac{(x-t)^n}{n!} f^{(n+1)}(t)\, dt.$$

True for $n = 0$, it is proved by induction after n using the formula

$$R_{n-1}(x) - R_n(x) = \int_a^x \left(\frac{(x-t)^{n-1}}{(n-1)!} f^{(n)}(t) - \frac{(x-t)^n}{n!} f^{(n+1)}(t) \right) dt$$

$$= -\int_a^x d\, \frac{(x-t)^n}{n!} f^{(n)}(t) = \frac{(x-a)^n}{n!} f^{(n)}(a).$$

As a second application, let us prove that the Gamma function (26) has the property that

$$\alpha > 0 \quad \Rightarrow \quad \Gamma(\alpha + 1) = \alpha \Gamma(\alpha).$$

In fact, the formula (33) and $dx^\alpha = \alpha x^{\alpha-1}\, dx$, $de^{-x} = -e^{-x}\, dx$ give

$$\alpha \int_a^b x^{\alpha-1} e^{-x}\, dx = \int_a^b e^{-x}\, dx^\alpha = \int_a^b x^\alpha e^{-x}\, dx + b^\alpha e^{-b} - a^\alpha e^{-a}$$

when $0 < a < b$. When b tends to ∞ and a tends to 0, the last two terms tend to zero and this gives the desired result. Since $\Gamma(1) = 1$, repeated use of the formula just proved shows that $\Gamma(\alpha) = (\alpha - 1)!$ when α is a positive integer.

An important application of integration by parts to Fourier transforms is given in the next section.

The Fourier inversion formula

At this stage it is convenient to introduce a class of very well-behaved functions, the S functions employed by L. Schwartz in the theory of distributions. An infinitely differentiable function $f(x)$ from the real line to the complex numbers is an S *function* when all products $x^j f^{(k)}(x)$ of powers of x and derivatives of f are bounded. Example: when P is a polynomial and $a > 0$, $f(x) = P(x)e^{-ax^2}$ is an S function. It is clear that linear combinations, derivatives, and products by polynomials of S functions are S functions. Since $x^2 f(x)$ is bounded, an S function f is absolutely integrable and a passage to the limit in (14) shows that $\int f'(x)\, dx = 0$ with integration over the real line.

An important property of S functions is that the Fourier transform of an S function is also one. In fact, let f be an S function. Since the functions $x^k f(x)$ are absolutely integrable for every k, its Fourier transform

$$F(t) = \int e^{-itx} f(x)\, dx$$

is a C^∞ function whose derivatives are computed according to (20), in particular

$$F'(t) = -i \int x e^{-itx} f(x)\, dx. \tag{35}$$

Since $ite^{-itx}\, dx = -de^{-itx}$, an integration by parts gives

$$it \int_a^b e^{-itx} f(x)\, dx = \int_a^b e^{-itx} f'(x)\, dx + e^{-ita} f(a) - e^{-itb} f(b)$$

so that, letting $a \to -\infty$ and $b \to \infty$,

$$itF(t) = \int e^{-itx} f'(x)\, dx. \tag{36}$$

Hence $tF(t)$ is bounded and since $xf(x)$ and $f'(x)$ are S functions, iterations of the formulas (35) and (36) show F to be an S function.

Applying them to the S function $g(x) = e^{-x^2/2}$ with the Fourier transform $G(t)$ shows that

$$G'(t) = -i \int xe^{-itx}g(x)\, dx,$$

$$itG(t) = \int e^{-ixt}g'(x)\, dx = -\int xe^{-ixt}g(x)\, dx,$$

for $g'(x) = -xg(x)$. Hence $G'(t) = -tG(t)$, i.e., $(e^{t^2/2}G(t))' = 0$, so that $G(t) = ce^{-t^2/2}$ where c is a constant. Hence

$$G(t) = \int e^{-ixt-x^2/2}\, dx = ce^{-t^2/2}, \qquad c = \int e^{-x^2/2}\, dx, \qquad (37)$$

giving us an explicit formula for G apart from the value of c. We shall see later that $c = (2\pi)^{1/2}$.

We now have everything needed to prove one of the most important results in mathematics, the Fourier inversion formula.

Theorem (The Fourier inversion formula). *If f and its Fourier transform*

$$F(t) = \int e^{-ixt}f(x)\, dx$$

are continuous and absolutely integrable, then

$$f(x) = (2\pi)^{-1} \int e^{ixt}F(t)\, dt. \qquad (38)$$

The requirements on f and F are convenient but far from necessary. They are fulfilled when f is an S function but also, for instance, when f is a C^2 function and f, f', f'' are absolutely integrable and tend to zero as $|x|$ tends to infinity, in particular when $x^2f(x)$, $x^2f'(x)$, and $x^2f''(x)$ are bounded. In fact, then two integrations by parts show that $t^2F(t)$ is bounded and hence F absolutely integrable.

Taking real and imaginary parts of both sides of (38) we get formulas that appeared for the first time in 1822 in Fourier's *Théorie analytique de la chaleur* with proofs that we now have to consider as purely formal. The computations are algebraically correct but at that time analysis was not so precise that it was considered necessary to be specific about the functions involved. But it is clear from the examples he gives that Fourier had a very clear idea of what he was doing.

To prove (38) note that (29) shows that it suffices to prove that

$$\int F(t)\, dt = 2\pi f(0). \qquad (39)$$

For applying this to the function $x \to f(y + x)$ whose Fourier transform is $e^{ity}F(t)$ gives (38) with x replaced by y. To prove (39) we shall make a suitable choice of the function g of the formula (23) extended above to continuous absolutely integrable functions f, g and their Fourier transforms. We shall put $g(x) = e^{-\varepsilon x^2/2}$ where $\varepsilon > 0$ is small. Replacing x by εx

in (37) and t by t/ε and dividing by ε shows that g has the Fourier transform $c\varepsilon^{-1}e^{-t^2/2\varepsilon}$. Hence, by (23),

$$\int F(t)e^{-t^2\varepsilon^2/2}\,dt = c\varepsilon^{-1}\int e^{-x^2/2\varepsilon^2}f(x)\,dx.$$

Letting $\varepsilon \to 0$, the left side tends to the left side of (39) by dominated convergence and, by virtue of (31), the right side tends to $c^2f(0)$, i.e., the right side of (39). This completes the proof.

When f is an S function, the proof of (39) can be simplified as follows. If $f(0) = 0$, then $h(x) = f(x)/x$ is an S function, for we have seen that it is infinitely differentiable in an interval around the origin, and, since $f(x) = xh(x)$, by (35), the Fourier transform $F(t)$ equals $-iH'(t)$ where H is the Fourier transform of h. Hence $\int F(t)\,dt = 0$ when $f(0) = 0$. Applying this to the linear combination $g(0)f(x) - f(0)g(x)$ of two S functions f and g with Fourier transforms F and G shows that

$$g(0)\int F(t)\,dt = f(0)\int G(t)\,dt.$$

Inserting e.g., $g(x) = e^{-x^2/2}$ proves (39).

Distributions

The theory of distributions, worked out by Laurent Schwarz in the 1940s is closely tied to integration by parts and we shall say a few words about it. We have already met with C_0 functions, i.e., continuous functions $g(x)$ from **R** which vanish for large $|x|$. If, in addition, g is infinitely differentiable we say that g is a C_0^∞ *function*. Via an integration, every continuous function f from **R** gives us a linear function (the equality sign is a definition of the symbol $f(g)$),

$$g \to f(g) = \int f(x)g(x)\,dx$$

from C_0^∞ functions to **R** or **C**. Such a linear function $g \to T(g)$, regardless of how it is defined, is called a *distribution* if it has the following continuity property: to every open bounded interval I there are numbers c and k such that $|T(g)| \leqslant c \max |g^{(k)}(x)|$ when g vanishes outside I. When $T(g) = f(g)$ with f as above, this holds with $k = 0$. When the inequality holds with $c = 0$ we say that $T = 0$ in the interval I. When f is a C^1 function, the formula for integration by parts can be written as $f'(g) = -f(g')$. The corresponding formula for distributions, $T'(g) = -T(g')$, is taken as a definition of T' which is immediately seen to be a distribution. In the same way, the formula (23), taken in the sense that $U(g) = T(G)$ where g is an S function and G its Fourier transform, defines the Fourier transform U of T when T is a so-called *tempered* distribution, i.e., extends in a continuous way to S functions. This extension of differential calculus and the Fourier transform to distributions is actually very useful, but we cannot give the details. Let us only note the important fact that the Fourier inversion formula extends to tempered distributions.

Convolutions

Let us call a function $f(x)$ from $\mathbf{R}$ a *C function* if it is continuous and vanishes for large negative x. We are going to study a product $f*g$ of two such functions and g called the *convolution* of f and g and defined by

$$(f*g)(x) = \int f(x-y)g(y)\, dy. \tag{40}$$

Note that the functions $y \to f(x-y)g(y)$ are C_0 functions, i.e., continuous and vanishing for large $|y|$, and that they vanish for all y when x is large negative. Hence the integral is defined, it vanishes for large and negative x and, by (15), it is a continuous function of x. Hence the convolution $f*g$ is another C function. A change of variables $y = x - z$ shows that convolution is commutative, $f*g = g*f$. It is trivially distributive but also associative, for if h is a third C function, then by (22),

$$f*(g*h)(x) = \int f(x-y)\left(\int g(y-z)h(z)\, dz\right) dy$$

equals

$$\int\left(\int f(x-y)g(y-z)\, dy\right)h(z)\, dz = (f*g)*h(x).$$

When f and g are C_0 functions so is $f*g$ and an integration of (40) gives

$$\int (f*g)(x)\, dx = \int\left(\int f(x-y)\, dx\right)g(y)\, dy = \int f(x)\, dx \int g(y)\, dy$$

for we can use (22), and $\int f(x-y)\, dx = \int f(x)\, dx$ is independent of y. The same equality, i.e.,

$$\int (f*g)(x)\, dx = \left(\int f(x)\, dx\right)\left(\int g(y)\, dy\right), \tag{41}$$

is also true for C functions which are $\geqslant 0$. This is seen by integrating (40) from $-\infty$ to b, changing the order of integration, and letting $b \to \infty$ afterwards. Both members of (41) can then be infinite. As an application let us find out what (41) means when f and g are C functions given by

$$f(x) = H(x)x^{\alpha-1}e^{-x}, \qquad g(x) = H(x)x^{\beta-1}e^{-x}.$$

Here $\alpha, \beta > 1$ and $H(x) = 1$ when $x \geqslant 0$ and $= 0$ otherwise is the so-called *Heaviside function*. In this case

$$(f*g)(x) = H(x)\int_0^x (x-y)^{\alpha-1}y^{\beta-1}\, dy.$$

The change of variable $y = tx$ where $0 \leqslant t \leqslant 1$ shows that

$$(f*g)(x) = B(\alpha, \beta)H(x)x^{\alpha+\beta-1}e^{-x}$$

where

$$B(\alpha, \beta) = \int_0^1 t^{\alpha-1}(1-t)^{\beta-1}\, dt \tag{42}$$

is called *Euler's Beta function*. Hence the definition (26) combined with the identity (37) proves the following formula by Euler (1731):

$$B(\alpha, \beta) = \Gamma(\alpha)\Gamma(\beta)/\Gamma(\alpha + \beta). \tag{43}$$

We have proved it when $\alpha, \beta > 1$ but a passage to the limit in (41) from C functions $\geqslant 0$ to functions $\geqslant 0$ with discontinuities in a finite number of points shows that we can use it also when $\alpha, \beta > 0$. Hence Euler's formula (43) holds with the same hypothesis. In particular, since $\Gamma(1) = 1$ and, by (30), $B(1/2, 1/2) = \pi$, (43) proves that $\Gamma(1/2) = \sqrt{\pi}$. This permits us to compute the constant c of (37). The change of variables $t^2 = 2x$, $dt = (2x)^{-1/2}\,dx$ gives

$$c = \int e^{-t^2/2}\,dt = 2\int_0^\infty e^{-t^2/2}\,dt = \sqrt{2}\int_0^\infty e^{-x}x^{-1/2}\,dx = \sqrt{2\pi}\,.$$

To see what convolution means for Fourier transforms, replace $f(x)$ and $g(x)$ in (40) and (41) by $e^{-ixt}f(x)$ and $e^{-ixt}g(x)$ and note that $(f*g)(x)$ then gets multiplied by e^{-itx}. Hence, restricting ourselves to C_0 functions, we have

$$\int e^{-ixt}\,(f*g)(x)\,dx = \left(\int e^{-ixt}f(x)\,dx\right)\left(\int e^{-ixt}g(x)\,dx\right).$$

Denoting the Fourier transform as an operator by $\mathcal{F}$, writing

$$\mathcal{F}f(t) = \int e^{-ixt}f(x)\,dx$$

this means that

$$\mathcal{F}(f*g) = \mathcal{F}f\,\mathcal{F}g. \tag{44}$$

In other words, the Fourier transform of a convolution is the product of the Fourier transforms of the factors or, if we want, the Fourier transform $\mathcal{F}$ turns convolution into multiplication. The formula (44), now proved for C_0 functions, extends to more general cases but we stop here.

Harmonic analysis

To arrive at the main properties of the Fourier transform we had to do a lot of work in a short time, but let us now relax and see what our results mean. Interchanging x and t, let us write the Fourier inversion formula as

$$f(t) = (2\pi)^{-1}\int F(x)e^{ixt}\,dx, \qquad F(x) = \int e^{-ixt}f(t)\,dt \tag{45}$$

and (44) as

$$(f*g)(t) = (2\pi)^{-1}\int e^{ixt}F(x)G(x) \tag{46}$$

without bothering about the kind of functions we are dealing with. We shall think of t as time and x as a frequency. The functions $t \to Ae^{ixt}$ and $t \to A\cos xt$ or $t \to A\sin xt$ are called *simple* or *harmonic* oscillations since they represent uniform movement along a circle and projections of such a movement on a line. All have the frequency of $x/2\pi$ periods per unit time. The number A is called the amplitude of the oscillation. Its absolute value measures the maximal deviation from a neutral position, here taken as the origin.

Let us now imagine that the function $f(t)$ measures something that changes with time and call it a *time process*. When f is complex we measure with two parameters, and when f is real by just one. In the latter case, taking the real part of the first formula (45), the simple oscillations $\cos xt$ and $\sin xt$ will then appear under the integral sign with amplitudes depending on x. Since every integral is a limit of Riemann sums, we can think of f as a linear combination of simple oscillations with varying amplitudes. The philosophical formulation of the inversion formula is then: *every* time process is a linear combination of simple oscillations. According to (45) the function F is a frequency resolution of f, i.e., it lists the amplitudes $F(x)$ of the simple oscillations e^{ixt} of which f is composed. If, e.g., F vanishes in an interval, the corresponding simple oscillations do not appear in f, and if F is large in an interval but small outside, f is dominated by the simple oscillations in that interval. Looking at it in this way, the function g of (46) becomes a frequency filter or a resonator. Convolution by g kills all the frequencies in f that do not occur in g. Frequency resolutions and filters are to be found both in nature and in the laboratory. The eye makes a frequency resolution of light, the ear of sound, and a band pass filter lets through electromagnetic waves with desired frequencies and absorbs the others. If f contains all frequencies and g varies, the formula (46) indicates that $f*g$ comes close to every time process h. This is in fact the content of a famous result of Norbert Wiener's from 1930: if f and h are absolutely integrable and $F(x) \neq 0$ for all x, then there are C_0^2 functions g such that $\int |f*g(t) - h(t)| \, dt$ is arbitrarily small.

One of the relations between a time process and its Fourier transform F is the Parseval formula $2\pi \int |f(t)|^2 \, dt = \int |F(x)|^2 \, dx$ gotten from (46) by putting $t = 0$ and $G = \bar{F}$, where $\bar{F}$ denotes the complex conjugate of F. If we normalize so that $\int |f(t)|^2 \, dt = 1$, there is the inequality

$$\left(\int t^2 |f(t)|^2 \, dt \right)\left(\int x^2 |F(x)|^2 \, dx \right) \geq \frac{1}{4}$$

which is not hard to prove. This is the uncertainty relation of quantum mechanics. It means that f is spread out as F is concentrated, i.e., gets small except in a small interval. This is not difficult to understand. When F is small far away, f has a small share of high frequencies and hence is a smooth and spread-out function. To concentrate f or to get it to develop singularities, its share of high frequencies must increase.

The Fourier transform extends to several variables, and since 1940 it also has existed in an abstract version. It applies to distributions which meant a big step forward both technically and conceptually and a vindication of Fourier's formal computations. The Fourier transform is closely connected with the theory of Fourier series dealt with in the next chapter. Both are part of Fourier analysis, also called *harmonic analysis* after the harmonic oscillations. It is a theoretically and practically important part of mathematics, full of interesting results, still growing, and an indispensable tool in technology, classical physics, and quantum mechanics.

8.3 Integration in $\mathbf{R}^n$ and measures

So far we have limited the Riemann integral to functions of one variable. Repeated integrations lead to the Riemann integral in $\mathbf{R}^n$. The integral of the function 1 over an interval in $\mathbf{R}^n$ then turns out to be the n-dimensional volume of the interval, and changes of variable will involve a determinant. Considered as a function from C_0 functions to numbers, the integral is linear and monotone increasing. We shall establish a theorem due to F. Riesz saying that every such function is a Stieltjes integral, i.e., a Riemann integral where the ordinary volume is replaced by a more general measure. Finally, there are a few words about the Lebesgue integral.

The Riemann integral in $\mathbf{R}^n$, change of variables

When f is a continuous function from some n-dimensional interval $I : a_1 \leqslant x_1 \leqslant b_1, \ldots, a_n \leqslant x_n \leqslant b_n$ we can integrate f over I by performing n repeated integrations, for instance

$$\int_I f(x)\, dx = \int_{a_1}^{b_1} \left[\int_{a_2}^{b_2} \left(\int_{a_3}^{b_3} f(x, x_2, x_3)\, dx_3 \right) dx_2 \right] dx_1 \qquad (47)$$

when $n = 3$ and analogously in the general case. The left side is here just a notation for the right side. It is well-motivated since repeated use of (22) proves the right side to be independent of the order in which the integrations are performed. When $f = 1$, the value of the integral is $(b_1 - a_1) \cdots (b_n - a_n)$, which is the n-dimensional volume of I (area when $n = 2$) provided the coordinate system is orthonormalized. Repeated use of (8) and (9) give

$$f \leqslant g \quad \Rightarrow \quad \int_I f(x)\, dx \leqslant \int_I g(x)\, dx \qquad (48)$$

$$\int_I (Af(x) + Bg(x))\, dx = A \int_I f(x)\, dx + B \int_I g(x)\, dx. \qquad (49)$$

In other words, the integral is a linear and monotone increasing function of the integrand f. When f is a C_0 function, i.e., when f is continuous and $f(x) = 0$ when the sum $|x_1| + \cdots + |x_n|$ is large enough, then the integral (47) is independent of I when I is so large that $f = 0$ outside I, and then we shall write it as $\int_{\mathbf{R}^n} f(x)\, dx$ or $\int f(x)\, dx$. The properties (48) and (49) then still hold.

Of course we have to integrate over other regions than just intervals. We want to integrate continuous functions over open subsets V of $\mathbf{R}^n$ and we shall define the corresponding integral in two steps. Let $C_0(V)$ be the set of C_0 functions vanishing outside compact subsets of V and put, by definition,

$$\int_V f(x)\, dx = \sup \int g(x)\, dx \quad \text{when} \quad g \in C_0(V) \quad \text{and} \quad 0 \leqslant g \leqslant f, \quad (50)$$

when $f \geqslant 0$ is continuous. When f is a $C_0(V)$ function, the supremum is attained when $g = f$ and we get nothing new. In the general case the right side is allowed to be plus infinity.

In the second step we shall integrate a function which is allowed to change its sign, but it must be absolutely integrable, i.e., $\int_V |f(x)|\, dx < \infty$. We then put

$$\int_V f(x)\, dx = \int_V f_+(x)\, dx - \int_V f_-(x)\, dx \quad (51)$$

where $f_+(x) = \max(f(x), 0)$ is the positive part of $f(x)$ and $f_-(x) = \max(-f(x), 0)$ the negative part with its sign reversed. Since $|f| = f_+ + f_-$ both terms of the right side are finite and since $f = f_+ - f_-$, (51) is true for $C_0(V)$ functions. It is not difficult to show that the integral (51) has the properties (48) and (49).

With our choice of notation, the formula for changing variables in (47) looks like the corresponding formula (33) for one variable except for an additional sign "det" for a determinant,

$$\int_{h(V)} f(x)\, dx = \int_V f \circ h(y) |\det h'(y)|\, dy \quad (52)$$

where h is a C^1 bijection from V to $h(V)$ with the Jacobian matrix $h'(y) = (\partial_k h_j(y))$. We can only sketch the rather laborious proof. If h changes only one variable, i.e., if $h_k(y) = y_k$ except for one k, e.g., $k = n$, and $V = \mathbf{R}^n$, we get the desired result by writing the right side as

$$\ldots \left(\int f \circ h(y) |\partial_n h_n(y)|\, dy_n \right) \ldots$$

and changing to the variable $x_n = h_n(y)$ in the inner integral. The formula (52) is also true when h just permutes the variables, and by repeated changes of variables of these two kinds, the general formula follows at least when $f \circ h$ vanishes outside a small part of V but arbitrarily located. Hence, letting the support of a function be the complement of the largest open region where it vanishes identically, we now know that (52) holds for functions with small supports. To proceed further we shall use *partitions of unity* defined as sequences $g_1, g_2, \ldots$ of $C_0(V)$ functions with the sum 1 such that every compact part of V meets at most a finite number of the supports of these functions. The point of this notion is that it can be combined with open coverings of V defined as families (W) of open sets W which together cover V. We say that a partition of unity is *subordinated* to such a covering if, to every g_k there is a W_k in the family such that g_k belongs to $C_0(W_k)$, i.e., g_k vanishes outside a compact part of W_k. See Figure 8.5.

189

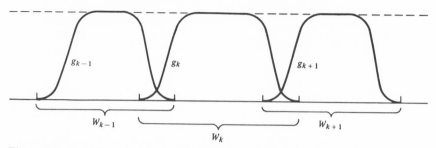

Figure 8.5   Partition of unity on **R**. The sum $g_1 + g_2 + \ldots$ equals 1 everywhere. Every g_k vanishes outside a compact part of a corresponding interval W_k.

We have

Theorem (Partitions of unity). *Every open covering of an open set in* $\mathbf{R}^n$ *has a subordinated partition of unity.*

The proof is not difficult but is so long that we cannot give it. To show how the theorem is used, suppose that we have proved that every point of V has an open neighborhood W such that (52) holds when $f \circ h$ is a $C_0(W)$ function. We then have an open covering of V. Let $g_1, g_2, \ldots$ be a subordinated partition of unity, let $f \in C_0(V)$ and consider (52). Replacing f by $f_k(x) = g_k(h^{-1}(x))f(x)$, then $f_k \circ h(y) = g_k(y)f \circ h(y)$ has so small support that (52) holds for f_k. Adding the corresponding formulas and noting that $f(x) = f_1(x) + f_2(x) + \cdots$ where at most a finite number of terms are different from zero proves (52) for f.

The formula (52) invites the use of differentials, for by the formula (32) of Chapter 7 we have

$$dx_1 \wedge \cdots \wedge dx_n = \det h'(y)\, dy_1 \wedge \cdots \wedge dy_n.$$

Taking symbolic absolute values and putting

$$dx = dx_1 \ldots dx_n = |dx_1 \wedge \cdots \wedge dx_n| \tag{53}$$

we get $dx = |\det h'(y)|\, dy$. This reduces (52) to a formal identity.

Let us now apply (52) when y denotes polar coordinates in two and three dimensions. The corresponding Jacobian determinants are given by the formulas (35) and (36) of Chapter 7. First let $y = r, \theta$ be polar coordinates in the plane, $x_1 = r \cos \theta$, $x_2 = r \sin \theta$ and V the interval $\theta_1 < \theta < \theta_2$. The region $U = h(V)$ is then a sector of the circular annulus $r_1^2 < x_1^2 + x_2^2 < r_2^2$ and we get

$$\int_U f(x_1, x_2)\, dx_1\, dx_2 = \int_V f(r \cos \theta, r \sin \theta)\, r\, dr\, d\theta.$$

When $f = 1$, the right side equals $(\theta_2 - \theta_1)(r_2^2 - r_1^2)/2$ which is the area of the sector. Polar coordinates in space according to (36) of Chapter 7 give a corresponding formula

$$\int_U f(x_1, x_i, x_3)\, dx_1\, dx_2\, dx_3$$

$$= \int_V f(r \cos \varphi \cos \theta, r \cos \theta \sin \varphi, r \sin \theta) r^2 \cos \theta\, dr\, d\theta\, d\varphi.$$

If V is the interval $r_1 < r < r_2$, $\theta_1 < \theta < \theta_2$, $\varphi_1 < \varphi < \varphi_2$ and $f = 1$, the right side is

$$3^{-1}(r_2^3 - r_1^3)(\sin \theta_2 - \sin \theta_1)(\varphi_2 - \varphi_1)$$

which is the volume of the corresponding piece of the spherical shell $r_1^2 < x_1^2 + x_2^2 + x_3^2 < r_2^2$. Finally, we shall compute the n-dimensional volume of the unit ball $E : x_1^2 + \cdots + x_n^2 < 1$ in $\mathbf{R}^n$, i.e., the integral

$$\int_E dx_1 \ldots dx_n = 2^n \int_V dx_1 \ldots dx_n$$

where V is the part of E where $x_1 > 0, \ldots, x_n > 0$. Changing variables to $t_k = x_k^2$ gives $2dx_k = t_k^{-1/2}$. Hence the right side equals

$$\int_U (t_1 \ldots t_n)^{-1/2}\, dt_1 \ldots dt_n$$

where U is the region $t_1 > 0, \ldots, t_n > 0$, $t_0 = 1 - t_1 + \cdots + t_n > 0$. This is a special case of the integral

$$\int_U t_0^{\alpha_0 - 1} t_1^{\alpha_1 - 1} \ldots t_n^{\alpha_n - 1}\, dt_1 \ldots dt_n = \Gamma(\alpha_0 + \cdots + \alpha_n)/\Gamma(\alpha_0) \ldots \Gamma(\alpha_n)$$

$$(54)$$

with $\alpha_0 > 0, \ldots, \alpha_n > 0$ generalizing Euler's formula (43). Hence the volume is

$$\Gamma\left(\frac{1}{2}\right)^n / \Gamma\left(\frac{n}{2} + 1\right).$$

To prove (54) by induction from (43), i.e., the case $n = 1$, change variables so that $t_1 = s(1 - t_2 - \cdots - t_n)$ and integrate with respect to s_1, using (43). The details are left to the reader.

The Stieltjes integral and measures

We have already remarked that the Riemann integral (47) considered as a map $f \to L(f)$ from C_0 functions to numbers is monotone increasing and linear. In other words,

$$f \geq 0 \quad \Rightarrow \quad L(f) \geq 0, \qquad L(Af + Bg) = AL(f) + BL(g) \quad (55)$$

for all numbers A, B and C_0 functions f and g. We shall get a good idea of what integration is by asking ourselves if anything interesting can be said about an arbitrary map L with these very general properties. The answer is yes provided we turn to the formula (50), now using it to let L define a measure. More specifically, let L be given with the properties (55) and put

$$m(E) = \sup L(f) \quad \text{when } 0 \leq f \leq \chi_E \text{ and } f \text{ is a } C_0 \text{ function.} \quad (56)$$

191

Here, E is any open set in R^n and χ_E is its *characteristic function* defined to be 1 on E and 0 outside E. This defines a function $E \to m(E)$ from open sets to numbers ≥ 0 including $+\infty$, called a measure, obviously with the following three properties (the arrow $\uparrow$ means "increases to the limit")

$$m(E) \geq 0, \quad E \text{ bounded} \quad \Rightarrow \quad m(E) < \infty \tag{57a}$$

$$E' \uparrow E \quad \Rightarrow \quad m(E') \uparrow m(E) \tag{57b}$$

$$E \cap E' = \phi \quad \Rightarrow \quad m(E \cup E') = m(E) + m(E'). \tag{57c}$$

For an arbitrary set F let us put

$$m(F) = \inf m(E) \quad \text{when } E \text{ is open and contains } F. \tag{58}$$

Suppose now, conversely, that we are given a measure m with the properties (57). We are going to define an integral of C_0 functions with respect to m. We then divide $\mathbf{R}^n$ into families (I) of disjoint bounded intervals $I_1, I_2, \ldots$ such that at most a finite number of them meets any given compact set and consider the corresponding Riemann sums

$$\sum f(\xi_k) m(I_k), \quad \xi_k \text{ belongs to } I_k \text{ for all } K. \tag{59}$$

Since f has compact support only a finite number of terms in them are different from zero. Precisely as for the Riemann integral one shows that these sums have a limit as the maximal size of the intervals in a family tends to zero. This limit, denoted by

$$\int f(x)\, dm(x)$$

is called the *Stieltjes integral* of f with respect to m. It is obvious that it has the properties (55), and we may then ask: if the measure is defined by (56), is this integral equal to the map L we started with? The answer is yes. The following result is essentially due to F. Riesz.

Theorem. *Every linear map L with the property* (55) *is the Stieltjes integral with respect to the measure* (56) *generated by L.*

The proof is not difficult, at least not when $n = 1$. Consider strictly increasing sequences of numbers x_k labelled by all integers and divide $\mathbf{R}$ into intervals I_k between x_k and x_{k+1}, the point x_k belonging to I_{k-1} or I_k. Then construct partitions of unity (g_k) such that g_k vanishes immediately outside I_k and at the endpoints not belonging to I_k while $g_k = 1$ at the other endpoints. The reader is asked to draw a figure of this similar to Figure 8.5. It follows from (56) and (58) that such partitions exist and that we can choose them so that $L(g_k)$ comes arbitrarily close to $m(I_k)$ as defined by (56) and (58) and this for all k. Hence the difference between the Riemann sum (59) and $\sum f(\xi_k) L(g_k)$ can be made as small as we please. On the other hand, since $L(f) = \sum L(fg_k)$,

$$\left| L(f) - \sum f(\xi_k) L(g_k) \right| \leq \sum \left| L(fg_k) - f(\xi_k) L(g_k) \right| \leq \sum' \delta_k L(g_k)$$

where δ_k is the difference between the maximum and minimum of f on the support of g_k, the last sum running over such k that the support of g_k intersects the support of f. Making a fine division and keeping the support of g_k sufficiently close to I_k, the maximal δ_k tends to zero while $\Sigma' L(g_k)$ remains bounded. This proves the theorem when $n = 1$ and the proof when $n > 1$ is analogous.

After this rather abstract piece of mathematics it is high time that we produce examples of measures and Stieltjes integrals. Let $g(x) \geq 0$ be continuous, let $(x^{(k)})$ be a sequence of points in $\mathbf{R}^n$ and (a_k) a sequence of numbers ≥ 0 such that all sums Σa_k with $x^{(k)}$ belonging to a compact set of $\mathbf{R}^n$ are finite. Then

$$L(f) = \int f(x) g(x) \, dx + \sum a_k f(x^{(k)})$$

is a Stieltjes integral. Clearly, the corresponding measure is

$$m(E) = \int_E g(x) \, dx + \sum' a_k$$

where the sum runs over all k such that $x^{(k)}$ lies in E. Without being completely general, this example is very representative. When $g = 0$ we say that the measure is *discrete*, and when all $a_k = 0$ that it has the *density* g. When $n = 1$, every measure m is represented by a nonincreasing function $h(x)$ defined by $h(x) = m(I_x)$ when $x \geq 0$ and $h(x) = - m(J_x)$ when $x < 0$ where I_x is the interval $0 < t \leq x$ and J_x the interval $x < t \leq 0$. The Stieltjes integral $\int f(x) \, dm(x)$ then has the Riemann sums

$$\sum f(\xi_j) \big(h(x_{j+1} - h(x_j)) \big)$$

and is also written as $\int f(x) \, dh(x)$, the form used by Stieltjes (1894). Riesz's theorem is from 1909. He was concerned with the linear space C of all continuous real functions f from an interval $a \leq x \leq b$ and linear maps $f \to L(f)$ to the real numbers such that $|L(f)| \leq c \max |f(x)|$ for all f and some c independent of f. He proved that every such L is the difference between two Stieltjes integrals

$$L(f) = \int_a^b f(x) \, dh_1(x) - \int_a^b f(x) \, dh_2(x) \tag{60}$$

where h_1 and h_2 do not decrease and h_2 can be taken $= 0$ when $L(f) \geq 0$ for $f \geq 0$. Let us note that in the latter case, the condition $|L(f)| \leq c|f|$ with $|f| = \max |f(x)|$ is satisfied with $c = L(1)$, for $|f| \geq f(x) \geq -|f|$ gives $|f| L(1) \geq L(f) \geq -|f| L(1)$. In modern terms Riesz's theorem says that every continuous linear functional on the Banach space C has the form (60) and conversely. Expressed in this way it is one of the earliest and also most basic results of functional analysis.

193

The Lebesgue integral and other integrals

So far we have, for simplicity mostly integrated continuous functions, but it is possible to go much further. Performing monotone passages to the limit in the Stieltjes integral we get the Lebesgue-Stieltjes integral or, if we start with the Riemann integral, the Lebesgue integral. More precisely, starting from a map $L(f)$ from C_0 functions with the properties (55) we extend the integral L by putting $L(f) = \lim L(f_k)$ when f is the pointwise limit of an increasing (decreasing) sequence (f_k) of C_0 functions. Such f's are said to be *lower (upper) semicontinuous*. An arbitrary function f is said to be integrable if it lies between two such functions (one of each kind, the larger being lower semicontinuous) whose integrals differ as little as we wish. The integral $L(f)$ is then naturally defined. The space of integrable functions turns out to be linear and L still has the properties (55) considered as a map from this space. The advantage of this kind of an integral, is another formulation due to Lebesgue (1900), is that the theorem on dominated convergence holds without locally uniform convergence. It suffices to have pointwise convergence.

There are also other aspects of integration. We may for instance start from the so-called fundamental theorem of calculus expressed as

$$f(b) - f(a) = \int_a^b f'(x)\, dx. \tag{61}$$

When $a = x_0 < x_1 < \cdots < x_{n+1} = b$ is a subdivision and f' exists, then by the mean value theorem,

$$f(b) - f(a) = \sum_0^n \left(f(x_{j+1}) - f(x_j) \right) = \sum_0^n f'(\xi_j)(x_{j+1} - x_j)$$

is a Riemann sum for the right side of (61) and hence this equality holds when f' is Riemann integrable. But the derivative f' may be a rather complicated function even if it is bounded and it need not be Riemann integrable. The right side of (61) is then meaningless. On the other hand, in Lebesgue's theory a bounded derivative is integrable and (61) holds. But even then not all is well. If, for instance, f is monotone and continuous, f' exists except on a very small set, a so-called *null set*, and is integrable according to Lebesgue, but the left side of (61) may be larger than the right side (when f is increasing). There are integrals where (61) holds or is taken as a definition, but the Lebesgue integral is the standard integral of analysis and will certainly keep this position in the future.

8.4 Integration on manifolds

Among the problems that led to infinitesimal calculus was the computation of the lengths of curved arcs and the areas of curved surfaces. Both were reduced to integrations. The formula for the length of the arc $y = f(x)$

between $x = a$ and $x = b$ turned out to be

$$\int_a^b ds \qquad (62)$$

where $ds = (dx^2 + dy^2)^{1/2} = (1 + f'(x)^2)^{1/2} \, dx$ is the so-called arc element. It is motivated by the idea that small parts of the arc are so close to straight lines that we can apply the theorem of Pythagoras to them. Leibniz's ingenious notation makes for a perfect fit between the formula and our intuition. Posterity has contributed nothing except for defining the length of the arc as the supremum of the lengths of piecewise linear arcs approaching the given one.

We shall extend the formula (62) to higher dimensions and to manifolds but only after preparing ourselves by integrating densities on manifolds and making a remark on Riemannian geometry. After that follows integration of differential forms on oriented manifolds and, as an application, proofs of the formulas of Green and Stokes. In spite of their formidable appearance they are just expanded versions of the fundamental theorem of calculus, the fact that

$$\int_a^b f'(x) \, dx = f(b) - f(a)$$

for C^1 functions. In order to follow the exposition it is necessary to have gone through what was said about differential calculus on manifolds in section 7 of Chapter 7.

Integration of densities

In connection with the Stieltjes integral we have defined the integral of a continuous density $g(x) \geqslant 0$ in $\mathbf{R}^n$ as the map $f \to L(fg) = \int f(x) g(x) \, dx$ from C_0 functions. We shall now have to do with integrals of continuous densities also on a p-manifold M. That f is a C_0 function on M means that f is continuous and has compact support, i.e., vanishes outside a compact part of M. We shall consider linear maps $f \to L(f)$ from C_0 functions such that, for every chart h, V of M,

$$L(f) = \int f_V(t) H_V(t) \, dt, \qquad dt = dt_1 \ldots dt_p,$$

when f vanishes outside a compact part of the charted region $h(V)$. Here $f_V(t) = f(x)$ when $x = h(t)$ and $H_V(t) \geqslant 0$ is a continuous function. When f has its support in the intersection $h(V) \cap h'(V')$ of two charted regions we get

$$\int f_V(t) H_V(t) \, dt = \int f_{V'}(t') H'_{V'}(t') \, dt'$$

where $f_V(t) = f_{V'}(t') = f(x)$ when $x = h(t) = h'(t')$. To make this equality fit the formulas (52) and (53) for changing variables in an integral, the functions H_V and $H'_{V'}$ must be connected by the formula

$$h(t) = h'(t') \quad \Rightarrow \quad H_V(t) |\, dt_1 \wedge \cdots \wedge dt_p\,| = H'_{V'}(t') |\, dt'_1 \wedge \cdots \wedge dt'_p\,|$$

(63)

for all charts h, V and h', V'. Hence a density on a manifold is given by a collection (H_V) of continuous functions $\geqslant 0$, one for every chart, such that (63) holds. In particular, if $\omega(x)$ is a p-form on M, we can put $H_V(t) = |g_V(t)|$ where $\omega(x) = g_V(t)\, dt_1 \wedge \cdots \wedge dt_p$ expresses the form in terms of the variables of the chart h, V.

It is easy to construct densities for they can be added, and an arbitrary $C_0(h(V))$ function H_V where h, V is a given chart gives us a density by way of (63) if we put $H'_{V'}(t') = 0$ when $h(V) \cap h'(V')$ is empty or when $h'(t') = h(t)$ and $H_V(t) = 0$. The same method, incidentally, was used to construct differential forms. In order to construct an integral $L(f)$ with a given density (H_V) we shall use partitions of unity, i.e., countable collections (g_k) of C_0 functions $\geqslant 0$ with the sum 1 such that every one of their supports is contained in a charted region and at most finitely many of them meet a given compact part of M. Our earlier theorem on partitions of unity in $\mathbf{R}^n$ holds also on manifolds, but we have to forego the proof. In particular, every atlas has a subordinated partition of unity. The integral of a given density is now defined by

$$L(f) = \sum L(g_k f)$$

where (g_k) is a partition of unity, the support of g_k being contained in some charted region $h_k(V_k)$, and

$$L(g_k f) = \int (g_k f)_{V_k}(t) H_{V_k}(t)\, dt.$$

Since f is supposed to have compact support, at most a finite number of terms of the sum are not zero. If (g'_j) is another partition of unity and $L'(f) = \sum L'(g'_j f)$ the corresponding integral, (63) proves that $L(g_k g'_j f) = L'(g_k g'_j f)$ for all j and k so that, by a summation on j, $L(g_k f) = L'(g_k f)$ for $\sum g'_j = 1$ and, by a summation on k, $L(f) = L'(f)$ for $\sum g_k = 1$. Hence the integral does not depend on the partition of unity that we use. Next there follow some classical examples of densities.

Arc element and area element

Let $x = (x_1, \ldots, x_n)$ be orthonormal coordinates in $\mathbf{R}^n$ so that the square of the distance between the points x and y is $|x - y|^2 = (x_1 - y_1)^2 + \cdots + (x_n - y_n)^2$. The arc element of a C^1 curve γ in $\mathbf{R}^n$ is a density defined by

$$ds = \left(dx_1{}^2 + \cdots + dx_n{}^2\right)^{1/2} = |x'(t)|\, dt = \left(x'_1(t)^2 + \cdots + x'_n(t)^2\right)^{1/2} dt.$$

(64)

Here the first equality is symbolic; the density $|x'(t)|$ relative to a chart

$t \to x(t)$ appears in the last member. Note that if $s \to y(s) = x(t)$ is another chart, then $|x'(t)| = |y'(s)| \, |ds/dt|$ so that $|x'(t)|$ is actually a density. To see that $\int_I ds$ is the length of the arc I of γ, note that if I is the straight line from y to z given by $x(t) = y + t(z - y), 0 \leqslant t \leqslant 1$, then $ds = |z - y| \, dt$ so that the integral equals $|z - y|$.

As an example, take the ellipse $(x/a)^2 + (y/b)^2 = 1$ in the x, y-plane. In terms of the chart $x = a(1 - t^2), y = bt, -1 < t < 1$, its arc element is the expression under the integral sign of

$$F(u) = \int_0^u \left((a^2 - b^2)t^2 + b^2\right)^{1/2}(1 - t^2)^{-1/2} \, dt$$

and the integral itself is the length of the arc where $0 < t < u$. Using an angular coordinate, $x = a \cos \theta, y = b \sin \theta$ this can be rewritten as

$$F(u) = \int_0^{\arcsin u} (a^2 \sin^2 \theta + b^2 \cos^2 \theta)^{1/2} \, d\theta.$$

When the ellipse is a circle, i.e., if $a = b$, we can of course compute the integral explicitly, getting $F(u) = a \arcsin u$. But for a proper ellipse, the arc length $F(u)$ is not expressible in terms of the traditional elementary functions, and unlike its area πab, the total length of the ellipse is not a simple expression in a, b, and π. Through work by Euler, Jacobi, and Abel, the study of functions like F has led to an entire branch of mathematics, the theory of elliptic functions.

The area element of a C^1 surface Γ in R^3 with orthonormal coordinates x, y, z turns out to be

$$dS = \left((dy \wedge dz)^2 + (dz \wedge dx)^2 + (dx \wedge dy)^2\right)^{1/2}. \tag{65}$$

Substituting a chart $x = x(u, v), y = y(u, v), z = z(u, v)$ gives

$$dS = |J| \, du \, dv, \qquad |J| = \left(J_x^2 + J_y^2 + J_z^2\right)^{1/2}.$$

Here J_x, J_y, J_z are Jacobians defined by $dy \wedge dz = J_x \, du \wedge dv$, etc. The vector $J = (J_x, J_y, J_z)$ is a normal of the surface (see the section on tangents and normals in the last part of Chapter 7). The properties of Jacobians shows that $|J|$ is a density. When Γ is a parallelogram parametrized by $(x, y, z) = (x_0 + ux_1 + vx_2, y_0 + uy_1 + vy_2, z_0 + uz_1 + vz_2), 0 \leqslant u, v \leqslant 1$, and hence with corners corresponding to $(u, v) = (0, 0), (1, 0), (0, 1), (1, 1)$, then $|J_x| = |y_2 z_3 - y_3 z_2|$ is the area of the projection of Γ on the y, z-plane and analogously for the other Jacobians. Hence, by elementary geometry, $|J| = \int_\Gamma dS$ is then the area of Γ. This is ample motivation to consider $\int_\Gamma dS$ the area of Γ also in the general case.

Let us compute dS when Γ is the unit sphere $x^2 + y^2 + z^2 = 1$. Choosing a chart with coordinates z and θ where

$$x = r \cos \theta, \qquad y = r \sin \theta, \qquad r = (1 - z^2)^{1/2}$$

and $|z| < 1$, $0 < \theta < 2\pi$, we get

$$dy \wedge dz = r \cos \theta \, d\theta \wedge dz, \qquad dz \wedge dx = r \sin \theta \, d\theta \wedge dz,$$
$$dx \wedge dy = z \, d\theta \wedge dz,$$

so that $dS = dz \, d\theta$. This is equivalent to Archimedes' result that the area of a part of the sphere between two parallel planes with the distance h is $2\pi h$. The reader is now asked to prove that $(1 + x^2 + y^2)^{1/2} \, dx \, dy$ is the area element of the paraboloid $2z = x^2 + y^2$ when x and y are chosen as coordinates, and to compute explicitly the area of the part of the paraboloid where $x^2 + y^2 < R^2$.

Let us remark, finally, that a C^1 manifold M in $\mathbf{R}^n$ of dimension $n - 1$ has an element of $(n - 1)$-dimensional area given by the following formula, analogous to (65)

$$dS = \left(\sigma_1(x)^2 + \cdots + \sigma_n(x)^2\right)^{1/2}. \tag{66}$$

Here $\sigma_1, \ldots, \sigma_n$ are those $(n - 1)$-forms given by

$$\sigma_k(x) = (-1)^k \, dx_1 \wedge \cdots \wedge dx_{k-1} \wedge dx_{k+1} \wedge \cdots \wedge dx_n$$

which we met in connection with the formula (44) of Chapter 7.

Riemannian geometry

Let $M \subset \mathbf{R}^n$ be a C^1 manifold of dimension p and let h, V be a chart. Substituting $dx_i = \sum_1^p \partial_j h_i(t) \, dt_j$ into the arc element $(dx_1^2 + \cdots + dx_n^2)^{1/2}$ of $\mathbf{R}^n$, writing out the squares, and summing and writing the result as a linear combination of the formal products $dt_j \, dt_k = dt_k \, dt_j$, we get a corresponding arc element on the manifold M, namely

$$ds = \left(\sum g_{jk}(t) \, dt_j \, dt_k\right)^{1/2} \tag{67}$$

where $(g_{jk}(t))$ is a symmetric positive definite matrix. A curve $u \to h(t(u))$ on M, lifted from a curve $u \to t(u)$ in V, then has an arc element obtained from ds by putting $dt_j = t_j'(u) \, du$. Riemann's idea in his famous paper from 1854, "On the Basic Hypotheses of Geometry," was to measure distances on manifolds by metrics or arc elements (67) given a priori. How things look inside the parenthesis on the right then depends on the chart used, but it is required that

$$\sum g_{jk}(t) \, dt_j \, dt_k = \sum g_{jk}'(t') \, dt_j' \, dt_k'$$

for corresponding points in overlapping charts. An easy argument using a partition of unity shows that every manifold has Riemannian metrics. It is easy to see that, e.g., $(\det g_{jk}(t))^{1/2} \, dt_1 \ldots dt_p$ is a density for any choice of the metric. One can also define entities measuring the curvature of the manifold M without regard to any embedding into some $\mathbf{R}^n$. These formulas work even when the metric is not definite, i.e., (g_{jk}) is just a symmetric matrix. This is the mathematical starting point of the theory of general relativity.

Orientation and integration of differential forms

By definition the integral

$$\int_a^b f(x)\, dx$$

changes its sign when a and b change places. So far we have supposed that $a_1 < b_1, \ldots, a_n < b_n$ in the formula (47) for integration over an interval and hence have not used this mechanism. Our densities have been supposed to be $\geqslant 0$ and have led to integrals $L(f)$ which are $\geqslant 0$ when $f \geqslant 0$. In the formula (53) this is marked by surrounding the vector product $dx_1 \wedge \cdots \wedge dx_n$ by the signs for absolute value. We shall now remove these signs and consider integrals

$$L(f) = \int_I f(x)\, dx_1 \wedge \cdots \wedge dx_n$$

over intervals $I \subset \mathbf{R}^n$. We must then have a convention fixing the sign of the integral. One way is to orient the interval by one or the other of the rules

$$dx_1 \wedge \cdots \wedge dx_n > 0 \quad \text{or} \quad < 0 \qquad \text{on } I. \tag{68}$$

This is to be interpreted so that, respectively, $L(f) > 0$ and $L(f) < 0$ for positive f. Changing in (68) the order of the factors or multiplying by a function $\neq 0$ shall be accompanied by a corresponding change of the sign of inequality. Example: the orientations $dx_1 \wedge dx_2 > 0$, $dx_2 \wedge dx_1 < 0$, and $-(1 + x_1^2)\, dx_1 \wedge dx_2 < 0$ are all the same.

A p-manifold M is oriented with the aid of p-forms τ on M which are nowhere equal to zero, i.e., if h, V is a chart and $\tau(x) = g(t)\, dt_1 \ldots dt_p$ then $g(t) \neq 0$ for all t. The rule $\tau(x) > 0$ then orients every chart and at the same time the whole manifold. When the manifold M has been oriented we can start integrating p-forms ω over it. If the support of ω is contained in a charted region $h(V)$ and $\omega = f(t)\, dt_1 \wedge \cdots \wedge dt_p$, we put

$$\int_M \omega(x) = \int_V f(t)\, dt_1 \wedge \cdots \wedge dt_p$$

where V is oriented by $\tau(x) > 0$, so that, with notations as above, $g(t)\, dt_1 \wedge \cdots > 0$. For p-forms with compact supports we use a partition of unity, reasoning as for densities. Then our integral gives us a linear functions $\omega \to \int_M \omega(x)$ from the space of p-forms on M with compact supports.

In one, two, and three dimensions, orientation is a very intuitive notion that can be expressed in many ways. We give some examples.

A C^1 curve γ is oriented by a rule $dg > 0$ where g is a real C^1 function from the curve such that $dg \neq 0$ everywhere. This may be illustrated by drawing arrows on the curve in the direction of growth of g. See Figure 8.6. Let $\omega(x) = f_1(x)\, dx_1 + \cdots + f_n(x)\, dx_n$ be a continuous 1-form on γ with compact support and suppose that γ has been oriented. Then the line

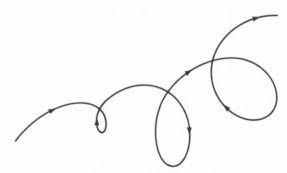

Figure 8.6   Orientation of a curve.

integral $\int_{\gamma}\omega$ is defined. Letting ds be the arc element and choosing coordinates $t \to x(t)$ with $dt > 0$, the equality $eds = dx = x'(t)\, dt$ defines a continuous tangent vector e to the curve of length 1, and we can also write the integral as an integral of a density,

$$\int_{\partial} f_1\, dx_1 + \cdots + f_n\, dx_n = \int_{\gamma} (f_1 e_1 + \cdots + f_n e_n)\, ds. \tag{69}$$

In $\mathbf{R}^2$ we can express a line integral as the integral of a density, using as a continuous normal vector $\nu = (\nu_1, \nu_2)$ of length 1. The formula is

$$\int_{\gamma} (P\, dy - Q\, dx) = \int_{\partial} (P\nu_1 + Q\nu_2)\, ds. \tag{70}$$

Since $e_1 = \nu_2, e_2 = -\nu_1$ defines a tangent vector of γ, it follows from (69) that (70) holds apart from the sign. Orienting γ by $\nu_1\, dy - \nu_2\, dx > 0$, the sign is correct for then both sides are $\geqslant 0$ when $P = h\nu_1$, $Q = h\nu_2$ and $h \geqslant 0$.

Integrals of 2-forms over a surface Γ in $\mathbf{R}^3$ can also be written as integrals of densities. The formula is

$$\int_{\Gamma} (P\, dy \wedge dz + Q\, dz \wedge dx + R\, dx \wedge dy) = \int_{\Gamma} (P\nu_1 + Q\nu_2 + R\nu_3)\, dS, \tag{71}$$

analogous to (70). Here dS is the area element, $\nu = (\nu_1, \nu_2, \nu_3)$ is a continuous unit normal of Γ, and Γ is oriented so that $\nu_1\, dy \wedge dz + \cdots \geqslant 0$. By a partition of unity it suffices to prove this when P, Q, R vanish outside a charted region $h(V)$. But then, by (65), both sides are equal to

$$\int_{V} (PJ_x + QJ_y + RJ_z)\, du\, dv$$

apart from the sign. Here $x = x(u, v)$ is a chart, $dy \wedge dz = J_x\, du\, dv$, etc., so that $J = (J_x, J_y, J_z)$ is a normal to Γ. That the sign of (71) is correct follows if we choose (P, Q, R) a nonnegative multiple of ν.

Finally we shall write down the analogue of (71) for a hypersurface M in $\mathbf{R}^n$. Putting $(f, \sigma) = f_1\sigma_1 + \cdots + f_n\sigma_n$ and $(f, \nu) = f_1\nu_1 + \cdots + f_n\nu_n$ it is

$$\int_{M} (f, \sigma) = \int_{M} (f, \nu)\, dS. \tag{72}$$

Here the components of $f = (f_1, \ldots, f_n)$ are C_0 functions, $\nu = (\nu_1, \ldots, \nu_n)$ is a continuous unit normal, dS is given by (66), and the components of $\sigma = (\sigma_1, \ldots, \sigma_n)$ are $(n-1)$-forms defined by the formula following (66). The orientation of M is $\nu_1\sigma_1 + \cdots + \nu_n\sigma_n > 0$. The proof runs exactly as in the case $n = 3$.

The formulas of Green and Stokes

In traditional notation, Green's formula (1827) (also called Gauss's formula), is written as

$$\int_V \operatorname{div} f \, dV = \int_S (f, \nu) \, dS. \tag{73}$$

Here V is an open bounded part of $\mathbf{R}^n$ whose boundary S is a C^1 manifold of dimension $n-1$, and dV and dS are the corresponding volume and area elements. The vector $\nu = (\nu_1, \ldots, \nu_n)$ is the exterior unit normal of S, the components of the vector $f = (f_1, \ldots, f_n)$ are C^1 functions, (f, ν) is the scalar product $f_1\nu_1 + \cdots + f_n\nu_n$, and $\operatorname{div} f$ is the function $\partial_1 f_1 + \cdots + \partial_n f_n$. The formula is used in all dimensions. See Figure 8.7.

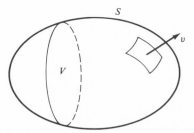

Figure 8.7 An illustration of Green's formula.

If we think of f as a flow of some kind of matter so that (f, ν) is the amount per unit time flowing through a piece of S of unit $(n-1)$-dimensional area in its normal direction ν, then the right side of (73) is the flow of matter per unit time out of V through S, and the left side the matter per unit time added to V by some kind of source. The density of the source per unit volume is then $\operatorname{div} f$. When $\operatorname{div} f = 0$ the flow is said to be source-free.

To prove (73), write both sides as integrals of differential forms. According to (72), the right side is the integral of the differential form

$$\omega(x) = f_1(x)\sigma_1(x) + \cdots + f_n(x)\sigma_n(x)$$

over S. Now a short computation shows that $d\omega(x) = \operatorname{div} f(x)\tau(x)$ where $\tau(x) = dx_1 \wedge \cdots \wedge dx_n$, and hence we can write Green's formula as

$$\int_V d\omega(x) = \int_S \omega(x) \tag{74}$$

where V and S are oriented by $\tau(x) > 0$ and $(\nu, \sigma(x)) > 0$ respectively.

201

Next make an atlas over an open set containing the compact region $V \cup S$ using charts $h(t)$, W of two kinds, one kind with $h(W) \subset V$ and the other kind such that $t_1 \leqslant 0$ on $h(W) \cap V$ and $t_1 = 0$ on S. Here W denotes open parts of $\mathbf{R}^n$. Using a partition of unity it is immediately clear that it suffices to prove (74) when ω vanishes outside such a charted region. Now, if $\omega = g_1 \sigma_1(t) + \cdots + g_n \sigma_n(t)$, then $d\omega = (\text{div } g)\tau(t)$ for, by Leibniz's lemma, d commutes with substitutions. Hence the whole thing reduces to proving that

$$\int \left(\partial_1 g_1(t) + \cdots + \partial_n g_n(t) \right) dt_1 \ldots dt_n = 0 \tag{75}$$

$$\int_{t_1 < 0} \left(\partial_1 g_1(t) + \cdots + \partial_n g_n(t) \right) dt_1 \ldots dt_n = \int_{t_1 = 0} g_1(t) \, dt_2 \ldots dt_n \tag{76}$$

for C^1 functions with compact supports in W, and in the part of W where $t_1 \leqslant 0$, respectively. Integrating $\partial_k g_k(t)$ with respect to t_k first, both formulas follow from the fact that $\int_a^b f'(x) \, dx = f(b) - f(a)$ for C^1 functions of one variable. This completes the proof apart from one detail, the sign in (76). But the sign is correct since both sides are $\geqslant 0$ when g_1 is $\geqslant 0$ and increases with t_1.

Green's formula in the plane is usually written as

$$\int_\Omega (Q_x - P_y) \, dx \, dy = \int_\gamma P \, dx + Q \, dy \tag{77}$$

where Ω is a bounded open region with a smooth boundary γ oriented by $e_1 \, dx + e_2 \, dy > 0$, where (e_1, e_2) is a tangent to γ chosen so that $(e_2, -e_1)$ is an exterior normal. Since $d(P \, dx + Q \, dy) = (Q_x - P_y) \, dx \wedge dy$ this formula follows from (74).

Stokes's formula (1840) is a generalization of (77) and has to do with integration of a flow $f(x) = (f_1(x), f_2(x), f_3(x))$ over a surface in $\mathbf{R}^3$ and its boundary. In traditional notation it reads as follows:

$$\int_\Omega (\text{rot } f, \nu) \, d\Omega = \int_\gamma (f, e) \, d\gamma$$

where $d\Omega$ and $d\gamma$ are area element and arc element, $\nu = (\nu_1, \nu_2, \nu_3)$ is a continuous unit normal to Ω, $e = (e_1, e_2, e_3)$ a continuous unit tangent to γ, and rot f, the rotation of f, is a vector with the components

$$(\partial_2 f_3 - \partial_3 f_2, \partial_3 f_1 - \partial_1 f_3, \partial_1 f_2 - \partial_2 f_1).$$

In addition, ν and e are chosen, so that their vector product $e \times \nu$ with components $(e_2 \nu_3 - e_3 \nu_2, e_3 \nu_1 - e_1 \nu_3, e_1 \nu_2 - e_2 \nu_1)$ points out from Ω. See Figure 8.8. Intuitively, the right side is a mass flow around γ and the left side the integral over Γ of the density of a more abstract vorticity. Since

$$\omega = f_1 \, dx_1 + f_2 \, dx_2 + f_3 \, dx_3 \quad \Rightarrow \quad d\omega = (\partial_2 f_3 - \partial_3 f_2) \, dx_2 \wedge dx_3 + \cdots$$

we can write Stokes's formula as

$$\int_\Omega d\omega = \int_\gamma \omega \tag{78}$$

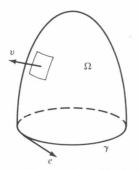

Figure 8.8   An illustration of Stokes' formula.

where Ω and γ are oriented by $\nu_1\, dx_2 \wedge dx_3 + \cdots > 0$ and $e_1\, dx_1 + \cdots >$ 0 respectively. The formula can be reduced to the special case (77) but it is also easy to prove the way we proved Green's formula using charts $x = h(t_1, t_2)$ of Ω and ending up with (75) and (76) for $n = 2$. In fact, this proof shows that (78) holds for $(p-1)$-forms when Ω is a p-manifold of class C^1 with a C^1 boundary γ, both suitably oriented. Taken in this generality it is also called Stokes's formula.

8.5   Documents

Green on Green's formula

From *An Essay on the Application of Mathematical Analysis to the Theory of Electricity and Magnetism*, by G. Green (1828). His δ is Laplace's operator $\partial_x^2 + \partial_y^2 + \partial_z^2$.

"Before proceeding to make known some relations which exist between the density of the electric fluid at the surfaces of bodies, and the corresponding values of the potential functions within and without those surfaces, the electric fluid being confined to them alone, we shall in the first place, lay down a general theorem which will afterwards be very useful to us. This theorem may be thus enunciated:

Let U and V be two continuous functions of rectangular co-ordinates x, y, z whose differential co-efficients do not become infinite at any point within a solid body of any form whatever; then will

$$\int dx\, dz\, dy\ U\delta V + \int d\sigma\ U\left(\frac{dV}{dw}\right) = \int dx\, dy\, dz\ V\delta U + \int d\sigma\ V\left(\frac{dU}{dw}\right);$$

the triple integrals extending over the whole interior of the body, and those relative to $d\sigma$ over its surface, of which $d\sigma$ represents an element: dw being an infinitely small line perpendicular to the surface, and measured from this surface towards the interior of the body."

Note that when $f = U\,\mathrm{grad}\,V - V\,\mathrm{grad}\,U$, $\mathrm{div}\,f = U\delta V - V\delta U$ and (f, ν) $= U(dV/dw) - V(dU/dw)$ where ν is the exterior normal. Hence, the

203

original Green's formula above is contained in (73). It is still a standard item of potential theory.

Riesz on linear functionals

From F. Riesz "On linear functional operators" (1909). His function $\alpha(x)$ of bounded variation is the difference between two increasing ones.

"To define what we mean by a linear operation we must first be precise about the functional field. We consider the class Ω of real and continuous functions defined between two fixed numbers, for instance 0 and 1. For this

class we define limit functions by uniform convergence. A functional operator $A[f(x)]$ which attaches a real number to every element of Ω is said to be continuous if $A(f)$ is the limit of $A(f_i)$ as $f_i(x)$ tends to $f(x)$. A distributive and continuous operation is said to be *linear*. It is easy to see that such an operator is also bounded, i.e., there is a number M_A such that, for every element $f(x)$, one has

$$|A[f(x)]| \leqslant M_A \max |f(x)|.$$

Hadamard has proved the remarkable fact that every such linear operation $A[f(x)]$ is of the form $\lim \int_0^1 k_n(x) f(x)\,dx$ where the k_n are continuous functions. In this note we shall give a new analytic expression of the linear operation containing only one generating form. To this end [here Riesz starts his arguments].... Hence we have the theorem: Given a linear operation $A[f(x)]$ we can define a function $\alpha(x)$ of bounded variation such that, for any continuous function, we have

Frédéric Riesz 1880–1955

$$A[f(x)] = \int_0^1 f(x)\,d\alpha(x)."$$

Literature

The bulk of the material in this chapter is in virtually every calculus book. *Smith's Primer of Modern Analysis*, by Kennan T. Smith (Bogden and Quigley, 1971) is close to our text both in terminology and spirit and contains a lot of integration. *An Introduction to Fourier Series and Integrals*, by R. T. Seeley (Benjamin, 1966), and *Distributions and Fourier Transforms*, by W. F. Donoghue, Jr. (Academic Press, 1969), give the elements of harmonic analysis. *Fourier Series and Integrals*, by H. Dym and H. P. McKean (Academic Press, 1972) is a more advanced text.

9
SERIES

9.1 *Convergence and divergence.* 9.2 *Power series and analytic functions.* Taylor's series. Formal power series. Functions defined by power series. Analytic functions. 9.3 *Approximation.* The Weierstrass approximation theorem. Approximation by trigonometric polynomials. Fourier series. Quantitative approximation. 9.4 *Documents.* Abel on the convergence of series. Dirichlet on Abel's theorem.

Measuring and counting, man meets infinity in the shape of distances that are too large to be measured and numbers too large to be counted. These experiences have been and always will be a source of speculation. Man's encounter with infinity has given us both the belief in eternal life with or without a beginning, and the opposite belief that the world was created once and will perish once. In the Bible, the sand of the sea stands for the uncountable: Jacob to God: "And thou saidst, I will surely do thee good, and make thy seed as the sand of the sea, which cannot be numbered for multitude." In *A Portrait of the Artist as a Young Man*, James Joyce wrote a very expressive passage about infinity. An Irish priest talks about eternal punishment:

> "What must it be then to bear the manifold tortures of hell for ever. For all eternity. Not for a year or for an age but for ever. Try to imagine the awful meaning of this. You have often seen the sand of the seashore.... Now imagine a mountain of sand, a million miles high, reaching from the earth to the farthest heavens...and imagine that at the end of every million years a little bird came to that mountain and carried away in its beak a tiny grain of sand. How many millions upon millions of centuries would pass before he had carried away even a square foot of that mountain, how many eons upon eons of ages before it has carried away all? Yet at the end of that immense stretch of time not even an instant of eternity could have said to have ended...after that eon of time the mere thought of which makes your brain reel dizzily, eternity would scarcely have begun."

Archimedes did something about the sand; he counted it. In the introduction to his book *The Sand Reckoner*, he says

> "There are people who think that the sand is uncountable and by sand I mean not only the sand of Syracuse and the rest of Sicily but that of every country, inhabited or not inhabited. There are also those who, without considering it infinite, do not believe that numbers have been named that are larger than the number of grains of sand . . . . But I shall

endeavor to prove with geometrical arguments that some of the numbers I
have named in a letter I sent to Zeuxippos are larger not only than the
number of grains of sand on the earth but also the number of grains of
sand in the universe."

The Greek philosophers were of course interested in infinity. One of
their great discoveries is that one can add arbitrarily many numbers
without their sum surpassing a given bound. This is one way of interpreting *Zeno's paradox*, which says that in order to go a certain distance, we
must first go half-way, and before that half-way to the midpoint, and so
on. From this Zeno drew the radical conclusion that motion is a theoretical
impossibility. Thought met infinity and stopped. We may, perhaps, hold
the view that the formula

$$2^{-1} + 2^{-2} + \cdots \text{ ad infinitum} = 1$$

gives an explanation, but it had certainly not made Zeno change his view.
He would, of course, understand the formula as well as we do but he
would probably reject it as a mere reformulation of his paradox. Anyway,
this has been the reaction of the many philosophers who after Zeno and up
to our time have thought and written about the paradox.

A more substantial result of the encounter with infinity is the theory of
infinite series, to be considered in this chapter. The first section is a quick
review of convergence and divergence. The second starts with Newton's
binomial series, continues with power series, and ends with a short presentation of analytic functions according to Weierstrass. The chapter
finishes with Lebesgue's proof from 1898 of the Weierstrass approximation
theorem and a look at Fourier series and quantitative approximation.

9.1 Convergence and divergence

Strictly speaking, an infinite series

$$\sum_0^\infty a_k = a_0 + a_1 + a_2 + a_3 + \cdots$$

with the terms $a_0, a_1, a_2, \ldots$ is just a signal to the reader to compute the
sequence $s_0, s_1, s_2, \ldots$ of the partial sums $s_n = \sum_0^n a_k$ and then wait. The
series is said to be *convergent* if the sequence $s_0, s_1, s_2, \ldots$ has a limit s,
and this limit is called the *sum* of the series. The terms of a convergent
series must tend to zero as $n \to \infty$ for $\lim a_n = \lim(s_n - s_{n-1}) = \lim s_n - \lim s_{n-1} = s - s = 0$. A series which is not convergent is said to be *divergent*. The simplest example is the geometric series $\sum_0^\infty a^k$ with the partial
sums $s_n = 1 + a + a^2 + \cdots + a^n = (1 - a^{n+1})/(1 - a)$ when $a \neq 1$ and
$s_n = n$ when $a = 1$. Hence the series diverges when $|a| \geq 1$ and converges to
the sum $(1 - a)^{-1}$ when $|a| < 1$. This is true for both real and complex a. It
will be convenient to talk about sections of the series (1) as sets
$(a_{m+1}, \ldots, a_n)$ of $n - m$ successive terms, the sum of the section being
$a_{m+1} + \cdots + a_n$.

The series (1) is said to be *positive* if all terms are $\geqslant 0$. The partial sums s_n then increase with n so that the series converges if and only if they are bounded. It may come as a surprise that the harmonic series

$$1 + 2^{-1} + 3^{-1} + 4^{-1} + \cdots$$

is divergent. In fact, if we divide it into sections with terms from 2^{-n} to 2^{-n-1}, the sum of every section is larger than the number $2^{n+1} - 2^n$ of its terms times the least term 2^{-n-1}, and hence larger than $1/2$. Hence the partial sums are not bounded.

Let $\Sigma_0^\infty a_k$ be a positive series with decreasing terms. The corresponding so-called alternating series

$$a_0 - a_1 + a_2 - a_3 + a_4 - a_5 + \cdots$$

may then converge and the original series diverge, for the alternating series converges when $a_n \to 0$ as $n \to \infty$. In fact,

$$s_{2n} = a_0 - (a_1 - a_2) - \cdots - (a_{2n-1} - a_{2n})$$

decreases and

$$s_{2n+1} = (a_0 - a_1) + (a_2 - a_3) + \cdots + (a_{2n+1} - a_{2n})$$

increases as n increases and, since $s_{2n} = s_{2n+1} + a_{2n} \geqslant s_1$ and $s_{2n+1} = s_{2n} - a_{2n+1} \leqslant s_0$, both $\lim s_{2n}$ and $\lim s_{2n+1}$ exist and they are equal if and only if $a_n \to 0$ as $n \to \infty$.

It is clear that if $0 \leqslant a_k \leqslant b_k$ for all k, then (we leave out the limits of summation)

$$\Sigma a_k \quad \text{divergent} \quad \Rightarrow \quad \Sigma b_k \quad \text{divergent}$$

$$\Sigma b_k \quad \text{convergent} \quad \Rightarrow \quad \Sigma a_k \quad \text{convergent.}$$

(2)

This simple fact may raise hopes that there is a largest convergent series Σa_k with positive terms such that $b_k \leqslant \text{constant} \cdot a_k$ for every positive convergent series Σb_k. This is not the case. In whatever way the first series is chosen, there is another series that converges more slowly. This can be seen if we divide Σa_k into successive sections $A_1, A_2, \ldots$ defined so that $a_k \in A_n$ when $(1 - 2^{-n})s < s_k \leqslant (1 - 2^{-n-1})s$. The sum over A_n then does not exceed $2^{-n}s$. Here, as above, $s_1, s_2, \ldots$ are the partial sums of the series and s the sum. Now let $B_1, B_2, \ldots$ be the corresponding sections of a series Σb_k chosen so that $b_k = (4/3)^n a_k$ when a_k belongs to A_n. Then b_k/a_k tends to ∞ as k tends to ∞, but the sum over B_n is at most $(4/3)^n 2^{-n}s = (2/3)^n s$ and hence Σb_k converges. In the same way, given a divergent series Σb_k with positive terms, it is possible to construct a divergent series Σa_k such that $a_k/b_k \to 0$ as $k \to \infty$. But this does not, of course, mean that a comparison between series according to (2) is meaningless. The series $\Sigma_1^\infty k^{-a}$, convergent when $a > 1$ and divergent when $a \leqslant 1$, is often used as a comparison series. It is clear that (2) does not require that $0 \leqslant a_k \leqslant b_k$ for all k. It suffices to assume this for all sufficiently large k.

207

A series Σa_k is said to be *absolutely convergent* if $\Sigma |a_k|$ is convergent. If s_n and σ_n are the partial sums of these series, then $|s_n - s_m| \leq |\sigma_n - \sigma_m|$ so that, by Cauchy's convergence principle, absolute convergence implies convergence. Rearranging a series means writing its terms in a different order. When the series is positive, this does not affect the sum, for any section of the original series is contained in a sufficiently large section of the rearranged series, and conversely. And it is not difficult to show that rearranging the terms also leaves unchanged the sum of an absolutely convergent series. But Riemann proved the striking result that a series which is convergent without being absolutely convergent can be re-arranged so that its sum becomes any number given in advance. It can also be made to diverge. Convergence without absolute convergence is a very delicate matter which we shall not go into, except for mentioning a theorem proved by Abel in a paper from 1826 where he remarks that nobody at the time really knew whether some of the most frequently used series of analysis converge or diverge. Abel's theorem is as follows: if Σa_k converges, then the series $\Sigma a_k x^k$ also converges when $-1 < x \leq 1$, and its sum $f(x) = \Sigma_0^\infty a_k x^k$ is a continuous function in the same interval.

We pass to the proof. Since the series Σa_k converges, the absolute values of its terms are bounded by some number M, and hence $|a_k x^k| \leq M|x|^k$ for all k. A comparison with the geometric series then shows that $\Sigma a_k x^k$ converges when $|x| < 1$. Hence the function $f(x) = \Sigma a_k x^k$ is defined when $-1 < x \leq 1$. To show that it is continuous, introduce the partial sum $f_n(x) = \Sigma_0^{n-1} a_k x^k$. Then

$$|f(x) - f_n(x)| \leq M \sum_n^\infty |x|^k = M|x|^n (1 - |x|)^{-1}$$

so that the convergence is uniform and hence f continuous in every interval $|x| \leq c < 1$. When Σa_k is absolutely convergent, we get, of course, directly

$$|f(x) - f_n(x)| \leq \sum_n^\infty |a_k|,$$

proving uniform convergence in the interval $|x| \leq 1$, so that f, if also defined by the series when $x = -1$, is continuous in the same interval. When Σa_k is convergent but not absolutely convergent, our simple estimates no longer suffice to show that f is continuous at the point $x = 1$, i.e., $f(x) \to \Sigma_0^\infty a_k$ as $x \to 1$. To see this, write (following Abel) the difference $f - f_n$ as

$$f(x) - f_n(x) = a_n x^n + a_{n+1} x^{n+1} + \cdots$$

$$= (s_n' - s_{n+1}')x^n + (s_{n+1}' - s_{n+2}')x^{n+1} + \cdots$$

where $s_n' = s - s_n = a_n + a_{n+1} + \cdots$ and rewrite it as

$$f(x) - f_n(x) = s_n' x^n - s_{n+1}'(x^n - x^{n+1}) - s_{n+2}'(x^{n+1} - x^{n+2}) - \cdots,$$

using the fact that $|x| < 1$. Here, if $x \geq 0$, all parentheses are ≥ 0 and hence

$$|f(x) - f_n(x)| \leqslant s_n^*(x^n + x^n - x^{n+1} + x^{n+1} - \cdots) = 2s_n^* x^n \leqslant 2s_n^*$$

where s_n^* is the supremum of the numbers $|s_n'|, |s_{n+1}'|, \ldots$. Since the same inequality is trivially true when $x = 1$ and since $s_n^* \to 0$ as $n \to \infty$, the convergence $f_n \to f$ is uniform and hence f continuous in the interval $0 \leqslant x \leqslant 1$.

So far we have assumed that the terms $a_0, a_1, \ldots$ of our series Σa_k are numbers, real or complex. But our definitions of partial sums, convergence, sums, divergence, and absolute convergence and our proof that absolute convergence implies convergence work equally well when the terms belong to some linear, normed, and complete space, in other words, a Banach space. An immediate application of this remark is the following: if U is a Banach space, $E : U \to U$ the identity map, and $A : U \to U$ a bounded linear operator whose norm is less than 1, $|A| < 1$, then $E - A$ is invertible, with the inverse

$$(E - A)^{-1} = \sum_0^\infty A^k = E + A + A^2 + A^3 + \cdots \tag{3}$$

where the series is absolutely convergent. All this is obvious. We remind the reader that if $|u|$ is the norm of U, all bounded linear $A : U \to U$ constitute a Banach space L with the norm

$$|A| = \sup |Au| \quad \text{when } |u| \leqslant 1.$$

Since $|A^k| \leqslant |A|^k$, the series (3) converges absolutely when $|A| < 1$. The partial sums $S_n = E + A + A^2 + \cdots + A^n$ have the property that $(E - A)S_n = S_n(E - A) = E - A^{n+1}$ and, letting $n \to \infty$, we get $(E - A)S = S(E - A) = E$ where $S \in L$ is the sum of the series. The series (3) is sometimes called the *Neumann series*, after Carl Neumann who used it around 1870 in a study of Dirichlet's problem.

9.2 Power series and analytic functions

Taylor's series

In 1669, when he was in his late twenties, Newton discovered that the function $(1 + x)^c$, with arbitrary real c, can be written as a power series in x, i.e., an infinite sum $\Sigma a_k x^k$. More precisely,

$$(1 + x)^c = 1 + cx + c(c - 1)x^2/2 + \cdots$$
$$+ c(c - 1) \ldots (c - n + 1)x^n/n! + \cdots. \tag{4}$$

The right side is called *the binomial series*. For positive integers c, (4) is an equality between polynomials, the classical *binomial formula*, which was known much earlier. In 1668, Mercator had found a power series for the logarithm,

$$\log(1 + x) = x - \frac{x^2}{2} + \frac{x^3}{3} - \frac{x^4}{4} + \cdots . \tag{5}$$

Both are special cases of Taylor's series (1715)

$$f(x) = f(0) + (x - a)f'(a) + (x - a)^2 f''(a)/2! + \cdots$$

$$= \sum_0^\infty f^{(n)}(a)(x - a)^n/n!.$$

According to Taylor's formula (15) of Chapter 7, the series converges and equality holds when f is infinitely differentiable in the closed interval between a and x and $n \to \infty \Rightarrow M_n(x - a)^n/n! \to 0$, where M_n is the maximum of $|f^{(n)}(t)|$ in the same interval. Applying this to $f(x) = (1 + x)^c$ and $\log(1 + x)$ and doing a bit of calculation proves (4) and (5) when $|x| < 1$. There are also the following well-known special cases

$$e^x = \sum_0^\infty x^n/n!, \qquad \cos x = \sum_0^\infty (-1)^n x^{2n}/(2n)!,$$

$$\sin x = \sum_0^\infty (-1)^{n-1} x^{2n+1}/(2n + 1)!$$

true for all x, and, for instance,

$$\text{arc tan } x = \sum_0^\infty (-1)^{n-1} x^{2n+1}/(2n + 1),$$

true when $|x| < 1$ and, by Abel's theorem, also when $x = 1$, which gives

$$\pi/4 = 1 - 3^{-1} + 5^{-1} - 7^{-1} + \cdots ,$$

a formula proved geometrically by Leibniz in the beginning of his mathematical career.

Formal power series

It is useful to consider power series $\Sigma a_k x^k$ as polynomials with an infinite number of terms which are determined by the sequences (a_k) of their coefficients. We can then forget about convergence and compute as with polynomials. Given two such formal power series, $f \sim a_0 + a_1 x + a_2 x^2 + \cdots$, and $g \sim b_0 + b_1 x + b_2 x^2 + \cdots$, we form linear combinations $af + bg$, products fg, and, when $b_0 = 0$, compositions $f \circ g$ according to the following rules, where the powers $x^0 = 1, x, x^2, \ldots$ have served just as a bookkeeping device,

$$af + bg \sim (aa_0 + bb_0) + (aa_1 + bb_1)x + (aa_2 + bb_2)x^2 + \cdots$$

$$fg \sim (a_0 + a_1 x + \cdots)(b_0 + b_1 x + \cdots) \sim a_0 b_0 + (a_0 b_1 + a_1 b_0)x + \cdots$$

$$f \circ g \sim a_0 + a_1(b_1 x + \cdots) + a_2(b_1 x + \cdots)^2 + \cdots$$

$$\sim a_0 + a_1 b_1 x + (a_1 b_2 + a_2 b_1{}^2)x^2 + \cdots .$$

There is also the formal derivative

$$f' \sim a_1 + 2a_2x + \cdots = \sum_1^\infty ka_k x^{k-1}.$$

The sign $\sim$ stands as a reminder that we are dealing with formal series. Once convergence is established everywhere for certain values of x, it can be replaced by an equality sign. All these formulas were used as a matter of course straight from the beginning. Newton observed, for instance, that a differential equation

$$y' = F(x, y), \qquad y = 0 \quad \text{when } x = 0,$$

where F is a polynomial (or power series) in two variables,

$$F(x, y) = \sum b_{jk} x^j y^k$$

can be solved by inserting a power series $y \sim c_1 x + c_2 x^2 + \cdots$ for y with unknown coefficients $c_1, c_2, \ldots$ and equating the coefficients of corresponding powers of x. Since, with y as above,

$$F = b_{00} + b_{10}x + b_{01}y + b_{20}x^2 + b_{11}xy + b_{02}y^2 + \cdots$$

$$\sim b_{00} + (b_{10} + b_{01}c_1)x + (b_{01}c_2 + b_{20} + b_{11}c_1)x^2 + \ldots,$$

and

$$y' \sim c_1 + 2c_2 x + \cdots,$$

this gives, for instance,

$$c_1 = b_{00}, \; 2c_2 = (b_{10} + b_{01}c_1),$$

and so on. Once $c_1, \ldots, c_n$ are known we can compute c_{n+1}. This successive determination of coefficients corresponds to successive approximations of the solution y by the partial sums

$$c_1 x + c_2 x^2 + \cdots + c_n x^n$$

of its power series. Newton felt of course that this series had to converge to give a sensible result, and he wrote, for instance, $(1 - x)^{-1}$ as $1 + x + x^2 + \cdots$ or $-x^{-1} - x^{-2} - \cdots$, according as $|x| < 1$ or $|x| > 1$.

The power series is the universal tool of the theory of perturbations, small deviations from a state of a physical or other system, often in equilibrium. In the simplest case the deviation is assumed to be controlled by just one parameter, the quantities one wants to know are assumed to be power series in the parameter, and their coefficients are computed. At least the first coefficients are, as a rule, easy to get and, with some luck, they may give good information about the states of the system. Ever since the seventeenth century this method has been used in all kinds of theoretical physics, and it is also indispensable in very delicate situations in quantum mechanics. The theory of interaction between electrons and the quantized electric field depends on a perturbation series whose coefficients are unbounded linear operators, computed, around 1950, with a great amount of ingenuity and hard work. One still does not know if the series converges, and we let this fact illustrate the difficulties of perturbation theory. A power series is just the first step. The second step is to find out what the partial sums really say about the system.

Functions defined by power series

A function defined by a power series turns out to have very nice properties. Replacing powers of x by powers of $x - a$, let us write

$$f(x) = \sum_{0}^{\infty} c_k (x - a)^k \tag{6}$$

and let r be the least upper bound of numbers $s \geqslant 0$ such that $\Sigma |c_k| s^k$ converges. Our arguments above in connection with Abel's theorem prove that the series (6) converges when $|x - a| < r$ and diverges when $|x - a| > r$. Here $r = 0$ and $r = \infty$ are possible cases. The interval $|x - a| < r$, where its sum $f(x)$ is always defined, is called the *interval of convergence* of the series, and r the *radius of convergence*. We shall prove

Theorem. *The sum of a convergent power series is an indefinitely differentiable function in the interval of convergence, and its derivatives are obtained by differentiating the series term by term, the resulting series having the same interval of convergence as the original one.*

PROOF. The function f being given by (6), the theorem says that its derivative is given by

$$f'(x) = \sum_{1}^{\infty} k c_k (x - a)^{k-1}. \tag{7}$$

The terms of this series having larger absolute values than the corresponding ones of (6) when k is large, the radius of convergence of the series of (7) is, at most, that of (6), but we shall see that it is actually the same. Note first that, multiplying term by term, we get

$$(1 - t)^2 = (1 + t + t^2 + \cdots)^2 = 1 + 2t + 3t^2 + 4t^3 + \cdots$$

where the last series has bounded partial sums and hence converges when $0 \leqslant t < 1$. In particular, (kt^{k-1}) is a bounded sequence. Let r be the radius of convergence of (6), let $|x - a| < r$, choose s strictly between $|x - a|$ and r, and put $t = |x - a|/s$. Then

$$\sum k |c_k| \, |x - a|^{k-1} = \sum k t^{k-1} |c_k| s^{k-1} \leqslant \left(\sup_k k t^{k-1} \right) \sum |c_k| s^{k-1},$$

proving that the series of (7) and hence also the series of (6) converges absolutely and uniformly when $|x - a| \leqslant s$ and s is any number $< r$. In particular, their sums are continuous functions in the interval of convergence of (6). Further,

$$f_h(x) = (f(x + h) - f(x))/h = \sum_{1}^{\infty} c_k g_k(x, h)$$

where

$$g_k(x, h) = \left((x - a + h)^k - (x - a)^k\right)/h = k \int_0^1 (x - a + th)^{k-1} \, dt$$

tends to $k(x - a)^{k-1}$ as $h \to 0$. At the same time, $k(|x - a| + |h|)^{k-1} \geqslant |g_k(x, h)|$, proving that if $|x - a| + |h| \leqslant s < r$, the terms of the series for f_h are majorized by the corresponding terms of the convergent series $\Sigma k|c_k|s^{k-1}$. Hence, by dominated convergence, precisely as for integrals, as $h \to 0$, $f_h(x)$ tends to $f'(x)$ as given by (7). Repeated applications of (7) show that

$$f^{(p)}(x) = \sum_p^\infty k(k - 1) \ldots (k - p + 1)c_k(x - a)^{k-p} \tag{8}$$

with absolute convergence when $|x - a| < r$. Note that (6) is necessarily the Taylor series of f with center a, for (8) shows that $c_p = f^{(p)}(a)/p!$

When a function f is the sum of a power series (6) we can, of course, expect that it is equal to its Taylor series

$$f(x) = \sum_0^\infty f^{(k)}(b)(x - b)^k/k! \tag{9}$$

at any point b of the interval of convergence $|x - a| < r$, and that this series converges, at least in the interval $|x - b| < r - |a - b|$, the largest possible with center b contained in the first one (see Figure 9.1). This is indeed the case, and depends on the estimates

$$|f^{(p)}(b)| \leqslant g(t)p!(1 - |b - a|/t)^{-p-1}, \tag{10}$$

true when $|b - a| < t < r$. Here $g(t) = \sup_k |c_k|t^k$ is finite when $t < r$. They show that $(x - b)^p f^{(p)}(b)/p! \to 0$ as $p \to \infty$ when

$$|(x - b)/t| < |1 - |b - a|/t|,$$

i.e., $|x - b| + |b - a| < t$. To prove (10), take absolute values in (8), replace

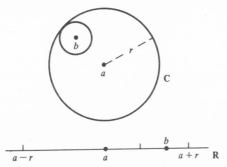

Figure 9.1   An interval of convergence $|x - a| < r$ and a disk of convergence $|z - a| < r$. Changing from powers of $x - a$ to powers of $x - b$, the series converges at least in the smaller interval with center b indicated in the figure. Similarly for the disk.

213

$|c_k|$ by $g(t)t^{-k}$, and sum the resulting series using (8) again with $a = 0$, $f(x) = (1 - x)^{-1}$, and $|x| < 1$ in which case $f^{(p)}(x) = p!(1 - x)^{-p-1}$.

Note that (6) may fail as an equality between a function f and its Taylor series with center a even if f is indefinitely differentiable. It does fail, for instance, when $a = 0$ and $f(x) = e^{-1/x}$ when $x > 0$ while $f(x) = 0$ when $x \leqslant 0$. This f is indeed indefinitely differentiable, but all its derivatives vanish when $x = 0$.

Analytic functions

Let us pass to power series in a complex variable with complex coefficients. This means replacing (6) by

$$f(z) = \sum_0^\infty c_k(z - a)^k$$

where now a is some complex number and z a complex variable. With r defined as above we find that the series converges when $|z - a| < r$ and diverges when $|z - a| > r$. Our interval of convergence has been replaced by a disk of convergence in the complex plane, but this is about the only difference. Word for word, our theorem and its proof still hold, the derivative f' being

$$f'(z) = \lim \frac{f(z + h) - f(z)}{h} \tag{11}$$

where $h \to 0$ through *complex* values. The arguments above also prove that (9) with x replaced by z holds in the disk $|z - b| < r - |a - b|$ with center b, contained in and tangent to the disk $|z - a| < r$ (see Figure 9.1).

A complex-valued function f from an open subset A of the complex plane is said to be *analytic* in A if, close to every point of A, f is equal to its Taylor series with center at this point. Examples: a function defined by a power series is analytic in the disk of convergence; polynomials are analytic everywhere; a function defined by (6) in an interval $|x - a| < r$ of the real axis extends to a function analytic in the disk $|z - a| < r$ simply by replacing x by z in (6); the function z^{-1} is analytic outside the origin for $z^{-1} = (z - a + a)^{-1} = \sum_0^\infty (-1)^k a^{-k-1}(z - a)^k$, the series converging when $|z - a| < |a|$.

The analytic functions, here defined according to Weierstrass ($\sim$1870), are of supreme importance in all branches of analysis and have a number of striking properties. Here are a few examples: linear combinations, products, quotients with denominators $\neq 0$, and compositions of analytic functions are analytic. A complex-valued function from an open subset A of the complex plane is analytic if its derivative (11) exists everywhere in the set. When A is connected, such a function f is uniquely determined by the coefficients of its Taylor series at any point of A. When C is a compact subset of A consisting of an open set Ω with a smooth boundary curve γ, the values of f in Ω can be computed from those on γ by Cauchy's integral formula

$$f(z) = (2\pi i)^{-1} \int_\gamma (w - z)^{-1} f(w)\, dw$$

where the integral is an ordinary line integral and γ is oriented so that Ω is to the left of γ.

The study of analytic functions usually comes next after basic analysis. It is therefore outside the scope of this book and we end the story of power series here.

9.3 Approximation

The Weierstrass approximation theorem

When $c = 1/2$, the binomial series (4) becomes

$$(1 - x)^{1/2} = 1 - 2^{-1}x - (2!)^{-1}(1 - 2^{-1})2^{-1}x^2 - \cdots - a_n x^n - \cdots$$

where

$$a_n = (n!)^{-1}(n - 1 - 2^{-1}) \ldots (1 - 2^{-1})2^{-1} > 0.$$

Letting x increase to 1 it follows that $\sum_1^n a_k < 1$ and hence the series is absolutely convergent for $|x| = 1$. According to what we have seen earlier, this implies that

$$n \to \infty \quad \Rightarrow \quad (1 - x)^{1/2} - s_n(x) \to 0 \quad \text{uniformly when} \quad |x| \leq 1 \quad (12)$$

where $s_n(x)$ are the partial sums of the series and hence polynomials. We express this by saying that the function $(1 - x)^{1/2}$ can be *approximated by polynomials*, uniformly in the interval $|x| \leq 1$. The Weierstrass approximation theorem says that this property is shared by all functions which are continuous in the interval. Hence, in careful formulation:

The Weierstrass approximation theorem (1885). *To every continuous function f from a compact interval of the real axis and every $\varepsilon > 0$ there is a polynomial P such that $|P(x) - f(x)| \leq \varepsilon$ for all x in the interval.*

Another way of formulating the theorem is to say that there exists a sequence $P_1, P_2, \ldots$ of polynomials such that $P_n - f$ tends to zero uniformly in the interval when $n \to \infty$. Just choose, for every integer $n > 0$, a polynomial P_n such that $|P_n(x) - f(x)| \leq n^{-1}$ in the interval. In the example above these polynomials happen to be partial sums of a power series, but this is no requirement of the theorem. We shall see that our example contains the clue to the proof.

Let I be a compact interval. Let us for simplicity say that a function from I is *approximable* when it has the property formulated in the theorem. Since, P_n and Q_n denoting polynomials and arrows uniform convergence,

$$P_n - f \to 0, \quad Q_n - g \to 0 \quad \Rightarrow \quad aP_n + bQ_n - af - bg \to 0$$

for all numbers a, b, all approximable functions form a linear space. In particular, if f and g are real and approximable by real polynomials, then $f + ig$ is approximable by complex polynomials. Hence we can restrict ourselves to real functions and real polynomials. Next, consider *polygonal* functions from I, i.e., real continuous functions whose graphs are broken straight lines with a finite number of corners. Figure 9.2 shows such a function.

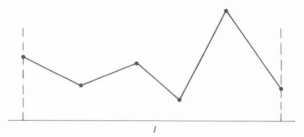

Figure 9.2   A polygonal function from a compact interval I.

It is clear that all polygonal functions from I constitute a real linear space. Also, if f is a real continuous function and f_n the polygonal function which is equal to f in n equally spaced corners, then, by the uniform continuity of f, the functions f_n tend uniformly to f as n tends to infinity. Hence, to prove the theorem it suffices to prove that every polygonal function is approximable. Now, clearly, every such function is a sum of polygonal functions, each one with just one corner. This reduces the proof to approximating these simple functions. The corner being at $x = c$, they are of the form

$$x \leqslant c \quad \Rightarrow \quad g(x) = A(x - c) + C, \qquad x > c \quad \Rightarrow \quad g(x) = B(x - c) + C$$

where A, B, C are real numbers. But this function can also be written as

$$g(x) = 2^{-1}(A - B)|x - c| + 2^{-1}(A + B)(x - c) + C$$

so that it is approximable if the function $x \to |x - c|$ is approximable. Now, if $a \neq 0$, $P(ax + b)$ is a polynomial at the same time as $P(x)$, and hence a simple change of scale reduces the entire proof to showing that the function $|x|$ is approximable in the interval $|x| \leqslant 1$. But this we know already, for changing x to $1 - x^2$ in (12) shows that

$$n \to \infty \quad \Rightarrow \quad |x| - s_n(1 - x^2) \to 0 \quad \text{uniformly when} \quad |x| \leqslant 1.$$

The Weierstrass approximation theorem has many variants, e.g., the theorem that if $h(x)$ is a strictly monotone function from a compact interval I, then every continuous function from I can be approximated uniformly by polynomials in h, i.e., finite sums $a_0 + a_1 h + a_2 h^2 + \cdots$. The proof is immediate, for h maps I continuously and bijectively onto the

216

compact interval $h(I)$, and hence the change of variable $y = h(x)$ reduces the theorem to the preceding one.

Approximation by trigonometric polynomials

Let us start playing a little with what we now know. A function f is said to be *even* when $f(-x) = f(x)$ and *odd* when $f(-x) = -f(x)$ for all x. Since $\cos x$ decreases steadily from 1 to -1 as x goes from 0 to π, every continuous function from that interval can be approximated by polynomials in $\cos x$, and the same goes for an even continuous function f from the interval $I : -\pi \leqslant x \leqslant \pi$. In fact, $\cos x$ is an even function. Next consider an odd polygonal function g from the same interval, and assume that $g(\pi) = g(-\pi) = 0$. Then $g(x)/\sin x$ is continuous and even and hence approximable by polynomials in $\cos x$. Hence, if f and g are continuous from I with the properties above, then $f + g$ can be approximated uniformly by functions $P(\cos x) + \sin x Q(\cos x)$ where P and Q are polynomials. Such functions are called *trigonometric polynomials*. Note now that if h is a continuous function from the real line which is periodic with the period 2π, i.e., such that $h(x + 2\pi) = h(x)$ for all x, then $h = f + g$ where $f(x) = (h(x) + h(-x))/2$ is even and $g(x) = (h(x) - h(-x))/2$ is odd and such that $g(-\pi) = g(\pi) = 0$. Hence we have proved the following variant of the Weierstrass approximation theorem, also due to Weierstrass:

Theorem. *Every continuous function with the period 2π can be approximated uniformly by trigonometric polynomials.*

It only remains to add that Euler's formulas

$$\cos x = \frac{e^{ix} + e^{-ix}}{2}, \quad \sin x = \frac{e^{ix} - e^{-ix}}{2i}$$

show that every trigonometric polynomial is a finite sum

$$\sum a_n e^{inx}$$

and conversely. Here n runs through all integers $0, \pm 1, \pm 2, \ldots$.

The two theorems by Weierstrass are also true for continuous functions of several variables, and an abstract variant was proved by Stone in 1932: every algebra A of continuous real functions from a compact topological space E that separates points (i.e., given x and $y \neq x$ in E there is an f in A such that $f(x) = 0$, $f(y) \neq 0$) has the property that every real continuous function from E can be approximated uniformly by functions in A. When E is the interval $0 \leqslant x \leqslant 2\pi$ with the points 0 and 2π identified (or, if we want, the unit circle), all real trigonometric polynomials constitute such an algebra, for already the linear combinations of $\cos x$ and $\sin x$ separate points.

Fourier series

Let f be a continuous function from the real line with the period 2π. The Fourier series of f is the series

$$\sum_{-\infty}^{+\infty} a_n e^{inx}$$

with coefficients

$$a_n = (2\pi)^{-1} \int_0^{2\pi} f(x) e^{-inx} \, dx, \qquad n = 0, \pm 1, \pm 2, \ldots, \tag{13}$$

called the *Fourier coefficients* of f. The series need not converge, but if it converges absolutely, i.e., if $\Sigma |a_n|$ converges, then

$$f(x) = \sum_{-\infty}^{+\infty} a_n e^{inx}$$

is equal to the sum of its Fourier series. This follows from the Weierstrass approximation theorem for trigonometric polynomials. In fact, the formula

$$\int_0^{2\pi} e^{inx} e^{-imx} \, dx = \begin{cases} 2\pi & \text{when } m = n \\ 0 & \text{otherwise} \end{cases}$$

and the uniform convergence of the partial sums

$$s_N(x) = \sum_{-N}^{+N} a_n e^{inx}$$

to a continuous function $s(x)$ show that s has the same Fourier coefficients a_n as f. Hence

$$\int_0^{2\pi} (f(x) - s(x)) e^{-inx} \, dx = 0$$

for all n so that

$$\int_0^{2\pi} (f(x) - s(x)) g(x) \, dx = 0$$

for all trigonometric polynomials and hence, by Weierstrass, also for every continuous function $g(x)$ with period 2π. Putting $g(x) = \overline{f(x) - s(x)}$ shows that $f(x) = s(x)$ everywhere. Let us remark, finally, that two integrations by parts in (13) show that $|a_n| \leqslant \text{constant} \cdot n^{-2}$ and hence that $\Sigma |a_n|$ converges when f is a C^2 function.

The fact proved here that rather arbitrary functions are represented by their Fourier series was first explored by Fourier in his book from 1822, *Théorie Analytique de la Chaleur*. Fourier series and their companions, the Fourier integrals, are the main objects of harmonic analysis, one of the most important branches of modern mathematics.

Quantitative approximation

Let f be a given continuous function from a compact interval I and put

$$\varepsilon_n = \inf_p \sup_I |f(x) - P(x)|$$

where P runs through the set of all polynomials of degree at most n. According to Weierstrass ε_n tends to zero as n tends to infinity; but we may well ask the question: how quickly? A result by Jackson from 1911 says that $\varepsilon_n \leqslant$ constant $\cdot n$ when f has a continuous derivative, or more generally, $|f(x) - f(y)| \leqslant$ constant $\cdot |x - y|$ for all x and y. This is just an example of the quantitative theory of approximation. There are many other similar results.

9.4 Documents

Abel on the convergence of series

Algebraic manipulations with power series were so successful that it took over 150 years before anyone had a serious look at the convergence of such series. Here is the introduction to a paper by Abel from 1826 about the convergence of the binomial series:

"Submitting the arguments habitually used in connection with infinite series to a close examination, we shall find that, on the whole, they are not satisfactory and that, consequently, the number of theorems about infinite series with a rigorous foundation is very limited. In general, the operations of analysis are applied to infinite series as if they were finite and this seems to me not permitted without proofs. If, for instance, we want to multiply two series with each other, we put

$$(u_0 + u_1 + u_2 + u_3 + \cdots)(v_0 + v_1 + v_2 + v_3 + \cdots)$$
$$= u_0 v_0 + (u_0 v_1 + u_1 v_0) + (u_0 v_2 + u_1 v_1 + u_3 v_0) + \cdots$$
$$+ (u_0 v_n + u_1 v_{n-1} + \cdots + u_n v_0) + \cdots.$$

Niels Henrik Abel   1802–1829

This equality is correct when the series $u_0 + u_1 + \cdots$ are finite. But if they are infinite, to begin with it is necessary that they converge for a divergent series has no sum; also the series of the right-hand side should converge. It is only with this restriction that the expression above is correct; but, if I am not mistaken, one has not paid attention to this circumstance. This is what I propose to do in this article. There are several other similar operations to justify, e.g. the ordinary procedure to divide a quantity by an infinite series, the determination of the powers of an infinite power series, its logarithm, its sine, its cosine etc....

"The divergent series may sometimes successfully serve as symbols of certain abridged propositions, but they should not take the place of determined quantities. By such procedures we can prove all we want, the possible as well as the impossible".

Dirichlet on Abel's theorem

Abel's theorem that the function $f(z) = \Sigma_0^\infty a_n x^n$ is continuous when $-1 < x \leqslant 1$ provided Σa_n converges is not easy to prove. In a note from 1863, apparently written in memory of Dirichlet, Liouville says that he himself had difficulties understanding Abel's proof but was helped by Dirichlet who, on the spot, wrote down the following easy variant. Note that "numerically" means "in absolute value," and that, at the end, the inequality $(1 - \varepsilon)^n > 1 - n\varepsilon$ is used as a matter of course. In the middle Dirichlet lets $n \to \infty$ in the identity

$$s_0 + (s_1 - s_0)x + \cdots + (s_n - s_{n-1})x^n - s_n x^{n+1}$$
$$= (1 - x)(s_0 + s_1 x + \cdots + s_n x^n).$$

"In view of the convergence of the series

$$A = a_0 + a_1 + a_2 + \cdots + a_n + \text{etc.}$$

the sum

$$s_n = a_0 + a_1 + \cdots + a_n$$

is numerically inferior to a certain constant k and converges to the limit A as n grows indefinitely. Let us consider the series

$$S = a_0 + a_1 x + a_2 x^2 + \cdots + a_n x^n + \text{etc.};$$

the quantity x is supposed to be positive and less than 1; replacing a_0, a_1, a_2, etc. by $s_0, s_1 - s_0, s_2 - s_1$, etc. it takes the form

$$S = s_0 + (s_1 - s_0)x$$
$$+ (s_2 - s_1)x^2 + \cdots + (s_n - s_{n-1})x^n + \text{etc.}$$

and then, in a different order,

$$S = (1 - x)$$
$$\times (s_0 + s_1 x + s_2 x^2 + \cdots + s_n x^n + \cdots),$$

a transposition which is without difficulty for it reduces to adding to the first $n + 1$ terms the quantity $-s_n x^{n+1}$ which vanishes when $n = \infty$.

"Let us now see to what limit S converges when the positive variable $\varepsilon = 1 - x$ tends to zero. To this end, decompose S into two parts, one coming from the n first terms and the other consisting of the following ones and let n increase as ε decreases but so slowly that the limit of $n\varepsilon$ is zero. The first term, being numerically less than $n\varepsilon k$, converges to zero. As for the other part,

Peter G. Leieune Dirichlet 1806–1859

zero. As for the other part,

$$(1-x)\left(s_n x^n + s_{n+1} x^{n+1} + \cdots\right)$$

it can be written as

$$P(1-x)(x^n + x^{n+1} + \cdots) = Px^n = P(1-\varepsilon)^n$$

where P is a number between the largest and the smallest of the quantities $s_n, s_{n+1}, \ldots$. But, the latter ones converging to A, it is the same for P and since, by hypothesis, the other factor $(1-\varepsilon)^n$ converges to 1, this proves that if the variable x tends to 1, the limit of S is A, the very sum of the series first considered."

Literature

Series are just a mathematical tool and do not form an independent part of the subject, but there are some books entirely devoted to the theory of series, the classic being *Theory and Application of Infinite Series*, by K. Knopp (London and Glasgow, 1951). The Taylor series of the elementary functions are in every calculus book. There are plenty of elementary texts dealing with analytic functions. Among them *Complex Analysis*, by Ahlfors (McGraw-Hill, 1966), and *Elementary Theory of Analytic Functions of One or Several Complex Variables* by H. Cartan (Addison-Wesley, 1963) are outstanding. For harmonic analysis see the literature for Chapter 8. The elements of approximation theory and also some more advanced material can be found in *Approximation of Functions*, by G. G. Lorentz (Holt, Rinehart and Winston, 1966).

221

10
PROBABILITY

The word probability has to do with probing the truth. We live in a world full of uncertainties which we try to master by making conjectures about the true state of affairs and about the future. This process is an important part of our analysis of the world around us. It is natural that we should prefer certainty. Normally, situations are classified as absolutely dangerous or absolutely harmless and danger is avoided. We move cautiously on rough ground and, as pedestrians and drivers, keep large margins of security. But this kind of classification involves risks. After two or three similar experiences of the same phenomenon we are inclined to think that it always occurs in the same way.

Insecurity is both a strain and a challenge. Forced to choose between alternatives whose consequences are not fully known, we may react with a feeling of pleasure and arousal provided the choice really means something. But the consequence of a wrong choice must not be too serious. Then the situation is critical and we may perhaps have to mobilize all our mental resources, intellectual and emotional, to meet it, and failure may be destructive. The fascination of the unknown is so great that man has invented innumerable games where he can play with it under orderly conditions and without danger to his life.

Probability theory is a mathematical model of chance. It started as an analysis of games of chance and is now an extensive mathematical theory with applications to the social sciences, biology, physics, and chemistry. We shall give a short review of its foundations with an eye to the law of large numbers and the central limit theorem, and then touch upon some applications.

10.1 Probability spaces

In a textbook on the art of computing printed in Italy in 1494, its author Paciuolo says that if 6 plays are needed to win a game and two players interrupt the game when one of them has 5 plays and the other 2, then the

sum at stake should be divided between the players in the proportion 5 to
2. This may seem reasonable, but the principle to share in proportion to
the number of plays won is certainly not reasonable. Suppose, for instance,
that we change the figures to 15 and 12 plays and that 16 are needed to
win. The players would then get almost equal shares—but there is something wrong with this. In fact, he who has 15 plays just needs one more
play to get everything while the other one needs 4 in a row. Some years
later Cardano treated a similar problem. He understood that it is the
remaining plays that have to be analyzed, not the ones already played. In
Paciuolo's problem one player needs one play to take everything and the
other one needs four. Hence the rest of the game has five possible
outcomes. The first player may win in the first, second, third, fourth play
or not at all. Cardano wanted the sum at stake to be divided in the
proportion $(1 + 2 + 3 + 4) : 1 = 10 : 1$. His motivation for this is obscure.
The correct result is $15 : 1$ and it follows by applying the principles of
probability theory as formulated by Pascal and Fermat 100 years later.
Both of them dealt with interrupted games, arriving at the same result but
with different methods. Fermat's solution of Paciuolo's problem was to
consider all possible outcomes of 4 plays. There are $2 \cdot 2 \cdot 2 \cdot 2 = 16$ of them
and the first player wins in all cases except one, namely, when all four
plays go to his opponent. This kind of reasoning was immediately criti-
cized on the ground that not all outcomes have to be played to the end, for
instance, not when the first player wins the first game. This was countered
by the argument that nothing changes if all outcomes are played to the
end.

These episodes from the birth of probability theory follow a pattern that
was to be repeated many times: a practical question is asked, wrong
answers are proposed, a simple mathematical model reduces the answer to
a triviality, and the discussion turns to the applicability of the model.

Fermat did not use the word probability, but he could have defined the
probability that the first player wins as $15/16$, i.e., the number of favorable
cases divided by the number of all possible cases. In this definition it is, of
course, assumed that all cases are equally possible. This condition is very
often satisfied in combinatorial problems, e.g., favorable and possible deals
in games of cards, throws of dice, and drawings from urns. There are
simple and also very complicated combinatorial problems of this kind, but
in principle they offer no difficulties. They fit beautifully into a universally
accepted mathematical model for probability, the *probability space*, proposed by Kolmogorov in 1933. The simplest probability space is a finite set
U equipped with a function $u \to P(u) \geqslant 0$ such that $\Sigma P(u) = 1$, where we
sum over all u in U. The elements u of U are called *elementary events*, and
P stands for probability. The *probability* $P(A)$ of a subset A of U is defined
as $\Sigma P(u)$, where u runs over A. The subset A is called an *event* and should
be thought of as the occurence of one of the elementary events of A. The

function P, now extended to all subsets of U, has the property that

$$P(A \cup B) = P(A) + P(B) - P(A \cap B). \qquad (1)$$

In fact, every $P(u)$ with u in $A \cap B$ appears precisely once in both $P(A)$ and $P(B)$. The empty set $\emptyset$ is supposed to have probability 0. We have supposed that U is finite, but there is nothing that prevents U from being infinite provided $P(u) > 0$ for at most countably many u.

A function $P(A) \geqslant 0$ from subsets of a set U with the property (1) is called a *measure* and, if $P(U) = 1$, a *probability measure*. The general definition of a probability space is now simply a set U equipped with a probability measure P. In this definition, the elementary events have disappeared and it may happen that $P(u) = 0$ for all u in U. An example of this is $U = \mathbf{R}$ equipped with the measure

$$P(A) = \int_A f(x)\, dx$$

where $f(x) \geqslant 0$ is such that $P(\mathbf{R}) = 1$. With this we have slid into integration theory where, unfortunately, not all functions f and sets A are permitted to appear. We really ought to have added some technical reservations to our definition of a probability space. But we did not since we trust the reader to stand for some fogginess at this point.

Most games of chance and many other things can be thought of as probability spaces:

Throws with a coin. U has two elements, heads and tails, each with the probability $1/2$.

Throws with a die. U has six elements $1, 2, 3, 4, 5, 6$ each with the probability $1/6$.

Win or lose. U has two elements, gain and loss, with probabilities p and $q, p + q = 1$.

Roulette. U has n elements $F_1, \ldots, F_n$ with the probabilities $p_1, \ldots, p_n$. We may think of $F_1, \ldots, F_n$ as sectors of a circular spinning disk, a roulette wheel, the area of F_k being p_k times the area of the disk.

Bernoulli sequences. Playing win or loose n times, we get a probability space U consisting of 2^n sequences $u = (u_1, \ldots, u_n)$, where each u_k is either a gain or a loss and $P(u) = p^r q^s$ where r is the number of gains and $s = n - r$ the number of losses of the sequence u. Note that the $P(u)$ are the 2^n terms of the product $(p + q)^n$ when written as a sum. It follows that, for all k, $P(u_k = \text{gain}) = p$ where the right side is the probability of the subset of U consisting of all u such that $u_k = \text{gain}$. We also get $P(u_k = \text{loss}) = q = 1 - p$. Here the number n has disappeared and one can show that these equalities also define a probability measure on the set of infinite sequences $u = (u_1, u_2, \ldots)$ of gains and losses. The objects which we have now introduced are called *Bernoulli sequences*, after Jacob Bernoulli, who studied them in a book, *Ars Conjectandi* (1713).

Weather. U has two elements, beautiful and foul, with the probabilities 0.1 and 0.9.

Lotteries. U consists of all possible drawings of n tickets from $N > n$ numbered ones, all with the same probability. If the order between the n tickets counts, this probability is $(N - n)!/N!$ for then U has $N!/(N - n)!$ elements. If the order does not count, U has $\binom{N}{n} = N!/(N - n)!n!$ elements, and the probability is 1 divided by this number.

Races. U consists of the outcomes of 10 horse races with one winning horse out of 5 in each race. Every outcome has the same probability, 5^{-10}.

Most of these examples are firmly anchored in the real world. Casinos and lotteries are stable enterprises built on reliable probability spaces. But the example with the weather is almost meaningless, and anyone betting on horse races according to the model above will go broke in no time.

The concept of a probability space is a radical axiomatization of our intuitive idea of probabilities. Finite probability spaces may seem unduly trivial from a purely mathematical point of view. But they become marvelous toys once we introduce stochastic variables and their expectation values.

10.2 Stochastic variables

When a game is played for money each player wins or loses a certain amount, depending on the outcome of a play. The sum that he gets (negative when he loses) is an example of a *stochastic variable*, defined simply as a real function $u \to \xi(u)$ from a probability space. In our list of examples we get stochastic variables by putting a price on heads, another one on tails, etc. Here, the Bernoulli sequences offer interesting possibilities. We can let $\xi(u)$ be the number of gains in the sequence u, the maximal number of gains in succession, or the number of changes between gains and losses. In lotteries we can choose the largest number, the least number, the largest square, etc., in a drawing.

Before entering a game, a cautious player would probably like to know something about his chances, for instance, the probability that the amount he gains lies in some interval I. For a general stochastic variable ξ the corresponding probability is usually written as $P(\xi \in I)$ and is equal to $P(A)$ where A consists of all u in U such that $\xi(u)$ lies in I. To take an example, let ξ be the number of gains of a Bernoulli sequence with n

elements and the probability p of a gain. Then $P(\xi = n) = p^n$, $P(\xi = 0) = q^n$, $q = 1 - p$, and, for any k,

$$P(\xi = k) = \binom{n}{k} p^k q^{n-k}$$

for there are $\binom{n}{k} = n!/k!(n-k)!$ sequences containing precisely k gains. Using this simple fact it is easy to handle the general form of the partition problem for interrupted games treated by Paciuolo, Cardano, Pascal, and Fermat. Let two players, A and B, play a game where the probabilities that they win a play are p and $q = 1 - p$ respectively. The game is interrupted when A needs another r and B another s plays to win the game. How should the sum at stake be divided? We choose as our probability space $n = r + s - 1$ plays, the maximal number needed for a decision, and let the stochastic variable ξ be the number of plays won by A. The sum at stake should then be divided as $P(\xi \geqslant r) : P(\xi < r)$, i.e., as

$$\sum_{k=r}^{n} \binom{n}{k} p^k q^{n-k} \quad \text{to} \quad \sum_{k=0}^{r-1} \binom{n}{k} p^k q^{n-k}.$$

In particular, if A is just one play from winning the game, the proportions are as $1 - q^s$ to q^s.

It is sometimes conceptually convenient to consider all functions from a probability space as stochastic variables and not only those with real values. We shall then use the term *general stochastic variable*. The color of a ball drawn from an urn filled with differently colored balls is a stochastic variable. The k-th component u_k of a Bernoulli sequence is a stochastic variable with two values, gain and loss.

Distribution and independence

Every general stochastic variable ξ from a probability space U to a set V gives us a probability measure on V called the *distribution* of ξ, namely the function

$$K \to P(\xi \in K)$$

from subsets K of V to numbers between 0 and 1. For, if K, J are subsets of V and $A, B \subset U$ are defined by $\xi(u) \in K$ and $\xi(u) \in J$ respectively, then $A \cup B$ consists of all u such that $\xi(u) \in K \cup J$ and $A \cap B$ of all u such that $\xi(u) \in K \cap J$. Hence, by the property (1) of probability measures, we have $Q(K \cup J) = Q(K) + Q(J) - Q(K \cap J)$ where $Q(I) = P(\xi \in I)$. It is also clear that $Q(\varnothing) = 0$, $Q(V) = 1$.

A number of general stochastic variables $\xi_1, \ldots, \xi_n$ from a probability space U to sets $V_1, \ldots, V_n$ are said to be *independent* if every joint probability $P(\xi_1 \in A_1, \ldots, \xi_n \in A_n)$ is a product as follows

$$P(\xi_1 \in A_1, \ldots, \xi_n \in A_n) = P(\xi_1 \in A_1) \ldots P(\xi_n \in A_n)$$

for all subsets $A_1, \ldots, A_n$ of $V_1, \ldots, V_n$. When the sets V are finite, this amounts to

$$P(\xi_1 = a_1, \ldots, \xi_n = a_n) = P(\xi_1 = a_1) \ldots P(\xi_n = a_n)$$

for all $a_1 \in V_1, \ldots, a_n \in V_n$. The Bernoulli sequences $u = (u_1, \ldots, u_n)$ and the stochastic variables $\xi_k(u) = u_k$ are an example of this. For then all V_k are equal and have two elements, gain and loss, and the equality above is nothing but the definition of the probability measure on the Bernoulli sequences with $P(u_k = \text{gain}) = p$, $P(u_k = \text{loss}) = q = 1 - p$.

In general, stochastic variables are, of course, not independent. Some may be functions of the others or be influenced by them. The way one stochastic variable ξ from a probability space U to a set V is influenced by another one η from U to a set W is expressed by the *relative probability*

$$P(\xi \in A | \eta \in B) = P(\xi \in A, \eta \in B) / P(\eta \in B)$$

that ξ lies in $A \subset V$ when η is assumed to be in $B \subset W$. It is easy to check that $A \to P(\xi \in A | \eta \in B)$ is indeed a probability measure on the subset of U defined by $\eta(u) \in B$ provided, of course, that $P(\eta \in B) > 0$. That ξ and η are independent can then be rewritten as $P(\xi \in A | \eta \in B) = P(\xi \in A)$ for all A and B. To give an example of relative probability, let ξ_1 and ξ_2 be the number of spots on two dice thrown at the same time. Then

$$P(\xi_1 + \xi_2 \leqslant 6 | \xi_2 \leqslant 3) = 12 \times 6^{-2}/2^{-1} = \tfrac{2}{3}$$

while $P(\xi_1 + \xi_2 \leqslant 6) = 5/12$

If ξ is a stochastic variable from U to V and h is a function from V to W, then $P(h(\xi) \in B) = P(\xi \in h^{-1}(B))$ where $h^{-1}(B)$ consists of all $v \in V$ such that $h(v) \in B$. It follows that functions $h_1(\xi_1), \ldots, h_n(\xi_n)$ of independent stochastic variables $\xi_1, \ldots, \xi_n$ are independent.

Distribution functions

Every real stochastic variable gives us a probability measure on the real line **R**. We shall take a closer look at such measures. The function

$$x \to F(x) = P(\xi \leqslant x)$$

has the property that $P(x_0 < \xi \leqslant x_1) = F(x_1) - F(x_0)$, and is called the *distribution function* of the stochastic variable ξ. As x increases from $-\infty$ to $+\infty$, F increases from 0 to 1. When ξ comes from a finite probability space, F is locally constant with a finite number of jumps occurring at points x where $P(\xi = x) > 0$, and $F(x)$ vanishes when x is large negative and $F(x) = 1$ when x is large positive. When U is infinite, it may happen that ξ has a frequency function, i.e., a function $f(x) \geqslant 0$ such that

$$P(x_0 < \xi \leqslant x_1) = F(x_1) - F(x_0) = \int_{x_0}^{x_1} f(x)\, dx.$$

In that case, the distribution function is continuous. Since $P(b\xi + a \leqslant x)$ $= P(\xi \leqslant (x - a)/b)$ when $b > 0$, the variable $a\xi + b$ has the distribution function $x \to F((x - a)/b)$ and the frequency function $b^{-1}f((x - a)/b)$

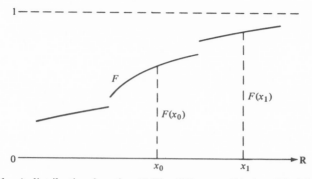

Figure 10.1   A distribution function F. The difference $F(x_1) - F(x_0)$ is the probability measure of the interval $x_0 < x \leqslant x_1$ of the real axis.

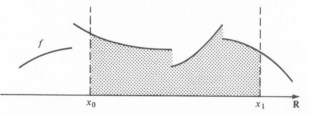

Figure 10.2   A frequency function f. The shaded area is the probability measure of the interval $x_0 < x \leqslant x_1$ on the real axis.

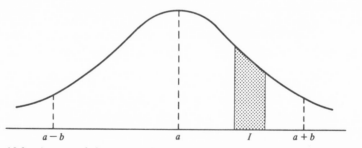

Figure 10.3   A normal frequency function. The shaded area is the probability measure of the interval I. When the parameter b decreases, the distribution concentrates around the point a.

when ξ has the distribution function $F(x)$ and the frequency function $f(x)$. The Figures 10.1 and 10.2 illustrate general distribution functions and frequency functions.

The normal distribution

A stochastic variable with a frequency function

$$(2\pi)^{-1/2}b^{-1}e^{-(x-a)^2/2b^2}$$

where $b > 0$ is said to be *normally distributed* and its frequency function is said to be *normal*. Later in this chapter we shall demonstrate the capital importance of these functions. Figure 10.3 illustrates their bell-shaped graphs.

10.3 Expectation and variance

Expectation values

Let ξ be a stochastic variable from a finite probability space U and let $h(\xi)$ be a real function of ξ. The *expectation value* $E(h(\xi))$ of $h(\xi)$ is the sum of its values multiplied with the corresponding probabilities, hence

$$E(h(\xi)) = \sum h(\xi(u))P(u),$$

the sum running over all u in U. Since all $P(u)$ are $\geqslant 0$ and their sum is 1, the expectation value lies between the largest and the least value of $h(\xi)$ and hence the expectation value of a constant is the constant itself. Collecting all terms with equal values of $\xi(u)$ we can write the expectation value as

$$E(h(\xi)) = \sum h(x)P(\xi = x)$$

with summation over all different values x of ξ. When the probability space is infinite, these sums should be replaced by integrals,

$$E(h(\xi)) = \int_U h(\xi(u))\, dP(u) = \int_R h(x)\, dP(\xi \leqslant x).$$

We must refrain from defining the first integral, but if h is continuous, the second one is a well-defined Riemann-Stieltjes integral when it converges. When $h \geqslant 0$ is continuous, the integral is also defined but may be infinite. We suppose in the sequel that the integral converges when $h(x)$ is not larger than a constant times x^2 for large x and restrict ourselves to such ξ and h. By the properties of the integral,

$$E(ah(\xi) + bg(\xi)) = aE(h(\xi)) + bE(g(\xi))$$

when a, b are numbers and h, g are continuous.

When ξ is real, the most important expectation values are the *mean* $m = E(\xi)$ and the *variance* $V(\xi) = E((\xi - m)^2)$. The square root of the variance is called the *standard deviation* and will be denoted by $\sigma(\xi)$. The variance appears in the important inequality

$$P(|\xi - m| \geqslant t) \leqslant t^{-2}V(\xi), \tag{2}$$

true for all $t > 0$ and called *Chebyshev's inequality*. It can also be written

$$\int_R (x - m)^2\, dP(\xi \leqslant x) \geqslant \int_{|x-m| > t} t^2\, dP(\xi \leqslant x) = t^2 P(|\xi - m| \geqslant t).$$

In this form it is an immediate consequence of the properties of the Riemann-Stieltjes integral. The inequality shows that large deviations

$|\xi - m|$ from the mean are not very probable when the variance is small. Distributions with a small variance are concentrated around the mean and when the variance vanishes, then $\xi = E(\xi)$ with probability 1. The stochastic variable $\xi - m$ has the expectation value 0 for $E(\xi - m) = E(\xi) - mE(1) = E(\xi) - m = 0$. Hence the variable $\eta = (\xi - m)/\sigma(\xi)$ has the mean 0 and the variance 1 for $E(\eta^2) = \sigma(\xi)^{-2}E((\xi - m)^2) = 1$.

EXAMPLE. When $\xi = 1$ and 0 with the probabilities p and $1 - p$, then $E(\xi) = 1 \cdot p + 0 \cdot (1 - p) = p$ and $E(\xi^2) = p$, too. Hence $V(\xi) = E((\xi - p)^2) = E(\xi^2 - 2p\xi + p^2) = E(\xi^2) - 2pE(\xi) + p^2 = p - p^2 = p(1 - p)$. When ξ is normally distributed with the frequency function

$$(2\pi)^{-1/2}b^{-1}e^{-(x-a)^2/2b}$$

then $\eta = (\xi - a)/b$ has the frequency function $(2\pi)^{-1/2}e^{-x^2/2}$. The formulas $\int_R te^{-t^2/2} \, dt = 0$ and (integrate by parts once)

$$\int_R t^2 e^{-t^2/2} \, dt = \int_R e^{-t^2/2} \, dt = (2\pi)^{1/2}$$

then prove that $E(\xi - a) = E(b\eta) = bE(\eta) = 0$ and $E((\xi - a)^2) = E(b^2\eta^2) = b^2 E(\eta^2) = b^2$ so that, finally, $E(\xi) = a$ and $V(\xi) = b^2$ and $\sigma(\xi) = b$. In other words, the parameters a and b equal, respectively, the mean and standard deviation of ξ.

Characteristic functions

The *characteristic function* of a real stochastic variable ξ is given by the formula

$$t \to E(e^{it\xi}) = E(\cos t\xi) + iE(\sin t\xi).$$

Since

$$E(e^{it\xi}) = \int_R e^{itx} \, dF(x)$$

where $F(x) = P(\xi \leq x)$, this is nothing but the Fourier transform of the measure given by the distribution function F of ξ. Here is a piece of information without proof: the characteristic function $\varphi = E(e^{it\xi})$ determines $F(x) = P(\xi \leq x)$ uniquely and, if $\varphi, \varphi_1, \varphi_2, \ldots$ are the characteristic functions of the stochastic variables $\xi, \xi_1, \xi_2, \ldots$ and $\varphi_n \to \varphi$ as $n \to \infty$, then also $\xi_n \to \xi$ as $n \to \infty$. The first convergence is taken in the sense that $\varphi_n(t) \to \varphi(t)$ for all t, and the second one in the sense that $P(\xi_n \leq x) \to P(\xi \leq x)$ for all x where F is continuous, i.e., where $P(\xi = x) = 0$.

EXAMPLES. When η is normally distributed with mean 0 and variance 1, its characteristic function is

$$E(e^{it\eta}) = (2\pi)^{-1/2} \int_R e^{itx - x^2/2} \, dx = e^{-t^2/2}.$$

(This formula is proved in section 2 of Chapter 8). Putting here $\sigma = \sigma(\xi)$ and $\eta = (\xi - m)/\sigma$, we get

$$E\left(e^{it\xi}\right) = e^{itm}E\left(e^{it(\xi-m)}\right) = e^{itm}E\left(e^{it\sigma\eta}\right) = e^{imt}e^{-t^2\sigma^2/2}$$

when ξ is normally distributed with mean m and standard deviation σ. From the estimate

$$e^{itx} = 1 + itx - 2^{-1}t^2x^2\left(1 + g(tx)\right)$$

where g is bounded and $g(s) \to 0$ as $s \to 0$ it follows that

$$E\left(e^{it\xi}\right) = 1 - 2^{-1}t^2V(\xi)\left(1 + h(t)\right), \qquad (h(t) \to 0 \text{ as } t \to 0) \qquad (3)$$

for every stochastic variable ξ with mean zero and finite variance $V(\xi)$.

10.4 Sums of stochastic variables, the law of large numbers, and the central limit theorem

Let $\xi_1, \ldots, \xi_n$ be stochastic variables from a probability space. We shall compute the expectation values of their sum and their product. The result is

$$E(\xi_1 + \cdots + \xi_n) = E(\xi_1) + \cdots + E(\xi_n) \qquad (4)$$

and, if the variables are independent,

$$E(\xi_1 \ldots \xi_n) = E(\xi_1) \ldots E(\xi_n). \qquad (5)$$

In fact, assuming that U is finite, (4) expresses the obvious equality

$$\sum(\xi_1(u) + \cdots + \xi_n(u))P(u) = \sum \xi_1(u)P(u) + \cdots + \sum \xi_n(u)P(u)$$

with summation over all u in U. The second formula follows from

$$E(\xi_1 \ldots \xi_n) = \sum \xi_1(u) \ldots \xi_n(u)P(u)$$

$$= \sum x_1 \ldots x_n P(\xi_1 = x_1, \ldots, \xi_n = x_n)$$

$$= \sum x_1 \ldots x_n P(\xi_1 = x_1) \ldots P(\xi_n = x_n)$$

$$= \left(\sum x_1 P(\xi_1 = x_1)\right) \ldots \left(\sum x_n P(\xi_n = x_n)\right)$$

where the sums run over, respectively, all u, all x_1 of the form $\xi_1(u), \ldots,$ all x_n of the form $\xi_n(u)$, and the independence is used once. The proofs in the general case are similar and use integration theory. From (4) and (5) follows the fundamental formula

$$V(\xi_1 + \cdots + \xi_n) = V(\xi_1) + \cdots + V(\xi_n) \qquad (6)$$

for the variance of a sum of independent stochastic variables. For, putting

$\xi = \xi_1 + \cdots + \xi_n$ and $\eta_k = \xi_k - E(\xi_k)$, the variables $\eta_1, \ldots, \eta_n$ are independent, $E(\eta_k) = 0$ and $\xi - E(\xi) = \eta_1 + \cdots + \eta_n$. Hence

$$V(\xi) = E\left((\eta_1 + \cdots + \eta_n)^2\right)$$

$$= E(\eta_1^2) + \cdots + E(\eta_n^2) + \sum_{j \neq k} E(\eta_j) E(\eta_k)$$

$$= V(\xi_1) + \cdots + V(\xi_n).$$

We shall also use the formula

$$E\left(e^{it(\xi_1 + \cdots + \xi_n)}\right) = E\left(e^{it\xi_1}\right) \ldots E\left(e^{it\xi_n}\right) \qquad (7)$$

where the variables are assumed to be independent. It is an immediate consequence of (5) and the properties of the exponential function.

Using Chebyshev's inequality and the formulas (3), (4), (6), and (7) we can now prove simple versions of two of the basic results of probability theory, the law of large numbers and the central limit theorem. To start with, we consider an infinite sequence $\xi_1, \xi_2, \ldots$ of independent stochastic variables and their successive sums $\xi_1, \xi_1 + \xi_2, \ldots, \xi_1 + \cdots + \xi_n, \ldots$. The variances $V(\xi_1 + \cdots + \xi_n)$ increase with n so that we can expect the sums to be widely distributed around their means. We now try to diminish these variances by considering small multiples of the sums,

$$\eta_n = a_n(\xi_1 + \cdots + \xi_n), \qquad a_n > 0,$$

for which

$$V(\eta_n) = a_n^2(V(\xi_1) + \cdots + V(\xi_n)).$$

Choosing a_n so small that the right side tends to zero as n tends to infinity, Chebyshev's inequality (2) shows that the variable $\eta_n - E(\eta_n)$ tends to zero, i.e., a stochastic variable which equals 0 with probability 1. Taking the mean, $\eta_n = (\xi_1 + \cdots + \xi_n)/n$, this certainly happens when the variances $V(\xi_1), V(\xi_2), \ldots$ are bounded, for then $V(\eta_n)$ is as small as $1/n$ and hence $P(|\eta_n - E(\eta_n)| > t^{-1})$ as small as t^2/n. Hence, if $\xi_1, \ldots, \xi_n, \ldots$ are independent stochastic variables with bounded variances we have proved, for instance, (taking $t = n^{(1-\varepsilon)/2}$) that

$$n \to \infty \quad \Rightarrow \quad P\left(|\xi_1 + \cdots + \xi_n - E(\xi_1 + \cdots + \xi_n)| < n^{(1+\varepsilon)/2}\right) \to 1 \quad (8)$$

for every $\varepsilon > 0$. This statement is part of a collection of theorems usually called the law of large numbers. The Bernoulli sequences $u = (u_1, \ldots, u_n)$ give us a famous special case. Let $\xi_k(u) = 1$ or 0 according as u_k is a gain or a loss. Then the means $E(\xi_k) = p$ and the variances $V(\xi_k) = p(1 - p)$ are independent of k. By virtue of (8), as n grows large, it is more and more improbable that the number of gains $\xi_1 + \cdots + \xi_n$ in a sequence u differs from its mean np by appreciably more than $\sqrt{n}$.

In the example just given, all the variables ξ have the same distribution. Let us assume that this is the case and let V be the common variance.

Then the variables

$$\eta_n = (Vn)^{-1/2}(\xi_1 + \cdots + \xi_n - E(\xi_1 + \cdots + \xi_n))$$

all have the mean 0 and the variance 1 and therefore are not likely to tend to zero. To see what happens for large n we shall compute the characteristic function of η_n. Using (7) we get

$$E(e^{it\eta_n}) = E(e^{it\bar{\xi}_1/\sqrt{n}}) \ldots E(e^{it\bar{\xi}_n/\sqrt{n}})$$

where $\bar{\xi}_k = (\xi_k - E(\xi_k))/\sqrt{V}$ has the mean 0 and the variance 1. The formula (3) shows that the right side equals

$$\left(1 - (2n)^{-1}t^2\left(1 + h(t/\sqrt{n})\right)\right)^n$$

where $h(s) \to 0$ as $s \to 0$. When n tends to infinity, this tends to $e^{-t^2/2}$, the characteristic function of a normally distributed stochastic variable with mean 0 and variance 1. Using the remark above dealing with characteristic functions, this means that

$$n \to \infty \quad \Rightarrow \quad P(\eta_n \leqslant x) \to (2\pi)^{-1/2} \int_{-\infty}^{x} e^{-s^2/2}\, ds$$

for all x. Expressed in terms of the variables $\xi_1, \xi_2, \ldots$ this is a sharpening of (8), namely that, as $n \to \infty$, the probability

$$P\left(\xi_1 + \cdots + \xi_n - E(\xi_1 + \cdots + \xi_n) \leqslant x(nV)^{1/2}\right)$$

tends to

$$(2\pi)^{-1/2} \int_{-\infty}^{x} e^{-s^2/2}\, ds \tag{9}$$

for all x. The assumption is that $\xi_1, \xi_2, \ldots$ are independent and that all $\xi_k - E(\xi_k)$ have the same distribution with the variance V. We have here a simple special case of the central limit theorem which, loosely formulated, says that sums of large numbers of independent stochastic variables have a tendency to be normally distributed.

The law of large numbers and the central limit theorem as presented here are more or less explicit in works by Jacob Bernoulli (1713), de Moivre (1733), and Laplace (1812). Later these results were generalized and refined, but to tell how would carry us too far afield.

10.5 Probability and statistics, sampling

What are the chances of DEP, the demopublican party, in the next election? The statistician B decides to find out precisely that. He asks 400 prospective voters and finds out that 80 of them are going to vote DEP. He then sells the following forecast to the National Television Company: DEP will get between 17 and 23 percent of the votes. Late at night after the election, when 40 million votes have been counted, it turns out that DEP has gotten 19.2 percent of the votes. The National DEP Committee chairman, a confirmed optimist, is not exactly happy. After many successful rallies he had expected much more.

That optimists can go wrong is clear, but how could B be so sure of himself? The answer is that he had used a stochastic model of the election. He thinks of his questions to voters as equally many drawings of black and white balls from an urn containing millions of black and white balls, letting white balls correspond to DEP votes and black ones to votes for other parties. He also assumes that the urn contains white and black balls in the proportions p to $1 - p$ and that it has been shaken so that all balls are thoroughly mixed with each other. Under these circumstances, taking a ball from the urn is a sample of a stochastic variable taking the value white with the probability p and the value black with the probability $1 - p$. Here p is a number between 0 and 1 which has to be estimated.

If the analogy with the urn is right, B can consider his sample when he asks n persons as a sample of n independent stochastic variables $\xi_1, \ldots, \xi_n$ taking the values 1 and 0 with the probabilities p and $1 - p$. From them he himself makes the variable $\xi = 100(\xi_1 + \cdots + \xi_n)/n$ representing the percentage of DEP votes in a sample of size n. This variable is distributed so that $P(\xi = 100k/n) = \binom{n}{k} p^k (1 - p)^{n-k}$ where $k = 0, \ldots, n$. When B was asking his questions he got a value x of ξ and he now decides to reject all values of the unknown p for which

$$P(|\xi - 100p| \geqslant |x - 100p|) \leqslant \frac{1}{20},$$

i.e., such p that the probability of getting a value of ξ farther away from the mean $E(\xi) = 100p$ than x is at most $1/20$. The remaining values of $100p$ then lie in a certain interval I around x (Figure 10.4).

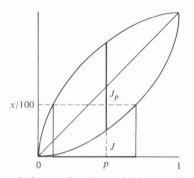

Figure 10.4 Sampling of the stochastic variable $\eta = (\xi_1 + \cdots + \xi_n)/n$ where $\xi_1, \ldots, \xi_n$ are independent and equal to 1 and 0 with the probabilities p and $1 - p$. The number p is not known. For every p, choose an interval J_p, symmetric around p, such that the samples of η fall in J_p with the probability $19/20$. All these form the cigar-shaped region. To a single sample $x/100$ construct an interval J as in the figure. The unknown p ought to be somewhere in J. The interval I of the text is equal to $100J$.

In practice B uses a shortcut to compute the interval I. He pretends that $100x$ is the true value of p and replaces I by an interval $x_0 < t < x_1$, symmetric around x, such that $P(x_0 < \xi < x_1)$ is close to $19/20$, where now the distribution of ξ is computed by putting $p = x/100$. Then $E(\xi) = 100p = x$ and $\sigma(\xi) = 100(p(1 - p)/n)^{1/2}$. When n is as large as 400 he can also assume that ξ is normally distributed, and in that case the choice $x_0 = x - 2\sigma(\xi)$, $x_1 = x + 2\sigma(\xi)$ corresponds well to the chosen probability $19/20$. Putting $n = 400$ and $x = 80$ gives $2\sigma(\xi) = 200(0.2 \times 0.8/400)^{1/2} = 4$. Trusting his theory, B is now convinced that between 16 and 24 percent will vote DEP in the coming election. Had he chosen a larger number than $1/20$ from the beginning, e.g., $1/10$, the same computations would have given him limits closer to 20 percent. After some hesitation he takes a chance to improve his image and delivers the figures 17 and 23 to the National Television Company, collecting a handsome payment.

After this we might well ask what B has done to earn his money. The theory is not his own, his computations are done in a minute, and asking 400 people the same simple question does not seem to require much work. All this is just a routine matter to B. His main contribution is to see to it that his mathematical model really applies. Here he faces many difficulties. If his business is to predict election outcomes he has to abandon the simple model presented here. The voters do not form a homogeneous community. B ought to take many things into account, e.g., age, social class, part of the country, results of previous elections, etc. B then has to use a more sophisticated stochastic model with several urns of various compositions. This complicates his computations a little but his main concern is the relation between reality and the model he employs. If his predictions fail he may loose his reputation and his customers. Nobody is the master of chance.

The desire to get reliable information from sampling is the origin of applied statistics. It employs a large collection of sometimes very ingenious statistical tests. Most of them use the fact that, under reasonable hypotheses, it is possible to compute the distributions of the stochastic variables represented by the sample. A typical one is Karl Pearson's χ^2-test from 1900. Its mathematical model is a roulette wheel with r sectors $F_1, \ldots, F_r$ occupying the fractions $p_1, \ldots, p_r$ of one turn. The roulette wheel is worked n times and if it stops ξ_s times in F_s, then $\xi_1, \ldots, \xi_r$ are stochastic variables with the sum n and a certain distribution which is simple to compute when n is small but rather complicated when n is large. But Pearson proved that, as n increases, the distribution of the stochastic variable

$$\chi^2 = (np_1)^{-1}(\xi_1 - np_1)^2 + \cdots + (np_r)^{-1}(\xi_r - np_r)^2$$

tends to the distribution of the sum of the squares of $r - 1$ independent normally distributed variables with mean 0 and variance 1. This distribution is easy to compute and tabulate. The miracle that takes place here is that it only depends on the number of sectors of the roulette wheel and not on their sizes. Pearson's test is used when certain observations are classified into a number of categories $C_1, \ldots, C_r$ and it is assumed that there is a certain probability p_s for one observation to fall into the category C_s. To test this hypothesis about the probabilities, take a large sample of n observations. If then $n_1, \ldots, n_r$ of them are in the corresponding categories, compute the corresponding value of χ^2 and look at the table for the probability of getting a larger value of χ^2. If this probability is small, for instance less than $1/100$, the observed numbers $n_1, \ldots, n_r$ are considered to deviate more from their expected values $np_1, \ldots, np_r$ than can be attributed to chance. The conclusion must be to reject the hypothesis about the probabilities.

But here is the risk of oversimplification. To use the test we just need two sets of numbers with the same number of elements. Then we look at the table to find out whether the two sets differ by chance or not. The background to this caricature is again the delicate connection between a stochastic model and the reality it is supposed to represent. It feels fine to use Pearson's test when throwing a die many times to find out whether it is loaded or not, but the same feeling of satisfaction would probably not be there if we used it to find out if a certain writer uses the word "she" significantly more than the word "he."

10.6 Probability in physics

In the eighteenth century, Euler and Lagrange invented a simple model for the movements of fluids and gases. It combines Newtonian mechanics with simple properties of pressure and density, and it has been very successful. Somewhat later Fourier made a model of heat flow based on the fact that it is proportional to the temperature gradient. In these models the medium —a gas, a fluid, or a heat conductor—is considered to be homogeneous and its equilibrium states are simple to describe. A fluid or a gas not influenced by outer forces is in equilibrium when pressure and density are constant, and there is no heat flow when the temperature is constant. But through the progress of chemistry in the beginning of the nineteenth century it became clear that gases, fluids, and solids consist of more or less free-moving molecules and that heat is a form of mechanical energy. The movements of molecules follow the laws of Newton but it is hopeless to keep track of them one by one. On the other hand, it is possible to study them statistically, for instance, the distribution of energy at various states of equilibrium and how this distribution may change with time. This is done in statistical mechanics, founded by Clausius, Maxwell, and Boltzmann. Combining the laws of Newton with various versions of the law of large numbers, they succeeded in deducing some known macroscopic laws, for instance Boyle's Law about the connection between pressure, temperature, and density of a gas and the laws of heat conduction. Statistical mechanics also has a branch in quantum mechanics which we cannot go into.

As an example of probability in physics we shall now describe a probability space with applications to heat conduction and Brownian motion, the motion of small particles in a fluid caused by impacts from the molecules of the fluid. This probability space consists of all continuous curves $t \to \xi(t) = (\xi_1(t), \xi_2(t), \xi_3(t))$ in three-dimensional space. Here $t \geq 0$ is time and we suppose that $\xi(0) = 0$. These curves correspond to all possible movements of a particle starting from the origin at time 0. The probability measure is such that the stochastic variables $\xi \to \xi_j(t_2) - \xi_j(t_1)$ where $j = 1, 2, 3$ are independent and normally distributed with means 0 and variances $c^2(t_2 - t_1)^2$ where $c > 0$ is a constant. Hence the frequency function of $\xi(t) = \xi(t) - \xi(0)$ is

$$f(t, x) = (2\pi ct)^{-3/2} e^{-|x|^2/2ct}$$

where $|x|^2 = x_1^2 + x_2^2 + x_3^2$ and we can interpret c as the velocity of the movement in every direction. Putting $dx = dx_1 \, dx_2 \, dx_3$, the number

$$P(\xi(t) \in A) = \int_A f(t, x) \, dx$$

is the probability that our stochastic particle shall be in the region A at time t. In the classical macromodel of heat conduction, $f(t, x)$ is the temperature at time t and at the point x of a three-dimensional heat conductor with the heat conduction coefficient c into which, at time $t = 0$ and at $x = 0$, one unit of heat was introduced. We can also connect our stochastic variable with potential theory, for it is not difficult to verify that

$$\int_0^\infty dt \int_A f(t, x) \, dx = (2\pi c)^{-1} \int_A |x|^{-1} \, dx.$$

Here the left side is the time which, on an average, the particle spends in A and the right side is the Newtonian potential at the origin of a uniform mass distribution on A. This kind of a connection has sometimes been used to guess and prove results in potential theory.

Finally, there is a loose connection between our probability space and quantum mechanics that tickles the imagination. The frequency function $f(t, x)$ is a solution of the equation of heat conduction,

$$2\partial_t f(t, x) = c(\partial_1^2 f(t, x) + \partial_2^2 f(t, x) + \partial_3^2 f(t, x))$$

where $\partial_t, \partial_1, \partial_2, \partial_3$ are the partial derivatives with respect to t, x_1, x_2, x_3. Changing t to it in the equation of heat conduction, we get one of the basic items of quantum mechanics, the Schrödinger equation. At the same time, the frequency function $f(t, x)$ turns into the complex function

$$(2\pi ict)^{-3/2} e^{-|x|^2/2ict}, \qquad (t > 0).$$

In this way our probability space of curves gets a complex measure which we can try to use to construct integrals. These so-called *history integrals* were invented by Feynman around 1950. So far they are not a bona fide mathematical tool, but they have played a considerable part in the intuitive arguments that are an indispensable part of quantum mechanics.

10.7 Document

Jacob Bernoulli on the law of large numbers

In his book *Ars Conjectandi* (1713), Jacob Bernoulli notes that probabilities are known a priori in games of dice or drawings from urns, but goes on to say:

"But, I ask you, who among mortals will ever be able to define as so many cases, the number, e.g., of the diseases which invade innumerable parts of the human body at any age and can cause our death? And who can say how much more easily one disease than another—plague than dropsy, dropsy than fever—can kill a man, to enable us to make conjectures as to what will be the future state of life or death? Who, again, can register the innumerable cases of changes to which the air is subject daily, to derive therefrom conjectures as to what will be its state after a month or even after a year? Again, who has sufficient knowledge of the nature of the human mind or of the admirable structure of the body to be able, in games depending on acuteness of mind or agility of body, to enumerate cases in which one or another of the participants will win? Since such and similar things depend upon completely hidden causes, which, besides, by reason of the innumerable variety of combinations will forever escape our efforts to detect them, it would plainly be an insane attempt to get any knowledge in this fashion."

Jacob Bernoulli 1654-1705

Noting that unknown probabilities can be determined by repeated experiments and that many repetitions seem to increase the precision, he continues as follows, announcing the law of large numbers and advocating the use of statistics in medicine and meteorology.

"Although this is naturally known to anyone, the proof based on scientific principles is by no means trivial, and it is our duty to explain it. However, I would consider it a small achievement if I could only prove what everybody knows anyway. There remains something else to be considered, which perhaps nobody has thought of. Namely, it remains to inquire, whether by thus augmenting the number of experiments the probability of getting a genuine ratio between numbers of cases, in which some event may occur or fail, also augments itself in such a manner as finally to surpass any given degree of certitude; or whether the problem, so to speak, has its own asymptote; that is, there exists a degree of certitude which never can be surpassed no matter how the observations are multiplied; for instance that it never is possible to have a probability greater than 1/2, 2/3 or 3/4 that the real ratio has been attained. To illustrate this by an example, suppose that, without your knowledge, 3000 white stones and 2000 black stones are concealed in a certain urn, and you try to discover their numbers by drawing one stone after another (each time putting back the stone drawn before taking the next one, in order to change the number of stones in the urn) and notice how often a white or black stone appears. The question is, can you make so many drawings as to make it 10 or 100 or 1000 etc. times more probable (that is, morally certain) that the ratio of the frequencies of the white and black stones will be 3 to 2, as is the case with the number of stones in the urn, than any ratio different from that? If this were not true, I confess nothing would be left of our attempt to explore the number of cases by experiments. But if this can be attained and moral certitude can finally be acquired (how that can be done I shall show in the next chapter) we shall have cases enumerated a posteriori with almost the same confidence as if they were known a priori. And that, for practical purposes, where "morally certain" is taken for "absolutely certain" by Axiom 9, Chap. II. is abundantly sufficient to direct our conjecture in any contingent matter not less scientifically than in games of chance.

For if instead of an urn we take air or the human body, that contain in themselves sources of various changes or diseases as the urn contains stones, we shall be able in the same manner to determine by observations how much likely one event is to happen than another in these subjects."

Literature

There is an overflow of textbooks on probability and statistics. Willy Feller's *An Introduction to Probability and Its Applications* (Wiley, 1968) faces mathematics squarely, is well written, and contains a lot of material.

11

APPLICATIONS

11.1 *Numerical computation.* History. Difference schemes. 11.2 *Construction of models.* Sound. The vocal tract. A mathematical model of sound. Vowels. The model and the real world.

Mathematics has two sides, the theoretical and the practical. The one attracting general attention is the practical side. The theoretical side is usually considered to be incomprehensible except to specialists. This dichotomy of the subject is as old as the subject itself. The mechanical inventions of Archimedes, for instance, his water screw and his water-driven model of the planetary system, made him famous, but he preferred theory. The historian Plutarch says that although his inventions had given him the reputation of superhuman wisdom, Archimedes did not want to write about these things but kept to the abstract world of mathematics. Gauss, who did a great deal of applied mathematics, took the same position. But if we leave the great innovators we shall find that the overwhelming majority of people who have devoted some of their time and energy to the subject have used it as a tool to get interesting numerical results and to understand more or less complex situations. Mathematics is used both for numerical calculation and the construction of models. We shall give a review of the history and practice of numerical analysis, look into some difference schemes, and construct a model of the vocal tract explaining how vowels are produced.

11.1 Numerical computation

History

To calculate, we need a notation for numbers and routines for computing sums, differences, products, and quotients. Man has shown great ingenuity in all these fields. The surviving notation for numbers, used the world over, is a positional system with the base 10, employing Arabic signs. Positional systems were invented more than 4500 years ago. One such system with the base 60 was used in the Babylonian empire and survives partly in our way of measuring time and angles. Clay tablets from this time exhibt the first multiplication tables. There are also computations with large numbers but as a rule no routines for performing, e.g., additions and multiplications. Only the results were worthy of being registered.

Computations were done in the head, with the aid of fingers, or else written in sand. There was also the counting board, the Latin abacus, a wooden board where figures were written in the dust or small stones were moved around. The very word *calculus* actually means "small stone". The modern abacus, a frame with sliding beads, is a Chinese invention. The routines that we use ourselves when doing arithmetic have the advantage that all small steps are registered and can be checked. They are described in an Arabic text from the tenth century. The name of its author, Al Kwarizmi, has become our word *algorithm*, meaning a systematic way of computing. Soon after the invention of the printing press there appeared many textbooks of elementary arithmetic, some of them treating also fractions and business mathematics, in particular equivalence of currencies, problems of partition, and rates of interest. The fact that $x = ac/b$ solves the equation $a/b = x/c$, the *regula de tri*, is found to be extremely useful. One writer calls it the golden rule, with the motivation that "it is so valuable that it surpasses other rules as gold surpasses other metals." Philip Melanchton, Luther's friend and an educational reformer, persuaded the university of Wittenberg to hire two professors of mathematics. One of them stressed in his inaugural lecture that "addition and subtraction are necessary in daily life and so easy that boys can learn to perform them. The rules for multiplication and division require more attention but, with some effort, they can readily be understood."

The first trigonometric tables appear in Almagest, the chief work of antique astronomy, compiled about 150 A.D. They were improved by the Arabs and began being used in warfare and navigation in the sixteenth century. One hundred years later, tables of logarithms were computed, and with this the tools of numerical calculation were fixed for at least 300 years. Improved measuring instruments made it more and more profitable to compute. The sextant and the chronometer, in combination with astronomical tables, made navigation safer. The interest of mathematicians in roots of algebraic equations made them invent routines for calculating them numerically. One of the most efficient ones, still used and taught, is due to Newton. It was often important to compute integrals numerically or to sum series. Requirements like these have led to a host of numerical methods, often very specialized. Since it works with a finite number of figures and a finite number of iterations, no numerical method gives, normally, the correct, ideal result. There is always an error: the difference between the number computed and the ideal result. The sign of a good method is that it results in a small error with a modest amount of work. Numerical methods without some kind of estimate of the error are worthless.

To get figures out of theoretical mathematics is an art in itself. From the seventeenth century until our time it was practiced on a small scale and with essentially the same tools, computation by hand and the use of tables. The counting board, which had been important earlier, was no longer used.

Slide rules and hand-operated calculating machines became common 100 years ago. Later, with electric machines, tables of logarithms became superfluous. This was the state of things as late as 30 years ago when a new abacus, the electronic computer, entered the scene. In a short time, this monster has revolutionized all kinds of bookkeeping and all steering processes of industry and technology. It has an enormous computing capacity, but it also requires a lot of foresight and ingenuity of its user. Old numerical methods were applied on a large scale and new ones were invented, some of them very refined. The universities needed no Melanchton to tell them to hire professors of numerical analysis and data processing. The computer itself required its own routines, its own languages, and a staff of technical mathematicians, the programmmers.

Difference schemes

With a computer it is possible to do extensive calculations. Complicated physical processes can be simulated in detail. Mathematical models that used to be just theory and general principles can be squeezed to yield numerical results. The old idea of replacing differential equations by difference equations is practiced on a large scale. Here are two examples of this procedure. Let us first consider the initial value problem for a first order differential equation $u' = f(t, u)$. Here we want to compute approximately a solution $u = u(t)$ such that $u = u_0$ is given when $t = 0$. Considering that the derivative $u'(t)$ is the limit of the difference quotient $(u(t + h) - u(t))/h$ as h tends to zero, it is not difficult to hit upon the idea of approximating the initial value problem with a problem about differences

$$\Delta v(t) = f(t, v(t))\Delta t, \qquad v(0) = u_0,$$

where $t = 0, \pm h, \pm 2h, \ldots$ with $h > 0$ small and fixed and the differences are $\Delta v(t) = v(t + h) - v(t)$ and $\Delta t = t + h - t = h$. It follows that $v(h) = u_0 + hf(0, u_0)$, that $v(2h) = v(h) + hf(h, v(h))$, and so on, and analogously for $v(-h), v(-2h), \ldots$. This determines v on a net consisting of the points $0, \pm h, \pm 2h, \ldots$ whose mesh width or, for simplicity, step, is h. When h approaches zero, we can expect the functions $v = u_h$ to converge to the solution u of the original problem. Under suitable assumptions about the function f, it is not difficult to estimate the difference $u - u_h$. We can also turn the problem around and, via a convergence of the functions u_h, prove that the original problem has a solution. The same method can of course also be used for systems of differential equations, or we can refine it by using a variable step or more sophisticated difference approximations to the derivative. But if f is well-behaved and we want to know u just in a small interval, our original approach goes a long way.

When partial differential equations in, e.g., two variables x and t are to be written as difference equations, we replace the partial derivatives $\partial u / \partial t$

and $\partial u/\partial x$ by difference quotients $\Delta_t u/\Delta t$ and $\Delta_x u/\Delta x$ where $\Delta_t u(t, x)$ $= u(t + h, x) - u(t, x)$, $\Delta t = t + h - t = h$, and $\Delta_x u(t, x) = u(t, x + k) -$ $u(t, x)$, $\Delta x = x + k - x = k$. The meshes of the corresponding net are rectangles with the step h in the t-direction and the step k in the x-direction. Even in simple cases, we encounter the fact that h and k are not always allowed to be small independently of each other. Take for instance the initial value problem.

$$\partial_t u + c\partial_x u = 0, \qquad u(0, x) = f(x)$$

whose solution is a progressing wave $u = f(x - ct)$ with the propagation velocity c. We assume that $c > 0$. A difference approximation is

$$\Delta_t v + c(\Delta t/\Delta x)\Delta_x v = 0, \qquad v(0, x) = f(x)$$

where $t = 0, \pm h, \pm 2h, \ldots$ and $x = 0, \pm k, \pm 2k, \ldots$. From this it follows, for instance, that

$$v(h, x) = v(0, x) - chk^{-1}(v(0, x + k) - (0, x))$$

for all x, giving v for $t = h$, and a similar formula expressing v for $t = 2h$ in terms of v for $t = h$. And so on. Hence we can compute the function $v = u_h$ successively on the whole net. While the value of the exact solution $u = f(x - ct)$ at the point (t, x) is equal to the value of f at the point $x - ct$, we see that the value of the function v at a point (t, x) of the net is a linear combination of the values of f in the part of the net on the x-axis lying between x and $x + th^{-1}k$. But it is clear that the difference approximation can only make sense when this interval contains the point $x - ct$. Hence we must have $k/h \leqslant - c$, i.e., $c\Delta t + \Delta x \leqslant 0$ when $t > 0$ and $h = \Delta t > 0$. It follows that we have to choose the step $\Delta x = k$ in the x-direction negative and at the same time see to it that the step Δt in the t-direction is not too large compared to the step in the x-direction. If we do not, the solutions of the difference problem have no connection with the exact solution. This shows that difference approximations may be unstable, i.e., may not give functions close to the exact solution when the steps are made small.

Difference approximations to Dirichlet's problem in the plane offer a reassuring contrast to instability. The problem is to find a function $u = u(x, y)$ satisfying Laplace's equation $\partial_x^2 u + \partial_y^2 u = 0$ in an open bounded set U of the plane, and equal to a given continuous function $f(x, y)$ on the boundary of U. We assume that the boundary is smooth and that f is defined and continuous also in a neighborhood of the boundary. Choosing $(g(x + h) - 2g(x) + g(x - h))/h^2$ as a difference approximation to the second derivative $g''(x)$ and taking equal steps $h > 0$ in the x- and y-directions, a little computation shows that the equation

$$v(x, y) = 4^{-1}(v(x + h, y) + v(x - h, y) + v(x, y + h) + v(x, y - h)) \quad (1)$$

is a difference approximation to Laplace's equation. We approximate

Dirichlet's problem by trying to find a solution $v = u_h$ of (1) defined in the part $V = U_h$ of the net $(x, y) = (rh, sh)$ with integral r, s, consisting of points (x, y) having at least one neighbor $(x \pm h, y)$ or $(x, y \pm h)$ in U. When (x, y) is a boundary point of U_h, i.e., has at least one neighbor outside U, we put $v(x, y) = f(x, y)$. This is possible when h is small enough, for f is defined in a neighborhood of the boundary of U. When (x, y) is an interior point of U_h, i.e., when all its neighbors are in U, we require (1). See Figure 11.1.

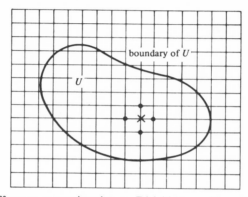

Figure 11.1   Difference approximation to Dirichlet's problem. The cross and the rings mark a point and its four neighbors.

Our difference equations have now become a system of linear equations (1), as many as there are interior points of $V = U_h$. Hence, the number of equations equals the number of unknowns. We claim that every solution assumes its least and largest values on the boundary of V. For let v be a solution and c its largest value. If $u = c$ at an interior point, then $u \leqslant c$ at all its neighbors. But then (1) shows that $u = c$ at all the neighbors, and then $u = c$ at a sequence of additional points which must end up at the boundary, for V has finitely many points. The same argument works for the least value of v. In particular, $v = 0$ is the only solution of (1) when all boundary values are put equal to zero. Hence, by the theory of systems of linear equations, the system (1) has a unique solution v for every choice of boundary values. The system is also well-behaved in other ways. It can be solved by successive approximation when the number of equations is very large, and it is easy to prove that the functions $v = u_h$ converge to the solution u of Dirichlet's problem when h tends to zero, and to obtain good estimates of the size of the difference $u_h - u$.

The two computing routines just described have a classical background and theoretically important features. Other routines may be uninteresting theoretically but are practically very important. A good example is linear programming. There the problem is to minimize (or maximize) a linear form $h(x) = c_1 x_1 + \cdots + c_n x_n$ in a large number of variables over a region of $\mathbf{R}^n$ determined by a number of linear inequalities of the form $a_1 x_1 + \cdots + a_n x_n + b \geqslant 0$. The practical economic background becomes clear if we interpret the variables as quantities of various utilities, $h(x)$ as cost of production, and the inequalities as limitations of the means of production at disposal. There are efficient routines giving the optimal value of h and at the same time the point x where it is attained. In practice, n is often a very large number, and then a computer is absolutely necessary for the calculations.

11.2   Construction of models

In the majority of the mathematical models of physics and technology, physical processes are described by differential equations which can be analyzed mathematically. As a rule, the differential equations just express simple physical laws, while every single process is determined by specific circumstances, the boundary conditions. Here follows an example which is typical on several counts. We shall try to see what happens to the air when we open our mouth and pronounce a vowel.

Sound

To begin with, some words about the physics of sound. What we hear as sounds are fast and small variations of air pressure. For sounds of moderate intensity, air pressure varies by about 10^{-3} millibars which is 10^{-6} times normal atmospheric pressure. Measuring the deviation $p(x, t)$ from atmospheric pressure along a line in the direction of the propagation of sound as a function of time t and position x on the line, we shall find that, approximately, $p(x, t)$ equals a function $f(x - ct)$ times a damping factor. This means that the pressure pattern along the line at time t, represented by the function $x \rightarrow f(x - ct)$, moves with the velocity c in the direction of propagation. Here $c \sim 340$ m/s is a constant, the velocity of sound. Sound propagates as a series of percussions of the mass of air. The movement of this mass itself is negligible. At pressure variations of 10^{-3} millibars, the maximal velocity of this movement has the order of magnitude of 10^{-8}m/s. There is a remarkable connection between the velocity of sound and the elasticity of air under pressure discovered by Newton and Laplace. We have $p = c^2 \rho$ where ρ is the deviation from the normal density of air. This equation holds with great precision when p and ρ are small.

Deviations of air pressure caused by outside forces add up when the forces act simultaneously. Hence two sound waves $p_1(x, t)$ and $p_2(x, t)$ add up to the wave $p_1(x, t) + p_2(x, t)$. A standing sound wave is, by definition, a product $p(x, t) = h(t)f(x)$ of an oscillating factor $h(t)$ and a function $f(x)$ of position. When h has the form $a \cos(2\pi\nu t + \alpha)$, it is said to be a *simple* or *harmonic oscillation*. It repeats itself every $1/\nu$ seconds, and is said to have the frequency of ν oscillations per second, abbreviated as ν Hz

(Hertz). Sound waves coming from, for instance, the oscillating columns of air of wind instruments are dominated by a few standing harmonic oscillations.

The vocal tract

When we speak we use our vocal tract. It consists of the pharynx and the mouth cavity and will be considered as a tube of varying cross section. The form of the vocal tract changes when the tongue, the jaw, and the lips assume different positions. The bottom of the vocal tract consists of the vocal chords enclosing a slit-like opening, the glottis. When the glottis is closed and under air pressure from below, it lets air through in pulses, normally with a frequency of 70 to 300 Hz, less for male than for female voices. Listening to this tone directly, it would be like a dry humming. What we hear when we listen to a vowel is the tone from the glottis modified by its passage through the vocal tract. In non-nasal vowels, the velum blocks the passage to the nasal tract. See Figure 11.2.

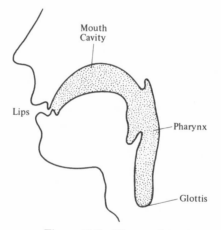

Figure 11.2   The vocal tract.

The human ear perceives sound by registering and processing the deviations $p(t)$ from atmospheric air pressure at the ear drum. It is shown in harmonic analysis that like any time process, $p(t)$ can be approximated as closely as we want by a sum of simple oscillations $a_k \cos(2\pi\nu_k t + \alpha_k)$, with the amplitudes a_k and the frequencies ν_k. Putting $c_k = a_k \exp i\alpha_k$ we can also write

$$p(t) \sim \text{Re}\left(\sum c_k \exp 2\pi\nu_k t \right)$$

with approximate equality. What the ear perceives of this process is, roughly speaking, its spectrum, i.e., the frequencies ν_k and the corresponding amplitudes. When all frequencies ν_k are multiples of a single frequency ν_0, the sound is periodic with the frequency ν_0, and will be heard as a tone with this frequency. Differences in the spectra of tones with the same frequency are heard as differences of timbre or acoustic quality. The spectrum of a sound can be registered electronically. Figure 11.3 exhibits schematic spectra of three vowels, /i/, /u/, /æ/. The curves give amplitude as a function of frequency. The frequencies where the curves have local maxima are called the *formants* of the vowels. Every vowel in the figure has three formants.

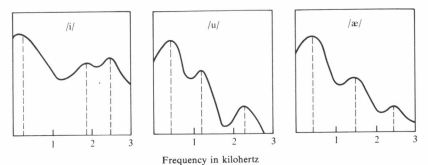

Frequency in kilohertz

Figure 11.3   Schematic spectra of /i/, /u/, /æ/. Pronounced with a glottis tone of, e.g., 100 Hz, their spectra are represented by points of the curves corresponding to 100, 200, ... Hz. The curve is actually the logarithm of the absolute value of the amplitude. The formants are marked by vertical broken lines. Within wide limits, vowels retain their identity when spoken with varying frequency of the glottis tone.

A mathematical model of sound

We shall now construct a mathematical model explaining qualitatively —and partly also quantitatively—what we have learned so far. Its ingredients are general laws of physics and mathematical analysis. We start by considering air enclosed in a tube with varying cross-section area $A(x)$, the cross-section being orthogonal to the axis of the tube and at the distance x along the axis from one of its points. We shall describe small movements of the air along the axis of the tube with the aid of a number of functions assumed to be constant on the cross-sections (see Figure 11.4). In the first place, there are the deviations $p = p(x, t)$ and $\rho = \rho(x, t)$ from atmospheric pressure p_0 and density ρ_0 connected by the equation $p = c^2\rho$.

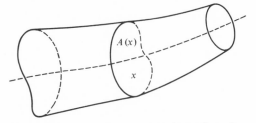

Figure 11.4   Axis and cross-sections of a tube.

There is also the velocity $v = v(x, t)$ of the air and the associated mass flow $u(x, t) = A(x)\bar{\rho}(x, t)v(x, t)$ per unit time. Here $\bar{\rho} = \rho_0 + \rho$ is the total density. Partial derivatives of such functions w will be denoted by $w_t = \partial w / \partial t$ and $w_x = \partial w / \partial x$ and when Δx is small, the difference $\Delta w = w(x + \Delta x, t) - w(x, t)$ will be approximated by $w_x(x, t)\Delta x$.

Now let D be a thin section of the tube between the cross-sections at x and $x + \Delta x$. The mass flow out from D is then $\Delta u \sim u_x \Delta x$ and must be compensated by the change of mass in D per unit time, $(A\bar{\rho})_t \Delta x$. Since $(A\bar{\rho})_t = Ac^{-2}p_t$, this gives

$$c^{-2}Ap_t + u_x = 0, \tag{2}$$

the so-called *equation of continuity*. The total outside pressure on D is $-A\,\Delta p$, which, according to Newton's law, equals the time derivative of the momentum $u\,\Delta x = A\bar{\rho}v\,\Delta x$ of D. This gives an equation of movement,

$$Ap_x + u_t = 0. \tag{3}$$

Differentiating (2) with respect to t and (3) with respect to x, we can eliminate the term u_{xt}, getting a differential equation for the deviation p from atmospheric pressure,

$$Ap_{tt} - c^2(Ap_x)_x = 0. \tag{4}$$

As a mathematical model of the states of the air in the tube we choose smooth functions p and u of x and t satisfying the equations (2), (3), and hence (4). Let us first see what happens when A is constant, i.e., when the area of the cross-section is the same all along the tube. The equation (4) is then equivalent to

$$p = f(x - ct) + g(x + ct)$$

where f and g are arbitrary. In other words, the pressure wave p is the sum of two pressure waves going in opposite directions and with the same velocity c. This confirms the interpretation of c as the velocity of sound. From (2) and (3) we can compute $u = A\bar{\rho}v$ and get

$$\bar{\rho}v = c^{-1}(f(x - ct) - g(x + ct)).$$

Here $\bar{\rho} \sim \rho_0 \sim 1$ kg/m^3. If the largest values of f and g are around 10^{-6} bar, i.e., 10^{-1} N/m^2, the largest value of v is of the order of magnitude of 10^{-8} m/s, which fits with direct observation. Sound travels fast but the air molecules are knocked about at a leisurely pace.

Vowels

After these general considerations let us see what happens when we pronounce a vowel. The vocal tract will be represented by our tube, and we choose coordinates so that $x = 0$ at the glottis and $x = a \sim 0.17$ m at the lips. The area function $x \to A(x)$ is different for different vowels and can be measured by radiography. What it looks like depends in an essential way on the positions of the tongue and the jaw, and on the size of the mouth opening. At the glottis, the pressure gradient p_x is determined by the movements of the vocal chords. At the same time they act as a reflecting wall. The pressure wave generated by the pulses at the vocal chords is travelling back and forth between the glottis and the lips where it meets and is reflected by the outer air. As a mathematical model of this process, we take solutions $p(x, t)$ of (4) such that the gradient $p_x(0, t)$ at the glottis is a given function of t while $p(a, t) = 0$, i.e., the total pressure at the lips equals atmospheric pressure. The pressure gradient $p_x(a, t)$ at the lips then determines a pressure wave, part of which, in practice, escapes at the lips and can be heard. The map

$$p_x(0, t) \rightarrow p_x(a, t) \tag{5}$$

between the two oscillations, one at the glottis, the other one at the lips, represents the action of the vocal tract on the vibrations of the vocal chords. We say that the vocal tract is an *acoustic filter*. To see how this filter acts note that (4) is a linear equation, i.e., any linear combination of solutions is again a solution. It follows that (5) is a linear map from functions to functions. Hence, since any function of time can be approximated by linear combinations of simple oscillations, it suffices to know the result of (5) when

$$p_x(0, t) = \mathrm{Re}\ e^{2\pi i \nu t} \tag{6}$$

is a simple oscillation. The corresponding solution of (4) turns out to be a standing wave

$$p(x, t) = \mathrm{Re}\ h(x)e^{2\pi i \nu t}$$

where $h(x) = h(x, \nu)$ is the solution of the system of equations

$$c^2(Ah_x)_x + (2\pi\nu)^2 Ah = 0, \qquad h_x(0) = 1, \qquad h(a) = 0 \tag{7}$$

and then (5) is given by

$$\mathrm{Re}\ be^{2\pi i \nu t} \rightarrow \mathrm{Re}\ bF(\nu)e^{2\pi i \nu t}$$

where $F(\nu) = h_x(a, \nu)$ and b is the amplitude at the glottis. Hence the frequency is the same at the lips as at the glottis, but the amplitude gets multiplied by $F(\nu)$, called the *transfer function* of the map (5) and of our idealized vocal tract. When A is constant we can compute it explicitly. The solution of (7) is then

$$h(x) = c(2\pi\nu)^{-1} \sin 2\pi\nu c^{-1}(x - a)/\cos 2\pi\nu c^{-1}a$$

so that

$$F(\nu) = 1/\cos 2\pi\nu c^{-1}a.$$

This function is infinite at the frequencies

$$\nu = \nu_0, 3\nu_0, 5\nu_0, \ldots \tag{8}$$

where $\nu_0 = c/4a$. Otherwise $|F(\nu)| \geq 1$ with equality sign only when ν equals $0, 2\nu_0, 4\nu_0, \ldots$. In practice it is, of course, impossible for the transfer function to be infinite. For the time being we hope that infinite quantities in the model correspond to very large ones in practice.

The frequencies (8) are the so-called *eigenfrequencies* or *resonance frequencies* of a tube of a with constant cross-sections. They are also identical with the values of ν for which the homogeneous system (7), obtained by replacing $h_x(0) = 1$ by $h_x(0) = 0$, has solutions $h \neq 0$. One of the most important features of our model is that every tube has an infinite sequence

$$f_1 < f_2 < f_3 < \cdots < f_n < \cdots$$

of resonance frequencies tending to infinity. They are also precisely the values of ν for which the transfer function $F(\nu)$ is infinite and for which the homogeneous system (7) has solutions $\neq 0$. All this can be proved directly, but also with the aid of the spectral theorem for linear self-adjoint compact operators (see Chapter 4, Section 4).

When we work with a sum

$$p_x(0, t) = \mathrm{Re} \sum c_k \exp 2\pi i \nu_k t \tag{9}$$

of simple oscillations (6), the right side of (5) comes out as

$$p_x(a, t) = \mathrm{Re} \sum c_k F(\nu_k) \exp 2\pi i \nu_k t, \tag{10}$$

showing the important fact that amplitudes of frequencies close to the resonance frequencies increase much more than amplitudes of other frequencies. If, e.g., (9) is periodic with the frequencies $\nu_0, 2\nu_0, \ldots$ and the corresponding amplitudes vary slowly we get the picture of Figure 11.5, comparing the spectra of the input (9) and the output (10).

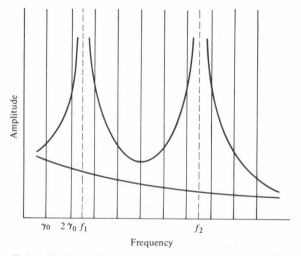

Figure 11.5 Tube filtering of a periodic sound source with the frequencies $\nu_0, 2\nu_0, \ldots$. The lower curve gives the amplitudes of the input (9), the upper curve the amplitudes of the output (10). The resonance frequencies $f_1 f_2, \ldots$ of the tube dominate in the spectrum of the output (10).

The model and the real world

We can now start comparing our model with observed vowel spectra according to Figure 11.3. Let us first take the vowel /æ/. It is pronounced with a relatively large mouth opening, the tongue resting on the bottom of the mouth cavity. The corresponding area function ought to be about constant, and the resonances as given by (8) close to 500, 1500, 2500, . . . , for $v_0 = c/4a = 340/4 \cdot 0.17 = 500$. This agrees rather well with the first three formants of /æ/. Even for the other vowels, computed and observed resonances are not far apart.

Having arrived at this point, we have had the same experience as did Fourier. In the introduction to *Théorie Analytique de la Chaleur*, he wrote:

> "I saw then that all phenomena having to do with heat depend on a very small number of general and simple facts in such a way that every problem of this kind can be stated in terms of mathematical analysis."

But our model is only a skeleton so far. We have not taken friction into account, nor that the walls of the vocal tract vibrate a little, nor the form of the vocal tract, nor the radiation from the lips. Neither have we accounted for the fact that at frequencies above 10^4 Hz it is not permitted to assume that all movements are parallel to the axis. To get better agreement with observed vowel spectra we have to complete the model. Fortunately, our work is facilitated by the fact that frequencies above approximately 15,000 Hz cannot be heard. On the whole, we have to improve the model by giving mathematical form to several physical facts known and analyzed before. It could be difficult enough and we might well ask if it is worth the effort. That the answer is yes and that the work has been done depends to a large extent on the requirements of linguistics. The languages of the world use a lot of different vowels. Descriptions of them should be so good that they can be used for analysis and reproduction, and take all linguistically important differences into account. Articulatory descriptions giving positions of the tongue and the lips have been used for a long time. Electronics has made acoustic descriptions possible, focusing on realistic transfer functions for vowels. It has turned out that they are quite adequately described by specifying the first four formants. Simplified descriptions use only the first two formants, making it possible to represent different vowels as points in a rectangle.

What we have said about vowel articulation is typical for the use of mathematics in technology and physics. Basic laws of physics are formulated mathematically to give a first generation of mathematical models which then are subject to various corrections, some of them empirical. When the results are processed for general use, they take the form of simple rules of thumb and numerical routines. In our example we got a simple and reasonably efficient model right from the start, but this is not always the case, in particular not in biology, chemistry, medicine, or the social sciences. But the success of mathematical models when predictability—either causal or stochastic—is high implies a constant temptation to also use them in less favorable situations.

Literature

Elements of Numerical Analysis, by P. Henrici (Wiley, 1964) is a good introduction. Virtually every book dealing with physics or technology is an applied mathematics text. A reference for just vowel articulation is *Speech Analysis, Synthesis and Perception*, by Flanagan (Springer-Verlag, 1972). There the vocal tract is thought of as an electrical transmission line, giving an application to articulation of the theory of electrical networks.

12
THE SOCIOLOGY, PSYCHOLOGY, AND TEACHING OF MATHEMATICS

12.1 *Three biographies*. 12.2 *The psychology of mathematics*. 12.3 *The teaching of mathematics*. How C reformed school mathematics. A fable.

To round off things we shall now give a short analysis of the role of mathematics in society. It starts with biographies of three persons or rather collectives A, B, and C, representing, in order, the public, the users of mathematics, and the professional mathematicians. In a section about psychology A meets C and C has a crisis. Then we are told how C fared when he wanted to reform the teaching of mathematics. The chapter ends with a morality.

12.1 Three biographies

A. Let us introduce the notation A for the more or less educated public as a mathematician. We then have to imagine a person who has lived and worked for thousands of years. He will even have different ages at the same time. The reader who is bothered by this phenomenon can conceive of A as sometimes split into several individuals. We shall sketch his development and his achievements in mathematics.

To start with, A is a child who learns to count. We assume that A lives in an environment where the numbers are important and that he has normal contact with grown-up people. Then, at the age of about six, A has a firm control of the first 10–20 integers and their order. He can also add small numbers and count with the aid of his fingers. He gets interested in larger numbers and asks for their names. Once he counts himself to sleep and comes to 100. A breathless moment. During his continued life as an illiterate A enlarges his mathematical territory to all numbers under 1000 and gets a good grasp of those fractions and geometrical figures that his language has names for. If he plays games of cards or dice, or otherwise is forced to make mental calculations, he makes them with accuracy. If A lives in a country where goods are bought and sold, where money is used, or where time is measured by the clock, he uses these talents every day.

This A has existed for many thousands of years, and still makes up over half of humanity. His almost daily contact with the integers and perhaps also the fractions makes him feel at home in the number model. He has an intuitive feeling for the kind of practical geometry that is used in, for instance, carpentry. But his knowledge is limited. To be able to handle large numbers and make long calculations, A has to go to school. There he ceases to be illiterate and can start absorbing parts of man's cultural heritage.

As a first task, A has to learn to read and write. On a large scale this did not happen until the nineteenth century. "Reading, writing, and 'rithmetic" were the three pillars of the education of the people. Mathematics came third, but the objective was very clear: to teach effective algorithms for the four arithmetic operations and to acquaint the pupils with the current systems of weight, volume, money, and time. A was first taught the names and signs of the integers and to make simple additions and subtractions. After that came the multiplication table, to be known by heart. Equipped with this indispensible tool, A learned the algorithms of multiplication and division, not without a considerable effort. There was a constant supply of practical applications. The teaching was very concrete with an emphasis on skill and the result was good. After school A had enough practice not to forget what he had learnt. In so far as A pondered what so-called higher mathematics could be about, he imagined a very comprehensive multiplication table or perhaps a fifth arithmetic operation.

Let us now follow A a bit further to high school where he studies Latin, history, languages, and some mathematics. The teaching is now less utilitarian and A meets scientific mathematics in two forms, geometry according to Euclid, and algebra including equations of the first and second degree. A is a good pupil and does all that is required of him but after school he has no use for his knowledge and it deteriorates very soon. As a public servant A now and then remembers the theorem of Pythagoras, but he has forgotten what a hypotenuse is and he has difficulties in computing percentages. As an editor A keeps the pages of his journal free from all kinds of formulas. If mathematics turns up as a subject of conversation, A wonders if there is a connection between musical and mathematical ability. This picture of A, true 100 years ago, is still true in all essentials. The difference depends on the fact that tedious calculations are no longer made by hand. A learns about the same things as before but he gets more explanation and less drill. The new things are: set theory to explain the number system, the binary system to explain computers, and the reading of simple diagrams to make A understand the complex society he lives in. As an adult A has very little reason to occupy himself with mathematics, and his view of the subject can be stated very briefly: it is something that has to do with computers, satellites, and nuclear power. He does not know any details but he believes that the binary system and sets somehow enter the picture. As an editor, A still keeps his pages free from formulas except for a short period when Einstein's formula $E = mc^2$, at times reproduced as $E = mc2$ or $F = m3^2$, was an accepted incantation. A treats figures produced by a computer with greater respect than those computed by hand. On social occasions A behaves as before.

B. Like A, B appears in many shapes at many times. We let B denote people who use mathematics as a tool without contributing to the subject. B stands between A and the professional mathematician, who is the subject of the next section. It is not so easy to differentiate between B and C, but let us say that the typical B nowadays is an engineer who designs a bridge or writes advanced programs for a computer. This B has had a brilliant career. It extends over thousands of years and we can only sketch it here.

B appeared for the first time 4000 years ago in the fertile valleys of the Nile and Euphrates. He did bookkeeping for princes, recording their stocks of grain, oil, wine, cattle, soldiers, and slaves. He wrote out the laws and built temples and palaces with the sign of civilization: straight lines and right angles. He measured distances, made maps, and registered the movements of the celestial bodies, and sometimes won the admiration of the crowd by predicting an eclipse of the sun. He ran a school where smart young boys learned arithmetic and bookkeeping. His life did not change much for 3000 years. In Italy, during the Renaissance, he still did bookkeeping, but now for a rich merchant. After the invention of gunpowder B devoted himself to the art of ballistics. After this first success on the battlefield, B started to build modern civilization.

He constructed, built, improved old processes and invented new ones. Among his achievements are the steam engine, big ships, railways, steel, long range guns, new explosives, cars, airplanes, telecommunication, synthetic materials, satellites, nuclear power, etc. In all his work, B uses mathematics as a tool, mostly just to do simple computations, but now and then he employs complicated mathematical models, for instance, when B discovered the planet Uranus or constructed optical lenses. As a rule, B got his models from C but sometimes B had to invent the model and its theory all by himself. This happened when B was Newton and thought about the movements of the planets. He then found the solution of the problem in the law of gravitation and invented the mathematical instrument, infinitesimal calculus, which made it possible for him to compute the orbits of the planets from this law. After his time as Newton, when B worked as a physicist, mathematics was absolutely necessary for him in order to formulate the fundamental laws that he discovered, e.g., the laws of hydrodynamics and electricity and the principles of atomic physics. Physics is a very difficult task for B and he often looks for new mathematical models. As a rule he no longer has sufficient force to construct them himself and it has happened that C has anticipated him. When B was Einstein and invented general relativity he needed a kind of geometry that Riemann had already put on the market. B had the same experience when he founded quantum mechanics. The mathematical tools, among others, group theory, were already available in stock. Today, when B works as a theoretical physicist, he is trying out lots of things in this stock, although infinitesimal calculus remains the main tool.

Otherwise, B is more active than ever. To his traditional fields of activity, engineering and physics, B has added numerical analysis on a large scale, data processing, and branches of the social sciences, biology, and medicine. Sometimes his pursuits are rather modest, for instance, when B is a doctor and wants to prove something with statistics. He then has to solicit the advice of another B, a specialist on the applications of mathematical statistics.

B's views on mathematics as a subject and his opinion of C varies between uncritical admiration and arrogant superiority. As a professor of physics or engineering B usually thinks that the teaching of mathematics should be done by physicists (engineers) and not be left to his colleague C, a hopeless theoretician. Lately, there have been signs of better relations between the two parties. B has gotten a better mathematical understanding and C has adjusted himself to a growing motley crew of students who want to use mathematics for very diverse purposes.

C. The mathematician C, the man who studies the subject for its own sake, appeared together with B for the first time 4000 years ago in the fertile valleys of the Nile and Euphrates. It is likely that B and C were the same person for some time and that B became C when he got more interested in the tool than in its use. The metamorphosis took place when B had plenty of time and found it to his pleasure to sit and think. C had his first great period in Greece from 500 to 0 B.C. B had then a time of stagnation after some initial successes, but C was busy thinking about mathematical problems, discussing them with others, and writing down what he found. His results are in Euclid's *Elements* and in the works of Archimedes and Apollonius. It was a brilliant debut and the *Elements* was a fantastic success in spite of its pedantic style and abstract reasoning. After this show of force C kept quiet for over 1000 years. His second appearance, with the solutions of third and fourth degree equations, was all right but did not measure up to his debut. But in the seventeenth century C created a sensation by inventing infinitesimal calculus, and he has worked hard and been rather successful ever since. He is also still respected by A, the general public. But his traditional prestige depends to a large extent on A's inability to distinguish between B and C. A is likely to give C undue credit for some of B's achievements, for instance, electronic computers.

We know already how A and B look at C and his subject and we shall now say a few words about the position of C in our society and his views of himself and others. Most of the time, C is a university teacher, but he can also be employed by industry or some research institute. He is a specialist in some branch of mathematics and he has written two or three articles that are known all over the world—if only in a small circle. At times he works very hard and sleeps badly. This is when he is trying to prove a new theorem. He is often unsuccessful but sometimes everything works out wonderfully, and then he is deeply satisfied. If he does not write good articles himself, there are others who do. C is proud of his collective self that has written and still writes so many wonderful things. He knows that his subject has unlimited possibilities and that it will always attract gifted young people. In his relations to A and B he is a realist in that he does not pretend to communicate on levels where it is not possible. But he sometimes finds B irritating, and it has happened that he thinks that A learns the wrong kind of mathematics and then decides to do something about it. We shall come back to his decision later on in this chapter.

12.2 The psychology of mathematics

Let us imagine A as an educated and intelligent person and C as an ordinary professor of mathematics, and let us assume that they are neighbors. A starts thinking about C. What does he do outside the daily routine of a citizen? What does he think about? How is it possible for him to think about such abstract things as mathematics? How is it possible to fill one's life with such dry stuff? C, on the other hand, has just read a new article by D, one of the geniuses of the subject. C had thought of doing something similar and had already come a bit of the way. But now he has to admit that he is thoroughly beaten. C feels depressed and cannot stop wondering why he did not see the problems in the same light as D. Then he would have gotten most of D's results.

To start with, there is nothing remarkable about the fact that C likes to occupy himself with mathematics. Millions of schoolchildren of all ages share his inclination, and there are many more peculiar ways of spending one's time, e.g., playing chess. Besides, C is only periodically trying to solve hard problems. Most of his time is devoted to teaching, and perhaps also to writing. C studied intensively to the age of 30. In one branch of mathematics he has read all the classics and all the recent literature and he has contributed a few theorems to it. Here he feels completely at home. To him the concept of, e.g., an analytic function, is like an old friend and provokes dozens of associations. Sometimes C is playing with the thought that he ought to be able to explain to A what an analytic function is. This has to be done by some physical imagery, for instance, a thin layer of water on a baking sheet flowing in from below in one hole and out through another. In fact, this flow is described by an analytic function. On second thought, C decides that this project is not realistic. But one day he gets an unexpected opportunity to try out his pedagogical talent. A is puzzled by what his daughter's mathematical textbook says about sets and asks C for explanations. Since A is an intelligent and polite listener, things work out reasonably well, but some suppressed questions can be read in A's face. Was this all? Why these strange symbols for such simple things? C gets the hint and explains that the algebra of sets is just a language used in mathematics where it can be useful conceptually and in complicated situations. To give A an idea what real mathematics is about, C now sketches the proofs that $\sqrt{2}$ is an irrational number, that there are infinitely many primes, and that an arc of a circle is seen under the same angle from all its points. The success is limited since A is getting tired and A and C part, assuring each other that they have had a nice time. A week later C has better luck beating A at chess and finding out one of his card tricks.

There is a very natural explanation for the difficulties experienced by C when he tried to explain the nature of mathematics to A. A has to work a lot to get accustomed to an unfamiliar mathematical model before he can appreciate the results of its theory. He must have a little of the enthusiasm and the patience that C had when he was a student. If, for instance, A had been interested in combinatorial problems, which would not be unusual, their mathematical contacts could have been much more fruitful, provided C adjusted to the situation and prepared himself carefully.

We now leave A and turn to the question of why D succeeded so much better than C with their common problem. C now thinks that he has a good answer. The problem had to do with an old established branch of mathematics that C knew very well. He had even contributed to it. Then a third mathematician E had given an expository lecture where he pointed out that a suitably generalized version of the problem was very significant in an entirely new context. When C heard this lecture he was almost frightened by the bold new approach. E had lifted his little circle of problems up to a higher level. Here was much to think about and much to do. C started at once with some special cases and made a number of guesses as to what the general answer would look like. The first ones had to be discarded, but just as a promising guess was emerging C made a visit to the library where he saw that D had found the right answer and would publish the detailed proofs in a few months. C could also see how D had been able to succeed. He had computed less and thought more. C visualized how D, an impatient and fast thinker, had picked out the most significant special cases and from there formulated the general result. It was almost uncanny to see how surefooted D was in unknown territory. After the first shock, C soon recovered and the story had a happy ending for him. C read D's final article very carefully, completed the theory in many details, and wrote an excellent book about the whole business that was considered a standard work for a long time.

The difference between C and D can be seen in every classroom where mathematics is taught. Most students like to do safe systematic computation but it sometimes conceals the true nature of things from them. In a new situation they try the old recipes and fail. A new idea is needed. Some get it by themselves, others need outside help. The one that succeeds has the power of perspective. If necessary he can compress the familiar material so that it does not dim the view of the unknown. This metaphor shall here serve as a modest explanation of why D succeeded so well.

12.3 The teaching of mathematics

The interplay between a mathematics teacher and his pupil can be a kind of artistry—the right question from the teacher, fast analysis of the pupil's thinking, followed by the right explanations at the right time. This kind of teaching cannot be systematized and shall not concern us further. This section is about the new mathematics in the schools, and ends with a fable entitled "The three ways."

How C reformed school mathematics

In order to understand old mathematical articles C must know the way people expressed themselves at the time. The contents of the theorems does not change, but terminology and ways of looking at things are forever changing. New terms are born and old ones disappear or change their meaning. In addition, there are perpetual dislocations within and between the branches of mathematics. The result is, of course, that the subject sometimes looks rather disorderly, and then C wants to clean up the mess. The middle of our century was such a period. Concepts like group, ring, field, and set had appeared and been generally accepted in the beginning of the century, but the big systematization began when C wrote a series of books called *Eléments des Mathématiques*, taking the pen name of Nicolas Bourbaki. His main purpose was to make the then relatively new theory of topological spaces the basis of mathematics, and he had a resounding success. He made a terminological revolution and created a new systematics founded on simple mathematical models like topological space, linear space, group, and so on.

But C was not satisfied with this; he also wanted to reform school mathematics. He thought, quite rightly, that it had too much routine computation, and, at the same time, too few of those concepts that are easy to understand and could lead to a better understanding of the nature of mathematics and human thinking. Such things were available—for instance, the concepts of a set, a relation, a function, and a group. C started putting forth propaganda for his program, and he succeeded in getting it adopted in many of the world's schools. Small children started studying sets before they could count the number of elements in them, older children occupied themselves with functions and relations. Parents went to special courses to enable them to understand what their children were doing.

After some years, C noticed that his initiative had not had the desired effect. It is true that the new concepts are simple, but they have to be handled with a linguistic precision that is beyond the majority of children. Besides, interesting applications were rare and the exercises were trivial and sterile. Other drawbacks were that arithmetic skill vanished, that elementary geometry was neglected, and that very little mathematics remained as active knowledge after school was finished. Briefly, the whole thing was, if not a complete failure, then not a success either. C now wants more emphasis on skill and less on theory, but he does not want to give up altogether. He believes that a reasonable compromise is possible. He wants to explain his failure with the following bit of analysis. The models and concepts that he wanted to introduce are simple and useful in mathematics, and there they contribute to law and order without being repressive in any way. They could do the same things in the school curriculum, too. But, unfortunately, the other side of their simplicity is their neutral, poor content: these models are simply not good playgrounds for children. There is quite a lot to see and learn in them—but very little to *do*. On the contrary, in the open country of numbers and geometry there are always nice and educational things to do for everybody. These models are a better everyday environment. In this up-to-date formulation C finds some consolation for his defeat.

A fable

Mathematics is an important basic subject and has to cater to many interests. It must serve not only its users and society in general, it has also to take care of its own interests. This means, among other things, that the subject should be taught in a mathematically meaningful way. Such a requirement has to be weighted against the realities of the teaching situation. To show how this can be done in three different ways we shall reproduce a fable where C expresses his teaching philosophy, leaving it to the reader to figure out the *sens moral*.

The three ways

The teacher says to his class: I will explain to you the important concept of direct proportionality. It is used in mathematics, in physics, in the social sciences, and in everyday life; it has to do with two variables, x and y, such that y depends on x. The definition is as follows (he turns to the blackboard and starts writing)

y is said to be proportional to x if there is a number a such that $y = ax$ for every value of x and the corresponding value of y.

Then he turns around and looks at the class. Only one or two have understood. The teacher tries again. Well, you see, what I just wrote means, for instance, that if we put $a = 2$ that (he turns to the blackboard and writes)

$$y = 2x \quad \text{for all } x.$$

He turns around again and looks at the class. Now almost everybody has understood, but there remain two empty faces. The teacher tries again. Well, you see, what I just wrote means, for instance, if we put $x = 3$, that $y = 6$ (he writes on the blackboard)

$$6 = 2 \times 3.$$

He turns around and looks at the class. Everybody has understood.

APPENDIX

Terminology and notation

Standard notation is used throughout, with one exception, "log" means the natural logarithm. The sign ∞ means "infinity". A double arrow, $\Rightarrow$, means "implies." A simple arrow as in "$a_n \to a$ when $n \to \infty$" means "tends to." When it is vertical as in "$b \uparrow \beta$" or "$a \downarrow \alpha$" (p. 177) it means "increases to" and "decreases to." A simple arrow is also used in connection with functions (see below). The general meaning of three dots in a formula is "etcetera." For instance, $a_1 + a_2 + \cdots + a_6$ stands for the sum $a_1 + a_2 + a_3 + a_4 + a_5 + a_6$ and $a_1 a_2 \cdots a_6$ for the product $a_1 a_2 a_3 a_4 a_5 a_6$.

Sets are usually described in words. The notation $(x; x \ldots)$ meaning "the set of x such that $x \ldots$" is used only twice, for the image and the kernel of a linear map (p. 83). Other notations of set algebra are kept at a minimum. The formula $x \in A$ means that the element x belongs to the set A, the union of two sets A and B is denoted by $A \cup B$, their intersection by $A \cap B$, the complement of B in A by $A \setminus B$, and the empty set by $\varnothing$: The product $A \times B$ of A and B stands for the set of pairs (x,y) where x belongs to A and y to B.

As in all mathematical texts, the concept of a function is expressed in many ways, both verbally and in formulas. The canonical definition is the following one: a function from a set A to another set B is a rule f assigning to every x in A an element $f(x)$ of B called the value of f at x. This is also expressed by saying that f maps or sends x to $f(x)$, or, using an arrow, $x \to f(x)$. In this formula, x is a variable, i.e. it stands for any element of A. Another way of looking at functions is to consider subsets F of the product $A \times B$ such that, for any given x in A, the pair (x,y) belongs to F for precisely one y in B. Putting $y = f(x)$, this produces a function f from A to B. Conversely, given such a function f, the set of all pairs $(x, f(x))$ of $A \times B$, called the graph of f, has the property above.

Depending on the context, as a rule the nature of A and B, functions are also called maps, mappings, operators, or transformations. The notation $f : A \to B$ indicates that f is a function from A to B and functions from A are also said to be functions on A. A real (complex) function is one with values in the real (complex) numbers.

When A, B, C are sets and $f : A \to B$ and $g : B \to C$ are functions, the composed function $g \circ f : A \to C$ is defined by putting $(g \circ f)(x) = g(f(x))$ for all x in A. Composition of functions is the basic operation of mathematics. It is associative, i.e. $(h \circ g) \circ f = h \circ (f \circ g)$ whenever the two sides are defined.

There are a number of conceptually very useful notions having to do with a function $f : A \to B$. In the first place, A is called its domain and B the

263

receiving space. The subset of B consisting of all $f(x)$ and conveniently denoted by $f(A)$, is called the image of f. When $f(A) = B$, the function f is said to be surjective (onto) or a surjection. When a function f maps no two elements of A to the same element of B, i.e. when $x_1 \neq x_2 \Rightarrow f(x_1) \neq f(x_2)$, it is said to be an injection. In this case, the map $f(x) \to x$ is a function g from $f(A)$ to A such that $(g \circ f)(x) = x$ for all x in A. It is called a left inverse of f. A function which is both injective and surjective is said to be bijective (one-to-one) or a bijection. When $f : A \to B$ is a bijection, its left inverse g is also a right inverse, i.e. $(f \circ g)(y) = y$ for all y in B and we say that f and g are inverses of each other. Surjections, injections, and bijections are illustrated by the figures below.

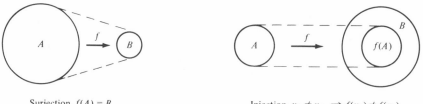

Surjection, $f(A) = B$.   Injection, $x_1 \neq x_2 \Rightarrow f(x_1) \neq f(x_2)$.

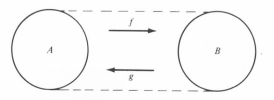

Bijection, $B = f(A)$, $A = g(B)$, g is the inverse of f.

How to read and choose a mathematical text

The main rule in the study of mathematics is to get a firm foothold somewhere by working through and really understanding some significant text. Without some such background it is virtually impossible to profit from a more passive study where difficult proofs are skipped. Elementary texts written for close and detailed study are as a rule less than rewarding for the general reader. He is more likely to find what he wants in advanced but not too technical books.

The best way to choose texts suited to one's interest is perhaps to browse in a good mathematics library with expert advice available on the side. Mathematical entertainment can be found in, e.g., James R. Newman's anthology *The World of Mathematics* (Touchstone, Simon and Schuster 1956) and in the journal *The American Mathematical Monthly* published by the Mathematical Association of America. For an outline of the history of mathematics see e.g. Struik's *A Concise History of Mathematics* (Dover 1948).

INDEX

QUEEN MARY
COLLEGE
LIBRARY